Nuclei Far from Stability and Astrophysics

NATO Science Series

A Series presenting the results of scientific meetings supported under the NATO Science Programme.

The Series is published by IOS Press, Amsterdam, and Kluwer Academic Publishers in conjunction with the NATO Scientific Affairs Division

Sub-Series

I. **Life and Behavioural Sciences**	IOS Press
II. **Mathematics, Physics and Chemistry**	Kluwer Academic Publishers
III. **Computer and Systems Science**	IOS Press
IV. **Earth and Environmental Sciences**	Kluwer Academic Publishers

The NATO Science Series continues the series of books published formerly as the NATO ASI Series.

The NATO Science Programme offers support for collaboration in civil science between scientists of countries of the Euro-Atlantic Partnership Council. The types of scientific meeting generally supported are "Advanced Study Institutes" and "Advanced Research Workshops", and the NATO Science Series collects together the results of these meetings. The meetings are co-organized bij scientists from NATO countries and scientists from NATO's Partner countries – countries of the CIS and Central and Eastern Europe.

Advanced Study Institutes are high-level tutorial courses offering in-depth study of latest advances in a field.
Advanced Research Workshops are expert meetings aimed at critical assessment of a field, and identification of directions for future action.

As a consequence of the restructuring of the NATO Science Programme in 1999, the NATO Science Series was re-organized to the four sub-series noted above. Please consult the following web sites for information on previous volumes published in the Series.

http://www.nato.int/science
http://www.wkap.nl
http://www.iospress.nl
http://www.wtv-books.de/nato-pco.htm

Nuclei Far from Stability and Astrophysics

edited by

Dorin N. Poenaru

National Institute of Physics and
Nuclear Engineering,
Bucharest, Romania

Heinigerd Rebel

and

Jürgen Wentz

Forschungszentrum Karlsruhe,
Karlsruhe, Germany

Kluwer Academic Publishers

Dordrecht / Boston / London

Published in cooperation with NATO Scientific Affairs Division

Proceedings of the NATO Advanced Study Institute on
Nuclei Far from Stability and Astrophysics
Predeal, Romania
August 28–September 8, 2000

A C.I.P. Catalogue record for this book is available from the Library of Congress.

ISBN 0-7923-6936-X (HB)
ISBN 0-7923-6937-8 (PB)

Published by Kluwer Academic Publishers,
P.O. Box 17, 3300 AA Dordrecht, The Netherlands.

Sold and distributed in North, Central and South America
by Kluwer Academic Publishers,
101 Philip Drive, Norwell, MA 02061, U.S.A.

In all other countries, sold and distributed
by Kluwer Academic Publishers,
P.O. Box 322, 3300 AH Dordrecht, The Netherlands.

Printed on acid-free paper

Printed in the Netherlands.

NATO Advanced Study Institute

Nuclei Far from Stability
and
Astrophysics

Organizing Committee

Directors

Heinigerd Rebel (Karlsruhe)
Dorin N. Poenaru (Bucharest)

Scientific Secretaries

Jürgen Wentz (Karlsruhe)
Bogdan Vulpescu (Bucharest)

International Advisory Board

James W. Cronin (Chicago)
Amand Fäßler (Tübingen)
Alex C. Mueller (Orsay)
Livius Trache (College Station)

Contents

Foreword

Science of our days realizes that the age-old scientific questions about the evolution of the Universe and the structure of the world, that we live in, unify the modern scientific horizons and puzzles, in their aspects of macroscopic and microscopic structure of the Cosmos and carried by the current research in nuclear physics and nuclear and particle astrophysics. Since particle interaction and nuclear processes are the driving forces for the birth of the chemical elements and the energy sources of stellar burning, there is an obvious interrelation between a detailed understanding of nuclear structure and a variety of astrophysical phenomena. These unified views base the recent interest in extending our knowledge about the atomic nucleus beyond the limits of stability to nuclei of exotic structure, far off stability. The interest is reflected by the enormous world-wide activities in Europe, USA and Japan to set up powerful research facilities providing beams of radioactive nuclei of various kinds. The recent report and recommendations of the NuPECC Working Group and various International Workshops on Science of Advanced Radioactive Beam Facilities illustrate the lively interest in this topic.

The NATO Advanced Study Institute "Nuclei Far from Stability and Astrophysics", held in Predeal, Romania (August 28 – September 8, 2000) reviewed the scientific goals and methodological aspects of studies of nuclei beyond the stability limits, in particular with relevance to astrophysical problems.

One aspect addresses questions of key importance to our basic understanding of nuclei as finite many-body systems. By studying nuclei far from stability we expect to discover exotic properties of loosely-bound quantum systems, with new geometries such as nuclei with neutron halos or skin of neutron matter, and to find new regions of nuclei with special structure and symmetries, new types of nuclear superfluidity, new shell structure, and new collective modes. One- and two-proton decay modes determine the proton-rich edge of the nuclear landscape. For very heavy nuclei the limits of stability are determined either by spontaneous fission or by alpha-decay. The heaviest nuclei synthesized up to now by fusion reactions are those with $Z = 110 - 118$. It is still not clear which number of protons from the predicted $Z = 114$, $Z = 120$ or $Z = 126$

1

D. N. Poenaru et al. (eds.), Nuclei Far from Stability and Astrophysics, 1–3.
© 2001 *Kluwer Academic Publishers. Printed in the Netherlands.*

is the magic one in the superheavies region. New fission modes, like alpha- and ^{10}Be accompanied cold-fission have been recently discovered by using the latest generation of large arrays of gamma-ray detectors (GAMMASPHERE).

The other aspect of interest originates from the impact for our understanding of the abundance of nuclei in the cosmos, synthesized at various astrophysical sites. The horizon of these topics experiences a considerable extension by the question, how nuclei are injected into the interstellar medium and with which energies they are propagating through the space. This is the question of the energy spectrum and the mass composition of high energy primary cosmic rays, which continuously penetrate our Earth's atmosphere and lead to secondary interactions there. Several large scale experiments (large area detector arrays) like KASCADE in Forschungszentrum Karlsruhe are just put in operation for addressing experimentally these questions and to provide experimental information about sources and propagation of cosmic rays. A particular interest is due to the fact that the energy spectrum of the "Nuclei from the Cosmos" exceeds the energies provided by man-made accelerators. Most recently the installation of an extremely large area detector array PIERRE AUGER has been committed by an international collaboration to study the highest particle energies in the Universe, whose existence are an enigma of science.

In order to span the large range of topics (proton-rich nuclei, proton decay, $Z = N$ nuclei, neutron-rich nuclei, halo nuclei, nuclear astrophysics, creation of nuclei at various astrophysical sites, Coulomb dissociation experiments of astrophysical relevance, high energy cosmic rays: energy spectrum, mass composition and interactions, exotic fission modes, superheavy nuclei, radioactive beams facilities) the scientific Organizing Committee had invited representative and active scientists of international reputation, known for noticeable contributions to the field, for lectures reviewing the status of our understanding various items of current interests.

The lecture program was supplemented by a number of special illustrative seminars of well-known scientists in the field and contributions of ASI-students, peer reviewed and selected by the invited lecturers. In fact the Advanced Study Institute, understood as meeting point of nuclear structure knowledge with astrophysical topics up to the problem of the origin of highest energies, induced a vivid interaction of aspects from different modern research lines. Thus the present publication of the lectures and seminars as Proceedings in the NATO Science series provides a textbook with an original view, covering exciting topics of current interest in a coherent approach.

One of the goals of the ASI was to mediate the personal contacts among scientists of various countries and to display in the Balkanic region the frontier topics of science as encouragement and motivation for young researchers. The exciting and stimulating atmosphere with excellent lectures and seminars, presented in a pedagogical manner and the lively discussion style of the presentations did strongly corroborate these intentions and lead to new ideas, in particular, since the nice environment of the Carpathian mountains did invite for fruitful meetings between youngsters and advanced researchers also outside the lecture hall.

Since 1964 the Romanian nuclear physicists based in Bucharest organized several Summer Schools in Predeal and Poiana Brasov. Two of them (in 1992 and 1994) have been organized as NATO Advanced Study Institutes. During the present NATO Advanced Study Institute it has been announced that the National Institute of Physics and Nuclear Engineering (NIPNE), after celebrating in 1999 fifty years of institutional research, has been selected as Center of Excellence (IDRANAP: Inter-Disciplinary Research and Applications based on Nuclear and Atomic Physics), by the European Commission. This marks the permanent efforts of the host institute NIPNE to join research on the frontier science, even under less fortunate conditions.

The present NATO-ASI was awarded and basically supported by the NATO Scientific and Environmental Affairs Division (Brussels). It was also supported by the UNESCO Office for Science and Technology for Europe (Venice), the Romanian National Agency for Science, Technology and Innovation (Bucharest), the Forschungszentrum Karlsruhe, and last not least by the German Ministry for Education and Research through the Romanian-German Governmental Collaboration Agreement. The Organizing Committee expresses its deep gratitude to these institutions.

The scientific secretaries Dr. Jürgen Wentz and Dr. Bogdan Vulpescu and the technical secretaries, Mrs. Alexandra Olteanu and Mrs. Carmen Tuca, have done an extremely valuable job and managed also difficult matters with great success. The ASI profitted very much from their engagement. Finally the editors of these proceedings thank Dipl.-Phys. Joachim Scholz for his efficient and substantial help in preparing this volume.

Heinigerd Rebel
Dorin Poenaru

I

Nuclear Physics

High-Power Accelerators and Radioactive Beams of the Future

Alex C. Mueller

Institut de Physique Nucléaire, F-91406 Orsay, France
mueller@ipno.in2p3.fr

Keywords: Radioactive beam production, use, applications and developments for high-power accelerators.

Abstract   The development of radioactive beam experiments is governed by the quest for ever increasing luminosities. The present lecture will provide some selected information on this topic, show examples of the relevant R&D in accelerator technology and address the issue of possible synergies with other fields.

1.   Introduction

Today, a century after the discovery of the phenomenon of radioactivity, about 3000 (particle-bound) combinations of protons and neutrons have been investigated, some in a very detailed way, whereas for others only their existence has been established. According to today's theoretical predictions, about as much nuclei await discovery and study, in particular very neutron-rich species (part of the nuclear landscape nowadays often referred to as "terra incognita"). The aim is to appreciate the full dimension of the nuclear many-body problem with the ultimate goal of a unified theoretical description. This vision is intimately related to a comprehension of nucleo-synthesis in astrophysical sites and thus of the origin of the elements we and our world are made of. This is the fascinating physics which is made at radioactive beam facilities and which is reported in the lectures of my colleagues in the present proceedings. It is my aim to discuss in the following some technical issues, with the underlying assumption that the performance of the experimental tools quite substantially conditions the advance of physics.

7

D. N. Poenaru et al. (eds.), Nuclei Far from Stability and Astrophysics, 7–17.
© 2001 *Kluwer Academic Publishers. Printed in the Netherlands.*

2.   Some Generalities about Present and Future Radioactive Beams

2.1.   Techniques to Make the Beams

2.1.1   The ISOL Method.

The historically first developed method for making radioactive beams is called ISOL (Isotopic Separation On-Line). In this technique, the unstable nuclei are produced by charged-particle beams or neutrons which bombard a target that in general is sufficiently thick (else a catcher is used) to stop the recoiling reaction products. The latter are then transported (diffusion, jet transport ...) into an ion source (surface ionization, plasma, laser ...) providing element separation through chemical selection. After extraction (of the desired charge-state), the wanted mass is then obtained through electromagnetic separation.

Thus the ISOL technique provides high-quality low-energy (a few tens to a few hundred of keV) beams of radioactive nuclei that are ideally suited for certain type of nuclear physics experiments (e.g.: nuclear decay spectroscopy, high precision mass measurements, optical spectroscopy ...), but also for solid state physics and other applications.

More general technical information on the use of ISOL beams as well as the demanding, but highly promising R&D efforts for the ISOL technique itself, can, e.g., be found in the proceedings of the EMIS conference series [1]. "Classical" ISOL separators can be found at many places in the world, e.g. at CERN, Jyväskylä (Finland), Louvain-la-Neuve (Belgium), Orsay (France), Warsaw (Poland), Oak Ridge (USA), TRIUMF (Canada), CIAE Beijing (China),... see, e.g. [1], and the reviews [3, 4, 5].

2.1.2   In-Flight Separation.

The second method is the in-flight separation technique relying on the forward focusing present in peripheral (and certain other) nuclear reactions. The concept of in-flight "fragment-separators" was pioneered with relativistic-energy and intermediate -energy heavy ion-beams at Berkeley, USA [6] and GANIL, France [7], respectively. A high-energy ($E > 30\ MeV/u$) heavy-ion beam impinges onto a relatively thin production target. The energy-loss of the wanted reaction products is generally kept below $\Delta E/E < 10\ \%$. Under these conditions the latter, generally called fragments, exhibit a narrow momentum distribution and a substantial part of them may be directly collected by means of the optics ($\Delta\Omega$ = a few msr) of a momentum selecting spectrometer. Designing this instrument doubly achromatic provides additional momentum-loss analysis after equipping the intermediate focal plane with a energy degrader. Pioneered with "LISE" (GANIL) [7], this feature introduces atomic number Z selection.

Consecutively, other large devices for fragment separation have been constructed and put into operation: "FRS" (GSI, Germany), "A1200" (MSU, USA), "RIPS" (RIKEN, Japan) and "SISSI" (GANIL, France), "COMBAS" and "ACCULINNA" (JINR Dubna, Russia), "RIBLL" (Lanzhou, China) and "SBL" (Chiba, Japan). Brief information of all these instruments and the main references can be found in [3, 4, 5].

One has however to note that the optical quality of fragment beams is somewhat limited, in particular while aiming at a high transmission, which means privileging the angular and momentum acceptance of the separator. The situation is best at high energy, where also contamination of incompletely-stripped charge states is minimal, with an upper limit of about 1 GeV/u due to the increasing probability for "destroying" the wanted exotic nucleus by reactions in the various materials it passes (production target, Z-selective degrader, detector systems).

Complementary to the ISOL technique, we stress here the following highly attractive features of In-Flight Separation: (i) the short-separation times, in the order of μs, without any dependence on the chemical nature of the transmitted nuclei, (ii) the possibility to inject into a storage ring and (iii) the high energy of the fragment beam that can be used for inducing secondary nuclear reactions.

2.1.3 Post-Acceleration of Separated Radioactive Beams.

Up to now, the reaction experiments at the fragment separators have essentially been made down to a minimum energy of, say, 25 MeV/u. Slowing down by further passage of matter excludes to conserve, simultaneously, reasonable optical properties and the beam intensity. Decceleration in a storage cooler-ring is intensity- and time-limited, in particular for short-lived species.

A technical solution for low-energy beams, of interest for many nuclear and astrophysical reactions, was developed since the mid-80's at Louvain-la-Neuve, Belgium [8]: Here, the secondary beams from an ISOL system, are injected into a second (cyclotron) accelerator for post-acceleration in the range 0.65 – 5 MeV/u. This pioneering development has given rise to several analogue projects, see, e.g. the overviews [3], [4] and [5] containing the main references for the Asian, European and North-American projects, respectively.

In the US "RIA" (Rare Isotope Accelerator) project [9] also a combination of ISOL and In-flight is foreseen as one of several possible production schemes. Here the in-flight separated projectile fragments would be stopped in a gas catcher feeding the ion-source of a post accelerator.

2.2.   Radioactive Beam Intensities

2.2.1   Efficiency Considerations.   The intensity I, available for an experiment is obviously a prime requirement for any future progress. Often quite inferior to the in-target production rate, it is determined by:

$$I = \sigma \phi N \, \varepsilon_1 \, \varepsilon_2 \, \varepsilon_3 \, \varepsilon_4 \, \varepsilon_5 \qquad (1)$$

which contains the following factors: σ is the cross section of the production reaction, ϕ the primary beam intensity, N the thickness of the production target, ε_1 the efficiency of release from the target and transfer to the ion source, ε_2 the efficiency of this latter, ε_3 the efficiency of the separator, ε_4 the delay transfer efficiency due to radioactive decay losses and ε_5 the efficiency of the post-accelerator.

The interdependence of these factors and their best combination is subject to intense debate among the specialists, e.g. [1, 10], but it is clear that maximizing the luminosity L, i.e. the product $L = \phi N$, is definitely the way of future progress. For the fragment separators ε_3 is in the order of 1 % to 80 %, $\varepsilon_1 \, \varepsilon_2 \, \varepsilon_4 \, \varepsilon_5 = 1$ ($\varepsilon_1 = 1$ due to direct recoil out of the target, $\varepsilon_2 = 1$ since there is no ion source, $\varepsilon_4 = 1$ because of the short flight-time, $\varepsilon_5 = 1$ since there is no post-acceleration). In contrast, for the ISOL method, the total efficiency is extremely case dependant and lies between, say, a few % to 10^{-8} (ε_5 being 5 % to 50 % depending on the type of post-accelerator. However, at least for not too short-lived species with "benign" release properties, the ISOL technique can overcompensate its often lower efficiency due to its luminosity advantage (see below).

2.2.2   Luminosity Considerations.   For the production one uses (essentially charged-particle induced) fragmentation/spallation, fusion, nucleon transfer, deep-inelastic and fission reactions. The luminosities are highest for proton-induced reactions, because of the high intensities of proton accelerators and the larger possible target thickness. Indeed, they presently reach $10^{13} \, b^{-1} s^{-1}$ at ISOLDE or more than $10^{14} \, b^{-1} s^{-1}$ at Louvain-la-Neuve whereas the luminosity four to six orders of magnitude lower for typical heavy-ion fragmentation reactions at GANIL or GSI, respectively.

One may note the great potential for substantial improvements at the heavy-ion facilities: GANIL is commissioning the acceleration of 95 MeV/u ions up to a beam power of 6 kW, GSI will have finished during the year 2000 its programme to boost the heavy-ion synchrotron up to its incoherent space-charge limit [11], RIKEN has started the construction of a new high-intensity facility [3], the MSU upgrade will considerably increase the performance of the present facility [12]. In the context of the NuPECC working group, a European fragmentation facility with a luminosity of more than 10^{12} $b^{-1}s^{-1}$ for up to 1 GeV/u uranium primary beams is considered. Presently discussed up-grade plans at GSI [13] meet this requirement while the US RIA proposal [9] is in the same luminosity class but at the lower energy of 400 MeV/u.

2.2.3 Selected R&D Topics in the Present Context.

A general limitation for RNB facilities, be it ISOL or in-flight, is given by the maximum heat-deposition in the target, due to the energy-loss of the primary charged-particle beam. Extensive R&D is under way at various places (CERN-ISOLDE, Rutherford Appleton Laboratory (RAL), IPN Orsay, GANIL-SPIRAL ...) for the design of targets with minimized local thermal overstress [14]. The SIRIUS-facility design study [15] of the RAL relies, e.g. on this R&D.

In contrast to charged particles, neutrons will heat the target only through the energy released by the "useful" nuclear reactions. In particular neutron induced fission is a very promising reaction since the produced, very neutron-rich nuclei either allow a direct access to the "terra incognita" or via secondary reactions, like transfer, fusion and fragmentation, of the post-accelerated beams. The fission cross sections for thermal neutrons on uranium (^{235}U) are extremely large in the top of the distribution. The success of this production method (for an overview of past experiments see [16]) has been at the origin the PIAFE project [17], now discontinued, and it is the basis for the MAFF project at Munich [18]. Nolen [19] recently proposed the use of fast neutrons on very thick ^{238}U targets where the luminosities may reach more than 10^{15} $b^{-1}s^{-1}$ despite of the smaller fission cross sections and difficulty of efficient release from such targets.

This important R&D issue, i.e. the optimization of $\varepsilon_1 \varepsilon_4$ is addressed by the PARRNe program at Orsay where the deuteron beam from the 15 MeV tandem is used to produce neutron fluxes of more than 10^8 s^{-1} with an energy around 10 MeV. The neutrons impinge on a ^{238}U target of the device PARRNe which allows extraction and collection of radioactive noble gases. Promising results have recently been obtained for targets of uranium carbide (containing up to 30 g of uranium) or molten uranium (up to 250 g) [20–23]. In order to investigate the best operational parameters for strong fission-fragment beams at the GANIL/SPIRAL facility based on the PARRNe developments, a European Union RTD Project has been started, with the collaborating laboratories GANIL, Jyväskylä, KVI Groningen, Louvain-la-Neuve and IPN Orsay [24]. The concomitant development of intense primary beams for the neutron generation will be discussed in some detail in the next chapter.

Another possibility for a fission based radioactive beam facility is the use of a low-energy $(25-50\ MeV)$ electron driver beam. The generated bremsstrahlung would induce photo-fission in the uranium targets [25], the Flerov laboratory in Dubna will include this in its upgrade programme [26]. Although this approach may not have the same ultimate luminosity potential than the use of fast neutrons (in addition, the fission cross section for photon induced reactions are an order of magnitude smaller and their mass and charge distribution considerably narrower), one may observe that the generation of intense electron beams in this energy range, compared to a hadron driver may be of economic interest; in particular one may benefit from the developments of the TTF collaboration [27].

For completeness, note that ultra-intense shot laser pulses can also be used to induce fission [28]. However, this technique is not (yet ?) practically applicable, because of the extremely low repetition rate of the pulses and the fact that such an intense laser shot vaporizes a great amount of the target material.

The experimental activities discussed above are complemented by an important effort made by several groups for predicting reliably cross-sections for the various envisaged production reactions. Among recent developments, the complementarity of different methods has been analyzed in calculations by Benlliure et al. [29] and modeling for fast particle induced fission is presently made by Rubchenya [30], Mirea [31] and Ridikas et al. [32].

3.   Applications of High Power Proton Accelerators and Related R&D

3.1.   The Domain of HPPA

Since the last years considerable efforts are devoted by accelerator teams from several countries to develop HPPA (high-power proton accelerator) technology. The aim is here to reach beams in the multi-megawatt class. Such accelerators allow real breakthroughs in several scientific domains. Here, only a very short description of selected aspects of these activities is been given, the interested reader can find up-to-date information and useful references e.g. in the proceedings of EPAC 2000 which just appeared [33]. Summarizing,

- condensed matter physics would greatly benefit from the generation of pulsed neutron beams that would allow time-of-flight techniques and have a pulse luminosity greatly exceeding the level of the best (continuous) high-flux reactor sources. In the US, the construction of the Spallation source SNS at Oak Ridge has been started, Japan plans to built the KEK-JAERI joint project from the next year on, and in Europe the project ESS is in the design phase. The power levels are in the $2-5\ MW$ range, to be compared to the presently strongest $200\ kW$ spallation source ISIS at the Rutherford Appleton Laboratory.

- subcritical hybrid reactor systems could be driven by means of the external neutrons provided by an HPPA, e.g. for the incineration of radioactive waste by transmutation. The KEK-JAERI joint project foresees a test experiment, although initially at very low power. In the US, DOE has recently released a report which proposes a roadmap for waste transmutation in which the development and construction of numerous $45\ MW$ HPPA is envisaged. In Europe, a technical study group (chair C. Rubbia), reporting to a number of governments, is presently investigating the specifications and the way to a European ADS (accelerator driven system) demonstrator. One important aspect of ADS-class accelerators is a reliability requirement exceeding conventional accelerator performance by orders of magnitude.

- neutron irradiation for technological purposes would e.g. have the advantage, compared to the classical use of nuclear reactors, that the fissile material inventory of the irradiation facility as such is zero. Convenient irradiation volumes would require a $20\ MW$ class HPPA.

- neutrino beams with fluxes exceeding $10^{20}\ y^{-1}$ could be generated by a $4\ MW$ pulsed HPPA like the one in the CERN ν-factory project (note that in addition to the HPPA, a lot of consecutive accelerator & storage ring equipment also needs to be developed). The projected long base-line neutrino-oscillation experiments (Gran Sasso in Italy, Soudan mine in the US, Kamioka mine in Japan), presently relying on much lower power accelerators, would greatly benefit from the HPPA developments.

- radioactive beams, as explained in more detail in chapter 2.2, not only quest for intense proton beams directly (with probably a practical limit of (a few ?) hundred kW of heat deposition in the production target), but intense spallation neutron fluxes also present great perspectives.

3.2. R&D for HPPA Technology

The following contains, as an example, a short summary of the R&D effort presently made in France, by a CEA/CNRS-IN2P3 team, in HPPA technology, more details and also the considerable efforts accomplished elsewhere can be found, e.g., in the proceedings of EPAC 2000 [33].

The construction of a 100 mA, 10 MeV proton injector accelerator has been started since 1997. This project, called IPHI (Injecteur de Protons de Haute Intensité) aims at (i) validating the beam dynamics codes of the low-energy section of an HPPA where the space-charge effects are worst, (ii) define the technological choices and assess the performance of the design tools, (iii) measure the spatial beam profile and its energy distribution in order to understand and control beam halo effects prior to injection into the high-energy part of the accelerator, (iv) collect, for this part of an HPPA, data on reliability, availability, component cost and manufacturers performance.

IPHI [34, 35] consists of a ECR source followed by the low-energy transport line to a 5 MeV RFQ, a consecutive drift tube linac and the high-energy line for beam analysis. The ECR source and low energy transport are operational with remarkable performance, for the RFQ real-size and for the DTL model construction are started. By the end of 2002 first beams from the RFQ should become available.

For the high-energy section of the HPPA, the R&D in France is concentrated on the development of (low-β) super-conducting radio-frequency (SCRF) cavities as proposed in the French reference design [6]. Indeed, this choice has many advantages compared to the classical approach, using copper cavities at room temperature, and it has also been adopted by the US spallation source SNS and the KEK-JAERI joint-project.

The efficiency of the power transfer from the radio-frequency transmitters to the beam is 1 for SCRF but only about 0.5 for copper cavities. Thus, at the MW power level of HPPA accelerators, the extreme amount of power lost in the copper walls augments the operating costs considerably. The power to be evacuated in the copper is actually so high, that it is very difficult to achieve sufficient cooling for CW operation at room temperature. For the SCRF case, the RF power-supplies and associated systems only need to provide half of the total power, which provides important investment savings. Of further advantage are the high electrical gradients. For $\beta = 0.5$ test cavities, accelerating fields well above 10 MeV/m have been obtained by the CEA/CNRS-IN2P3 collaboration [37], to be compared to the typical 1.6 MeV/m for the room temperature case. Consequently the HPPA overall shortened length by about a factor of three, important for civil engineering and site considerations. It also means that the accelerator can be operated at a very safe performance level (e.g. of paramount importance for the reliability needs of ADS). Finally, the quite large opening for the beam of SCRF cavities helps to reduce drastically beam halo problems.

As a next step, the French team has started testing optimized multi-cell cavities in the new built horizontal tank CRYHOLAB [27], and as a future step, it is planned, in collaboration with the INFN (Italy) to design and build a fully equipped cryo-module, containing cavities, high-power couplers and cryogenic connections in order to make a fully realistic test of the feasible accelerating fields and necessary cryogenic power.

4. Concluding Remarks

The NuPECC study group on radioactive beams recommended, among other items, to consider a next generation European ISOL facility [38]. In fact, from this work was launched recently the EURISOL design study [39], funded by the EU. This two year study should "thoroughly investigate the scientific and technical challenges posed by a next generation ISOL facility, identify this R&D required before a full engineering design can be undertaken, and establish a cost-estimate of capital investment and running costs. Possible synergies with other European installations and projects will also be considered". The preceding chapters of the present paper have presented (selected) information and consideration on these issues.

The synergy issue is also to be investigated by the European Technical Working on ADS (see section on the domain of HPPA). Finally, the ESS council and the French research organization CEA have jointly launched the CONCERT study on multi-purpose HPPA facilities [40].

References

[1] G. Münzenberg (ed.), 13th Int. Conf. on Electromagnetic Isotope Separators and their Applications 1996, published in Nucl. Inst. Methods B126 (1996)

[2] B.M. Sherrill, D.J. Morrissey, C.N. Davids (eds.), Int. Conf. on Exotic Nuclei and Atomic Masses 1998, AIP Conference Proceedings 455, Woodbury NY, 1998

[3] Isao Tanihata in [2], p. 943

[4] Alex C. Mueller in [2], p. 933

[5] Jerry A. Nolen in [2], p. 953

[6] T.J.M. Symons et al., Phys. Rev. Lett. 42 (1979) 40

[7] Alex C. Mueller and Rmy Anne, Nucl. Inst Meth. B56/57 (1991) 559 and references therein

[8] J. Vervier, Nucl. Phys. A616 (1997) 97c

[9] H. Grunder, in Proc. Radioactive Nuclear Beams 2000, 1–6 April 2000, Divonne-les-Bains, France, eds. U. Köster et al., in press

[10] H.L. Ravn et al., Nucl. Inst. Methods B88 (1994) 44

[11] Beam intensity upgrade of the GSI accelerator facility, Report GSI-95-05, Darmstadt, FR Germany, 1995

[12] D.J. Morrissey, Nucl. Phys. A616 (1997) 55c

[13] N. Angert and W.F. Henning, 2000, private communication

[14] see, e.g. in [1]: J.R.J. Bennet et al., p. 105 and p. 117; J.C. Putaux et al., p. 113; P. Drumm et al., p. 121

[15] htpp://www.dl.ac.uk/ASD/NPSG/sirius.html

[16] U. Köster et al. in International Workshop on Research with Fission Fragments, T. v.Egidy et al. (eds.), World Scientific, Singapore 1997, p. 29

[17] J.A. Pinston in [1], p. 22

[18] http/www.ha.physik.uni-muenchen.de/maff

[19] J.A. Nolen, in 3rd International Conference on Radioactive Nuclear Beams, D.J. Morrissey (ed.), Editions Frontires, Gif-sur-Yvette, 1993, p. 111

[20] F. Clapier, A.C. Mueller, J. Obert et al., Phys. Rev. ST — Acc. & Beams 1 (1998) 013501

[21] S. Kandri-Rody et al., Nucl. Inst Methods B160 (2000) 1

[22] E. Cottereau, in proceedings quoted in [9]

[23] E. Cottereau, J. Obert et al., private communication (2000) and to be published

[24] SPIRAL Phase II, http://ganila.in2p3.fr/spiral2

[25] W.T. Diamond, Nucl. Inst. Meth. A432 (1999) 471

[26] Yu. Ts. Oganessian in proceedings quoted in [9]

[27] Th. Junquera and H. Safa, 2000, private communication

[28] T.E. Cowan et al. Phys. Rev. Lett. 84 (2000) 903

[29] J. Benlliure, F. Farget et al. in [2], p. 960

[30] V. Rubchenya et al., private communication and to be published

[31] M. Mirea, F. Clapier, N. Pauwels et al., Il Nuovo Cimento 111A No 3 (1998) 267

[32] D. Ridikas and W. Mittig in [2], p. 1003

[33] Proceedings of the European Conference on Particle Accelerators EPAC 2000, Vienna, Austria, June 2000, http://www.accelconf.cern.ch/accelconf/e00

[34] J.M. Lagniel, S. Joly, J.L. Lemaire and A.C. Mueller, in 1997 Particle Accelerator Conference http://www.triumf.ca/pac97/papers/index.html and Inst. Electr. Eng. (ed.), Piscataway, NJ, p. 1022

[35] P.Y. Beauvais et al., in [33] p. 283

[36] H.Safa, J.M. Lagniel, Th. Junquera, A.C. Mueller, Proc. Int Conf. on ADTT, Praha, Czech Republic, April 1999

[37] H. Safa in [33], p. 197

[38] see: http://www.nupecc.org

[39] see: http://ganila.in2p3.fr/eurisol

[40] J.M. Lagniel, in [33], p. 945

Initial Studies of Neutron-Rich Nuclei with Next-Generation Radioactive Beam Facilities

R. F. Casten

Wright Nuclear Structure Laboratory, Yale University
*New Haven, CT 06520-8124, USA**
rick@riviera.physics.yale.edu

Keywords: Exotic nuclei, advanced RNB facilities, signatures of structure

Abstract   Unprecedented opportunities to extend the horizons of nuclear physics towards the "terra incognita" of neutron rich nuclei await the advent of next generation radioactive nuclear beam (RNB) machines, such as the Rare Isotope Accelerator (RIA) project. However, beam intensities, especially at the boundaries of accessibility, will be many orders of magnitude weaker than we are accustomed to. To overcome this obstacle and exploit the potential opportunities requires significant improvements in detectors systems and parallel improvements in our ability to extract more nuclear structure information from less data.

    The discussion focuses on structural evolution far from stability in neutron rich nuclei, different experimental approaches and advances in detector systems to study these nuclei, and new ideas on the more efficient use of the sparse data to understand how structure evolves. The discussion will cover nuclei ranging from those near stability to the most neutron rich species available.

1. Introduction

There have been innumerable discussions, White Papers, and review articles on the physics opportunities with radioactive beams over the last decade. As a result of the confluence of these opportunities with the technological capability to produce beams of exotic nuclei, this field has burgeoned. New generations of facilities, such as the coupled cyclotron fragmentation project at MSU in the US, or ISAC-1 in Canada, SPIRAL

*Work supported by the U.S. DOE under Grant number DE-FG02-91ER-40609.

D. N. Poenaru et al. (eds.), Nuclei Far from Stability and Astrophysics, 19–29.
© 2001 *Kluwer Academic Publishers. Printed in the Netherlands.*

at GANIL and the RB factory at RIKEN are starting operation, about to, or in the process of construction. In the offing are "ultimate" (as far as we can conceive at the moment) next generation facilities, such as the Rare Isotope Accelerator (RIA) in the US. These will open up tremendous vistas of the nuclear chart for exploration and will offer great discovery potential. One way to classify the physics opportunities embodies the rubrics:

- Exploration of the limits of nuclear existence, of the interactions that hold nucleons together, and of new exotic forms of nuclear matter.

- Study of energy generation and structural evolution in stars and of the origin of the elements.

- Tests of fundamental theories of matter and basic conservation laws.

The emphasis in the present paper is on the first of these, focusing especially on the evolution of structure in new exotic regions of nuclei.

The capabilities that are technologically possible may be illustrated by recent calculations (of course approximate) of the beam intensities produced by the RIA driver accelerator, a linac capable of accelerating all nuclear species from protons to uranium to energies at or above 400 MeV/A. These intensities are shown in Fig. 1, taken from calculations [1] by the ANL group, using input from a wide variety of sources.

Any of several production mechanisms can produce the optimum yield of a given isotope. For example, in the medium mass region of neutron rich nuclei, in-flight fission is the most prolific, while fragmentation dominates in many other regions. The extraordinary capabilities are exemplified by noting that most of the nuclei along the r-process path are accessible or nearly so at levels that allow at least some experiments (masses and half-lives).

In the rest of this paper, we will discuss the kind of experiments and the kinds of structural information that will become available with these beams.

2.   Spectroscopy Near and Far from Stability

For concreteness, let us consider a single (generic) isotopic chain of new nuclei, as illustrated in Fig. 2. Our emphasis is on neutron rich nuclei, which we therefore illustrate, since this side of stability offers the most discovery potential, but it should be realized that critically important questions abound on the proton rich side. As just four examples we

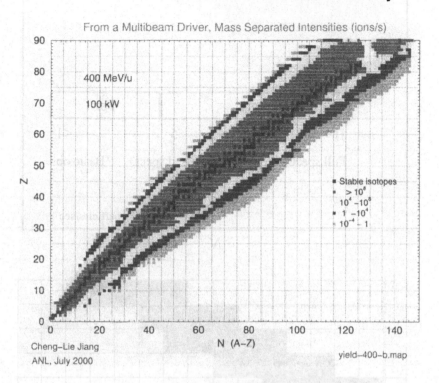

Figure 1 Projected RIA beam intensities, calculated in Ref [1]. This figure is a simplified version of a color-coded figure available from Argonne National Laboratory or the RIA Workshop website.

can cite: the exploration of the rp-process; the study of proton radioactivity; the study of the $T = 0$ interaction and possible evidence for a new $T = 0$ pairing phase in $N = Z$ nuclei; and the use of $N = Z$ nuclei as a laboratory for studies of the Standard Model by exploiting superallowed $0^+ \to 0^+$ β-decay processes.

Returning to the neutron-rich side of stability, Fig. 2 (top) makes an important point: RNB interactions will vary from the nanoamp ($10^9 - 10^{10}$ p/s) range near stability to well under 1 p/s at extremes of neutron richness. *At each stage*, significant experiments can be performed. Clearly, fewer will be possible far from stability, and the structural information will be less extensive, but crucial data on structure and structural evolution are accessible all along each iso-chain.

For illustrative purposes, Fig. 2 (top) rather artificially divides the iso-chain illustrated into "near" and "far" regions, roughly distinguished by intensities above and below 10^5 p/s. In the near region, rather sophis-

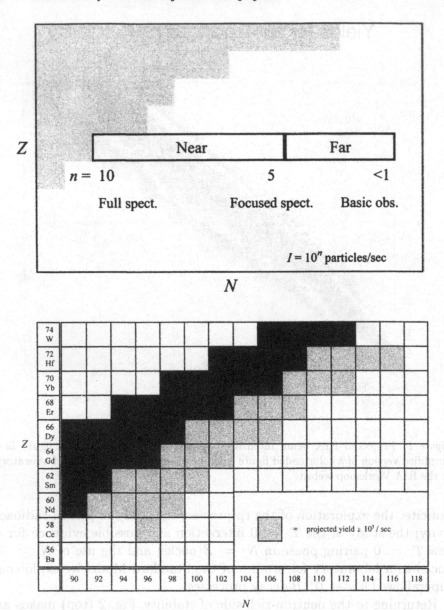

Figure 2 (Top) A hypothetical isotopic chain. The symbol n indicates beam intensity according to 10^n p/s. (Bottom) Contours of projected RIA intensities $> 10^3$ p/s for the even-even rare earth nuclei, based on the calculations in Fig. 1. The black squares are stable nuclei.

ticated spectroscopy is possible, similar to the kinds of studies carried out presently with stable beams leading to multi-band structures and a wealth of information on levels, transitions, level lifetimes, and transition rates.

At somewhat lower beam intensities, around $10^4 - 10^6$ p/s, abundant data are still available. In particular, single nucleon transfer reactions to extract spectroscopic factors for magic and near-magic nuclei are feasible as is two-step Coulomb excitation to excite, for example (in even-even nuclei), the 4^+ state and low lying collective modes. At these source intensities, extensive level schemes from β-decay can study difficult but important questions in nuclear structure such as the existence of multi-phonon states. Fig. 2 (bottom) shows, for example, for the deformed rare earth region, the even-even nuclei with expected RIA intensities of 10^3 or greater with which it should be possible, for example, to test the selection rules characteristic of 2-phonon γ vibrations.

When the RNB intensities fall to $\sim 10^3$ p/s or under, one still has access to mass and half-life measurements, β-decay studies (including $\gamma - \gamma$ and $\gamma - \beta$ coincidence measurements), and single step Coulomb excitation, either with fast beams or low-energy beams well below the Coulomb barrier. Both these types of Coulomb excitation experiments are specifically designed to *explicitly avoid* multiple excitation. Relying on the properties of the electromagnetic interaction, the excitation is limited to single step processes because (crudely speaking) the interaction is either too fast, or too weak, respectively.

Finally, at the lowest intensities, ranging down even to 10 s to 100 s of particles per day, limited information is still available, for mass measurements, β-decay lifetime determinations (both crucial input for astrophysics) and even limited spectroscopic information [e.g., $E(2_1^+)$, $E(4_1^+)$].

In discussing the data available at each intensity level, it is important to realize that these estimates are based on current detectors or current concepts for future systems. By definition, these estimates cannot take account of unforeseen technical developments. Yet these are sure to come, if recent history is any guide. Coulomb excitation provides a nice case study for the improvements possible in experimental sensitivity. Ten to fifteen years ago, Coulomb excitation was conceptually linked to the idea of beam intensities at the nA level ($10^9 - 10^{10}$ p/s). Nowadays, with detector advances, enhanced computer capabilities to handle complex data, and inverse kinematics, significant studies can be done at the 10^2 p/s level [2, 3]. It is hard to believe that this will happen to be the ultimate limit, or that other techniques will not similarly benefit from instrumentation developments.

3.   Signatures of Structure From Sparse Data

Let us now focus on the far region and consider what structural information may be deduced from the sparse data that will be available. We will assume, for the sake of discussion, that we have the minimal data shown in Fig. 3, namely $E(2_1^+)$, $E(4_1^+)$, their ratio $R_{4/2} \equiv E(4_1^+)/E(2_1^+)$, $B(E2 : 2_1^+ \to 0_1^+)$, $B(E2 : 4_1^+ \to 2_1^+)$, and the mass (from which we determine the two neutron separation energy, S_{2n}).

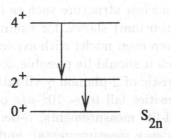

Figure 3 Simple observables discussed in the text and in Fig. 5 below. The label S_{2n} at the ground state refers to 2-neutron separation energies deduced from measured masses.

The dual ideas of maximizing the data rate per individual ion and of extracting a maximum of structural information from a nevertheless small amount of data, are schematically illustrated in Fig. 4 which emphasizes a fundamental point about the study of exotic nuclei: namely, as beam intensities drop, maintaining the level of physics understanding must involve an increase *either* in detector efficiency or "signature" efficiency (or, in fact, in both). By "signature efficiency" we refer to the ability to extract more physics from less data. Just as with instrumentation, enormous advances have been made in this area in recent years, as we now illustrate.

For example, for decades we have thought of the energy ratio $R_{4/2} \equiv E(4_1^+)/E(2_1^+)$ in terms of the simple structural paradigms of singly magic nuclei with 2-nucleon configurations of the type $|j^2 J >$, for which $R_{4/2} \sim 1.2$, harmonic vibrators ($R_{4/2} \sim 2.0$), and good symmetric rotors ($R_{4/2} \sim 3.33$). Yet, as we shall see, $R_{4/2}$, or $R_{4/2}$ in combination with other simple observables, can provide much, much more, and more subtle, structural information.

To illustrate this, consider Fig. 5 which gives several examples of the nuclear structure information available just from these simple observables. The top right panel shows the correlation of $B(E2 : 2_1^+ \to 0_1^+)$ values with $R_{4/2}$ for the region $Z = 50-82$ [4]. Traditionally, both of these observables are conceived to increase with the onset and devel-

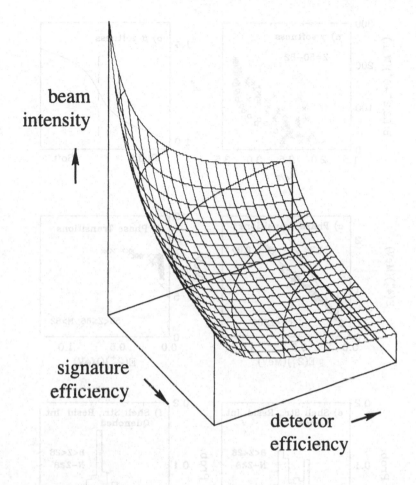

Figure 4 Schematic indication of the need for better instruments and better signatures of structure for exotic beam experiments at low beam intensities or source
strengths.

opment of collectivity and deformation. However, the data in Fig. 5a
shows that, in fact, there are two distinct trajectories for $R_{4/2} \geq 2.5$.
The upper trajectory corresponds to the $Nd - Gd$ spherical-deformed
transition region in which the nuclei are (essentially) axially symmetric.
The lower trajectory comprises the $Hg - Pt - Os$ region of nuclei from
γ-soft to deformed shapes. Clearly, there is a nice distinction based on
whether the nuclei evolve through axially symmetric or γ-soft shapes.

While there are other ways of distinguishing rigid triaxiality from γ-
softness (admittedly involving more difficult to measure observables —
namely γ-band energies), the concept of β-softness has always been much
much more difficult to assess. There have been, until now, essentially no
simple observables that yield such information. However, recent work

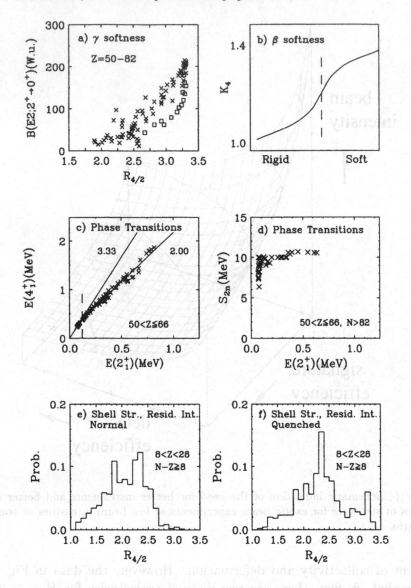

Figure 5 The panels show various signatures of structure based on simple data. The short text descriptions within each panel indicate the kind of structural information available from that plot. See text for discussion and applicable references.

has led to the development and testing of just such a signature, emerging from the concept of Q-invariants [5, 6].

These model independent quantities, involve, in principle, an infinite number of $E2$ matrix elements, and, as a consequence, have seldom been used. However, recently Brentano, Jolos, and colleagues have developed [7] approximate expressions for them. One of these,

$q_4 \sim\, <0^+|Q^4|0^+>$ is related to the fluctuations in β, i.e., β-softness. Specifically, the β-softness is given by the expression $\sigma_4 = [<\beta^4> - <\beta^2>^2]/ <\beta^2>^2 = K_4 - 1$ where $K_4 = <\beta^4> / <\beta^2>^2$. In Ref. 7 it was shown that

$$K_4 \sim \frac{7}{10} \frac{B(E2:4^+ \to 2^+)}{B(E2:2^+ \to 0^+)} \tag{1}$$

The derivation of Equ. 1 is model dependent, however, and hence needs empirical validation. This has recently been provided. New studies [8] of level lifetimes in ^{152}Sm have shown that $K_4(\sum) = 1.08(1)$ while $K_4(approx) = 1.02(3)$, where $K_4(\sum)$ is obtained by summing appropriate extensive products of E2 matrix elements, and $K_4(approx)$ is given by Equ. 1 above. K_4 varies from 1.0 for a well-deformed symmetric rotor to $\sim$1.4 for a β soft spherical vibrator. The general trend is illustrated in Fig. 5b.

That such simple data can reveal such subtlety of information is extraordinary. We will see momentarily that Fig. 5b reveals even more information than this.

Consider now the correlation of $E(4_1^+)$ against $E(2_1^+)$ in Fig. 5c. This result has been discussed a number of times, both in terms of an anharmonic vibrator interpretation of structural evolution [9] and in terms of evidence for phase transitional behavior in nuclei [10]. We focus on the latter implication here. We note the sharp change in slope from 3.33 to 2.00 at a particular 2_1^+ energy (marked by the dashed line), which we denote E_c because it simulates a critical point where the structure changes abruptly (as abruptly as possible in a finite-body system) from spherical to deformed.

Note that the evidence for a phase transition comes, not from a single nucleus, or even a single isotopic chain, but rather from the behavior of a region. This emphasizes two points: the ability to extract sophisticated information on phase/shape transitions from the simplest observables, and the need to obtain these data systematically over whole regions of nuclei.

There is an interesting link between panels b) and c) in Fig. 5. The dashed line represents the critical point E_c in Fig. 5c as just discussed. Fig. 5b also isolates the critical point (again denoted by the dashed line), namely the point (nucleus) where the second derivative of K_4 changes sign [11].

Masses (binding energies) provide another key observable. Masses embody the full set of interactions experienced by the nucleons in the nucleus. Mass differences of nearby nuclei give separation energies which are sensitive to structure and structural changes. Double differences of masses (differences in separation energies) can act as interaction filters giving, for example, the proton-neutron interaction of the last nucleons.

Fig. 5d illustrates the use of masses (S_{2n} values) to identify transitional regions. The panel again exploits the use of $E(2_1^+)$ as a kind of continuous control parameter for the phase transition, as a proxy for the equilibrium deformation β (which is not directly an observable), and as an observable whose potential regional continuity skirts the inherent integer nucleon number problem of nuclei [10]. The figure gives the empirical behavior of S_{2n} values in the $50 < Z \leq 66$ region, vividly showing the sharp kink at the spherical-deformed boundary.

Figs. 5e,f show distributions of the relative numbers of $R_{4/2}$ values as a function of $R_{4/2}$. It has been recently shown [12], in a comparison of robust nuclear data with the results of calculations with random interactions, that such $R_{4/2}$ abundance distributions are sensitive to the locus or distribution of valence nucleons for the nuclei considered and the nature of the residual interactions. Both can change in going from nuclei near stability to the neutron drip line: the valence nucleon number is sensitive to the location of the magic numbers and therefore to changes in the mean field potential. Residual interactions are also related to this potential and are particularly sensitive to effects of the nearby continuum as well. Thus, a distribution of $R_{4/2}$ values in a near-drip line region can give information on two of the most interesting and critical aspects of weakly bound nuclei, namely shell structure and residual interactions.

We have tried to show in this paper the extensive physics opportunities presented by beams of exotic nuclei, the range of issues that can be studied as one moves from intense sources of nuclei near stability to the weakest beams at the neutron drip line, and the power of simple signatures of structure to give extraordinarily subtle insights into the physics of nuclei in these new regions.

Acknowledgments

I am grateful to Witek Nazarewicz, Victor Zamfir, R.V. Jolos, P. von Brentano, V. Werner, and T. Klug for discussions and collaborations involved in this work, and to G. Savard and Cheng-Lie Jiang for providing the calculated RIA intensities in Fig. 1.

References

[1] Cheng-Lie Jiang et al., private communication and to be published.

[2] C.J. Barton et al., Nucl. Inst. and Meth. A391 (1997) 289

[3] T. Glasmacher et al., Nucl. Phys. A630 (1998) 278c

[4] R.F. Casten and N.V. Zamfir, Phys. Rev. Lett. 70 (1993) 402

[5] K. Kumar, Phys. Rev. Lett. 28 (1972) 249

[6] D. Cline, Annu. Rev. Nucl. Part. Sci. 36 (1986) 683

[7] R.V. Jolos et al., Nucl. Phys. A618 (1997) 126

[8] T. Klug et al., to be published

[9] R.F. Casten, N.V. Zamfir and D.S. Brenner, Phys. Rev. Lett. 71 (1993) 227

[10] R.F. Casten, D. Kusnezov and N.V. Zamfir, Phys. Rev. Lett. 82 (1999) 5000

[11] V. Werner et al., Phys. Rev. C61 (2000) 021301(R)

[12] D. Kusnezov, N.V. Zamfir and R.F. Casten, Phys. Rev. Lett. 85 (2000), to be published

Nuclear Structure Physics at GSI, Challenges and Perspectives

Gottfried Münzenberg

*Gesellschaft für Schwerionenforschung (GSI) mbH, 64291 Darmstadt, Planckstr. 1
and Johannes Gutenberg Universität Mainz, Germany*
G.Muenzenberg@gsi.de

Keywords: Exotic nuclear beams, halo nuclei, masses, superheavy elements

Abstract   Some characteristic examples from the ongoing GSI nuclear structure
research programme are presented such as recent experimental results
from nuclear reactions with exotic beams to explore the structure of halo
nuclei, direct mass measurements in the storage ring, and the structure
of heavy-elements. A brief outline of a next generation exotic beam
facility will be given.

1.   Introduction

Significant progress has been made in recent years to access the limits
of nuclear stability towards the proton- and neutron driplines and the
upper end of the nuclear chart [1, 2].

Reaction studies with light nuclei at the neutron dripline like ^{11}Li led
to the discovery of the nuclear halo [3, 4]: The weakly bound neutrons
spread out from the nuclear core and surround it as an extended halo of
diluted neutron matter. This unexpected discovery, made on nuclei well
known by their decay properties at this time, initiated a whole field of
research giving new access to nuclear structure which is complementary
to the conventional decay studies.

Experiments to explore shell structure far-off stability, yielded first
results on the weakening of the shell strength along the neutron-rich
mass 130 nuclei. First hints on a weakening of this shell came from
attempts to explain the r-process abundances near closed shells [5].

At the upper end of the nuclear chart, beyond the limit of liquid
drop stability, a region of deformed shell nuclei which exists only by

31

D. N. Poenaru et al. (eds.), Nuclei Far from Stability and Astrophysics, 31–42.
© *2001 Kluwer Academic Publishers. Printed in the Netherlands.*

shell effects has been discovered [6]. This paved the way for the present heavy and superheavy element research.

Such experiments became possible due to the development of new experimental techniques: accelerators delivering heavy ion beams of high intensity or with relativistic energies, the separation in-flight and single-atom identification [7] of exotic nuclei to access the rare and shortlived species at the very limits of stability. Recent detector developments include 4π-detection systems for charged particles, neutrons, and gammas. Heavy-ion storage rings opened-up new research possibilities.

Presently the prospects for a next generation GSI facility are explored. To evaluate the perspectives for a European Next Generation Radioactive Beam Facility a working group has been installed by NuPECC. The report is just being published [8].

2.   The GSI Accelerator System

GSI accelerators provide heavy-ion beams of all stable isotopes from all elements up to uranium with energies from near Coulomb-barrier to 1 $AGeV$, far above the Fermi domain. Typical beam intensities are $5 \cdot 10^{12}/s$ in the low-energy regime and $10^9/s$ to $10^{10}/s$ at relativistic energies. For the production of exotic nuclei we use three typical reactions:

— Complete fusion of heavy ions to access the elements beyond uranium
— Projectile fragmentation in peripheral collisions
— Fission in-flight

Whereas the fusion leads to recoils of low energy, the projectile — as well as the fission fragments — have energies far above Coulomb barrier. They can be used for reaction studies.

With these unique possibilities the GSI nuclear structure research program covers:
— Reaction studies of skin and halo nuclei
— The investigation of basic properties of nuclei at the limits of β-stability including precision experiments in the heavy-ion storage and cooler ring
— Fission studies with unstable nuclides
— The synthesis and investigation of heavy and superheavy elements.

3.   Nuclear Halos — Looking inside the Nucleus

Nuclear reactions with energetic beams of exotic nuclei give new insights into nuclear structure. Specific for the halo nuclei is that the reaction cross section for the weakly bound halo part of the nucleus is much larger than the cross section for the nuclear core. This permits the direct observation for the halo nucleons in nuclear reactions, a separate treatment of halo and core, and even more, a direct insight into the halo structure e.g. wavefunctions, correlations, and transition after nuclear of electromagnetic excitation. A new phenomenon for the weakly bound systems with the Fermi surface close to the unbound region is the interaction with the continuum (see also contribution by G. Schrieder).

The experimental technique applied in general for unstable nuclei is reversed kinematics: target and projectile are interchanged as compared to conventional experiments, the nuclide of interest is used as the projectile. The advantage of this technique, especially if applied at high energies, is that all reaction products, are kinematic focused in forward direction. This allows easily for 4π geometry. Moreover A and Z of projectile and reaction products can be identified event-by-event. Such experiments are carried out with the combination of the ALADIN magnet for charged particle identification, and the LAND large area neutron detector. Optionally a target calorimeter is used to detect gammas in coincidence with the reaction products to identify the involvement of excited nuclear states. As all kinematic parameters are measured, excitation energies of the involved nuclear states including resonances, correlation between break-up nucleons and the nuclear core etc. can be obtained.

Specifically projectile fragmentation and in-flight fission permit experiments with beam cocktails, e.g. the investigation of several nuclides at the same time, which is the ideal tool for survey experiments to explore unknown regions of the nuclear landscape.

3.1.   Momentum Widths of the Carbon Isotopes

Since the early experiments on halo nuclei it is well known that one-neutron separation energies, nuclear interaction cross sections, and the widths of the break-up momentum distributions of halo nucleons (or the core) are correlated. The local extension of the density distribution and the momentum spread are connected by the uncertainty relation. The data [9] for the carbon isotopes displayed in Fig. 1 clearly characterize ^{19}C as a halo nucleus by a low binding energy of the last neutron (upper panel), a narrow momentum distribution of 68 ± 3 MeV/c (central panel), and a large one-neutron removal cross section of 233 ± 51 mb (lower panel).

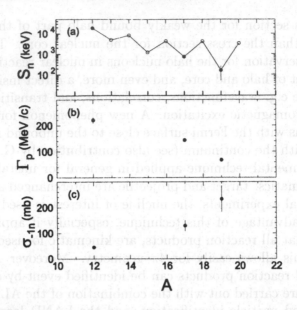

Figure 1 Systematic evolution [9] of the neutron separation energies (upper panel), momentum distributions (middle panel), and one-neutron removal cross sections (lower panel) for the neutron rich carbon isotopes ^{12}C to ^{19}C.

3.2.   Invariant Mass Spectroscopy

Fig. 2 displays the relative energy spectra of the $\alpha - n$ (upper panel) and $^6He - n$ system after break-up of ^6He and ^8He, in a carbon target [10]. The solid lines are the result of Monte Carlo calculations in a sequential fragmentation model. For the ^5He a p-wave resonance with $I^\pi = 3/2^-$ and $E_r = 0.77\ MeV$ was assumed which gives a perfect description of the data. The calculation for ^7He was performed taking two p-wave resonances with $I^\pi = 3/2^-$, $E_r = 0.44\ MeV$ and $I^\pi = 1/2^-$, $E_r = 1.2\ MeV$. The dashed and dotted lines display the separate contributions of the two resonances in ^7He. The $3/2^-$ groundstate resonance gives the main contribution and describes the maximum of the measured distribution. A contribution from its $1/2^-$ spin-orbit partner dominates the high-energy part of the spectrum.

For the ^6He the 2^+ state at 1.797 MeV is represented by a solid line, the broad distribution at high energies originates from non-resonant dipole and quadrupole excitations. For the ^8He two resonances were observed, a 2^+ resonance at 2.9 MeV (dashed line) and a resonance of unknown origin at 4.15 MeV (dotted line). The ^8He data were constructed from the momenta of ^6He and two neutrons.

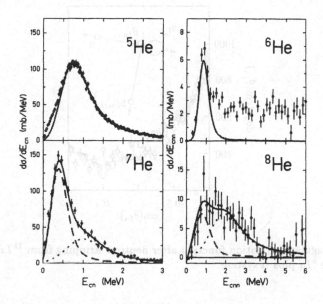

Figure 2 Left hand side: energy spectra of unbound systems: $\alpha - n$ system ^5He and $^6He - n$ system ^7He(lower panel), right hand side: energy spectra of ^6He, and ^8He from the interactions with a carbon target [10].

3.3.   Spin Alignment

As the reaction process is very fast the polarization of the remaining fragment is not affected by the sudden knockout of a halo neutron. Specifically for the Borromean two-neutron halo nuclei where the one-neutron knockout creates an unstable nucleus, the shape of the angular correlation gives information about the spin structure of the unstable nucleus[11]. Fig 3 shows the angular correlation for ^{11}Li for the one-neutron knockout leading to the unstable ^{10}Li which then emits a prompt neutron. The skewness of the correlation is due to a strong $s - p$-wave interference.

3.4.   Direct Mass Measurements — Mapping Driplines and Shells

A new and powerful tool for large-scale direct mass measurements are heavy-ion storage and cooler rings. As an advantage in comparison to ion traps, storage rings can accumulate in-flight separated atomic nuclei at their full energy and with large phase space, e.g. many masses can be measured simultaneously.

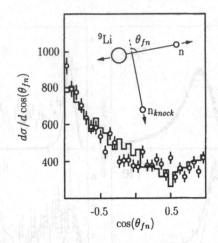

Figure 3 Angular correlation function after neutron stripping from ^{11}Li for the neutron and the ^9Li plus n system.

The nuclides of interest are produced by projectile fragmentation (in the experiment presented here from bismuth projectiles), separated in-flight, injected into the storage ring with their full energy, and stored and cooled. The masses (or more precise, the mass over charge ratios) are obtained from the Fourier-transformed noise signal of the coasting ions. This method, called Schottky-mass-spectroscopy, is very precise and extremely sensitive. It permits the measurement even of single ions. In our first experiments 104 new masses of proton rich nuclides from tellurium to plutonium[12] were obtained by direct mass measurement of the stored projectile fragments (for the isotopes of bismuth and below) or, for the heavier ones, by combining the known Q_α values along the α-decay chains of the trans-bismuth isotopes to masses at their endpoints determined directly in our experiment. So the proton dripline could be fixed between bismuth and protactinium. The precision is about 100 keV. In recent experiments we could improve our precision by a factor of about three to five.

Fig. 4 shows the persistence of the lead shell in dependence of the neutron number [13]. The data show that the lead-shell gap decreases fast towards the proton rich side. None of the displayed models predicts the experimental trend sufficiently [14, 15, 16].

Schottky-mass-spectroscopy is applicable to nuclei with halflives longer than seconds. For the short-lived species the storage ring is tuned to the isochronous mode, then the revolution time of the stored ions is only determined by their mass. This method is applicable to nuclei with half-lives down to microseconds. First experiments to determine the hitherto unknown masses of ^{39}Ca, ^{41}Se, ^{43}Ti, ^{45}V, and ^{47}Cr have been performed successfully.

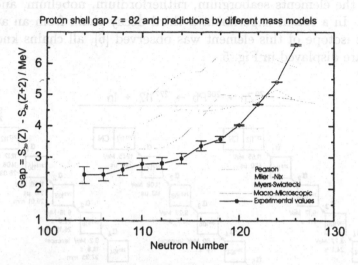

Figure 4 Experimental proton shell gap (circles with error bars) for neutron deficient lead isotopes compared to recent predictions [13].

4. Heavy and Superheavy Elements — Towards the Limits of the Nuclear Chart

Towards the heaviest elements the strong Coulomb forces dominate atomic and nuclear properties. They also limit the possibility for the synthesis. In terms of the liquid drop model all nuclei beyond $Z = 104$ are unstable against fission. The exciting discovery in heavy-element research is the enhancement of stability [17] for the elements beyond rutherfordium ($Z = 104$). Theory explained this in terms of a shell stabilization with a maximum at $Z = 108$ and $N = 162$, created by a hexadecapole deformation [14, 18]. The discovery of the new region of shell nuclei interconnecting the transuranium- and the superheavy elements created the basis for present trans-actinide research in physics and chemistry.

The heaviest known elements are produced in amounts of few atoms in experiments of typically one months duration. As the dominant decay mode for the heaviest elements is α-decay, the disintegration of the individual nuclei, implanted in silicon detectors, can be easily observed in situ and can be used for their unambiguous identification by the well established parent-daughter correlation technique [6]. The heaviest element presently identified with this method [19] is element 112. It was synthesized by complete fusion of ^{208}Pb with ^{70}Zn and identified by the observation of two atoms followed by long α-decay chains to known isotopes of the elements seaborgium, rutherfordium, nobelium, and even fermium. In a recent experiment a third decay chain from an atom of the same isotope of this element was observed [6], all chains known at present are displayed in Fig. 5.

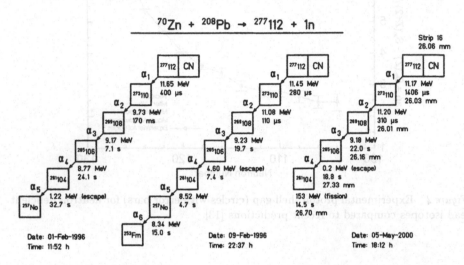

Figure 5 The alpha chains observed for element 112, the third chain observed in the recent experiment is terminated by fission at element 104 [6].

The location of the new shell is directly proven by the α-chains from element 112 (see Fig. 6) passing by the maximum of the shell stabilization [14] (lower panel). Above $N = 162$ the decays are accelerated, whereas below they are slowed down (upper panel).

Results on the synthesis of $Z = 114$ by complete fusion of $^{244,242}Pu$ with ^{48}Ca have been reported from Dubna. Very recently the discovery of a correlated decay sequence in the reaction ^{248}Cm plus ^{48}Ca leading to the previously known 114 and hence assigned to the decay of element 116 has been announced [20]. The production of $Z = 118$ in the reaction ^{208}Pb with ^{86}Kr has been reported from Berkeley [6]. Several laboratories including GSI, GANIL, and RIKEN have tried to confirm this experiment but without positive result. The decay properties observed for these nuclides as they have been assigned are generally consistent with expectations from theory, they form however a separate island located far in the neutron rich region not connected to the known region of the nuclear chart. New strategies for SHE-identification need being developed such as: direct determination of A by direct mass measurements

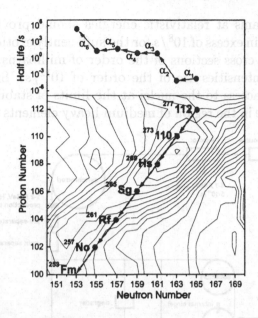

Figure 6 The calculated microscopic corrections of the heaviest elements [14]. The path of the decay chains from elements 112 are indicated. The upper panel displays the measured correlation times for one of the chains.

in traps, by time-of-flight spectrometers, or with calorimetric detectors. For the determination of Z chemical methods are being developed.

The still open question is the location of the spherical superheavy shell closure [21]. The macroscopic-microscopic method predicts $Z = 114$, self consistent calculations predict the location of the proton shell at $Z = 120$, or $Z = 126$. The neutron shell is predicted generally for 184. All of the recently reported superheavy elements including element 112 are formed with cross sections of the order of one picobarn, which is the limit of the sensitivity of present experiments.

5. The Next Generation Facility

The prerequisite to proceed to more exotic species are powerful heavy-ion accelerators which can deliver intense beams of all stable isotopes — including the rare ones like e.g. ^{48}Ca of all elements.

Strong beams at Coulomb-barrier energies are the prerequisite for the production of superheavy elements which will most probably be formed with picobarn cross-sections or even below. The required beam intensities should be at least by a factor of ten higher than available at present, e.g. of the order of $5 \cdot 10^{13}/s$.

Heavy ion beams at relativistic energies should provide secondary-beam intensities in excess of $10^8/s$ for the sufficiently exotic species which have production cross sections of the order of millibarns. The required primary beam intensities are of the order of $10^{12}/s$. Such intensities would also give access to the nuclei at the limits of stability and to the neutron drip-line in the region of medium heavy elements close to nickel.

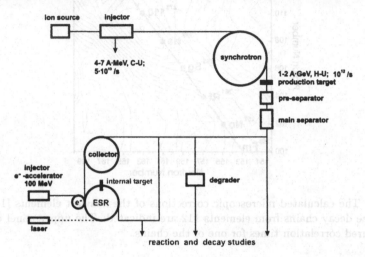

Figure 7 A possible Next Generation Exotic Beam Facility [8].

A possible future facility as outlined in Fig. 7. comprises a low-energy branch for studies of fusion products, preferably heavy and superheavy elements, and a high energy branch providing beams up to uranium at energies of up to 1 *AGeV*. The high beam energy is the prerequisite for clean separation of heavy elements and fission products. To optimize the high energy branch for the injection into the storage-ring system, the heavy ions are accelerated by a synchrotron. An improved multi-stage fragment separator system of large acceptance will separate projectile fragments and fission products for decay studies with high transmission and efficient background suppression.

Apart from efficient detection systems for decay studies the facility should include a set-up for reaction-experiments at high count-rates and with improved momentum and A-resolution in combination with an advanced $4\pi\gamma$-array with high efficiency and tracking capability. Trapping of in-flight separated projectile fragments will allow e.g. for fundamental physics experiments.

The storage-ring system requires an accumulator for the projectile- and fission fragments with fast stochastic cooling, an experimental storage ring, and a small electron collider for structure research. The magnetic rigidity of the ring system is 13 Tm, corresponding to 740 $AMeV$ for uranium. The electron ring should have 500 MeV electron energy.

For the next generation facility nuclear reactions at highest precision using a thin internal target to minimize atomic effects and cooled beams will play an important role. Elastic and inelastic scattering and Coulomb dissociation will allow for studies of matter distributions, nuclear structure, and break up as the reversed capture process. The latter playing a key role in the cosmic nucleo-synthesis.

The electron-heavy-ion ($e - A$) collider is a completely new experimental tool for structure studies with unstable nuclei to investigate e.g. charge radii or electromagnetic excitation. Electrons provide a clean electromagnetic probe. Such a mini electron-heavy-ion collider, first discussed at Dubna, is also part of the RIKEN project. The luminosities achievable with a next generation facility should be sufficient [22] for elastic scattering and reaction studies.

6.   Summary

The next generation exotic beam facilities will go for more intensity and more precision. They will enable us to proceed further to the limits of the existence of nuclear matter and open up the possibilities to apply new experimental methods for structure research including the exploration of the structure of skins and halos in heavy systems, correlations, continuum interactions, the influence of in-medium effects, the proton-neutron interaction, shells far-off stability, charge radii and diffuseness of the nuclear core, and the structure of the shell stabilized superheavy nuclei, to give a few examples. This will give us a more consistent picture of the atomic nuclei as the building blocks of elementary matter.

Acknowledgments

The author wishes to thank H. Emling, H. Geissel, S. Hofmann, B. Jonson, and Yu. Novikov for substantial discussions.

References

[1] A. Richter, Nucl. Phys. A553 (1993) 417c

[2] H. Geissel, G. Münzenberg, and K. Riisager, Ann. Rev. Nucl. Part. Sci. 45 (1995) 163

[3] I. Tanihata et al., Phys. Lett. B160 (1985) 380

[4] P. G. Hansen, B. Jonson, Europhys. Lett. 4 (1987) 409

[5] B. Pfeiffer et al., Z. Phys. A357 (1997) 235

[6] S. Hofmann, G. Münzenberg, Rev. Mod. Phys. 72 (2000) 733

[7] G. Münzenberg et al., Z. Phys. A315 (1984) 145

[8] NuPECC Report, A next Generation European Radioactive Beam Facility, B. Jonson Ed, (2000)

[9] D. Cortina Gil et al., submitted to Europhys. Journ.,

[10] K. Markenroth et al., Nucl. Phys., in press

[11] L. Chulkov et al., Phys. Rev. Lett. 79 (1997) 201

[12] Th. Radon et al., Nucl. Phys. A677 (2000) 75

[13] Yu. Novikov et al., to be published

[14] P. Möller et al., Atomic and Nuclear Data Tables 59 (1995) 185

[15] W. D. Myers et al., Phys. Rev. C58 (1998) 3368

[16] J. M. Pearsson et al., Nucl. Phys. A528 (1991) 1; At. Data, Nucl. Data Tables 61 (1995) 127

[17] G. Münzenberg et al., Z. Phys. A 322 (1985) 227

[18] Z. Patyk and A. Sobiczewski, Nucl. Phys. A533 (1991) 132

[19] S. Hofmann et al., Z. Phys. A354 (1996) 229

[20] Yu. Ts. Oganessian, priv. comm, 2000

[21] K. Rutz et al., Phys. Rev. C56 (1997) 238

[22] G. Münzenberg, G. Schrieder, in press

Study of Halo Nuclei

G. Schrieder

Institut für Kernphysik, TU Darmstadt, D-64289 Darmstadt, Germany

schrieder@ikp.tu-darmstadt.de

Keywords: Light nuclei near dripline, halo nuclei, structure of halos

Abstract   Studies of nuclei far from the valley of beta stability are one of the main topics in nuclear structure physics today. The search for states with spatially extended — by far larger than the range of nuclear forces — and low-dense matter distributions, so-called "halo" nuclei, found predominantly in light neutron-rich nuclei is of special importance. Results of experimental investigations that probe the structures of halo nuclei are presented and consequences for nuclear models are drawn. Some examples of experiments performed at GSI with relativistic secondary beams are given where cross sections, momentum distributions, angular, energy and momentum correlations after breakup into core fragments and valence nucleons have been measured. These data permit the extraction of cluster and single-particle ground-state properties and demonstrate the presence of large low lying multipole strengths near the threshold for particle emission.

1.   Introduction

With the advent of rare isotope beams, the study of exotic nuclei — far from the valley of β-stability — has become one of the main topics in modern nuclear structure physics. Close to, at and beyond the neutron and proton driplines new aspects of interactions among nucleons become important leading to a large variety of former unknown nuclear configurations [1, 2, 3]. Nuclear matter distributions, nuclear symmetries, and the structure of excitations may be quite different from those of nuclei situated near stability. For instance, a novel structure feature with spatially far extended valence nucleons ("halo"[1]) has been found

---

[1] Halo was introduced initially in physics to describe a luminous circle surrounding the sun or the moon. It is caused by ice crystals in the atmosphere producing reflection and refraction of the light.

D. N. Poenaru et al. (eds.), Nuclei Far from Stability and Astrophysics, 43–54.

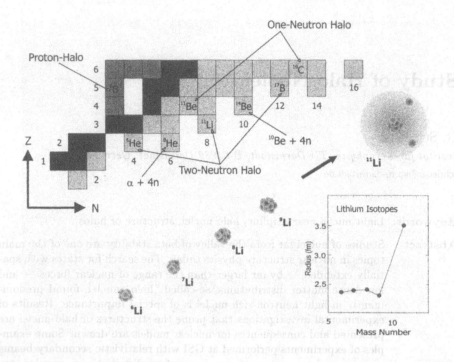

Figure 1 Nuclear chart with well established halo nuclei. To illustrate a halo a schematic sketch of sizes together with radii [5] of *Li* isotopes is shown in addition.

predominantly in extremely neutron-rich light nuclei. Proton halos are not so pronounced and may occur only in light elements ($Z < 20$) and the best condition for halo like configurations was found in systems with low valence orbital momentum $l \leq 1$ [4]. The reasons are long-range repulsive interactions like Coulomb and centrifugal interactions which hinder halo formation.

Since separation energies of the valence nucleon(s) are low (on the nuclear scale) down to 100 keV, quantum mechanics allow them to tunnel at distances far from the remnant system (the core) and the wave function of halo nucleon(s) gives a probability for being outside the core much larger than 50 %. As a result the spatial and energy-momentum structure of the valence particles are very different from the rest of the system. The valence and the core subsystem are to a large extent separable. Therefore, halo nuclei can be viewed to consist of a core with normal nuclear density surrounded by a low dense halo of valence nucleon(s) as illustrated in Fig. 1.

Halos have been observed experimentally in several light nuclei near the driplines: One-neutron halo nuclei such as ^{11}Be and ^{19}C, one-proton halo nuclei such as ^8B and two-neutron halo nuclei such as ^6He and ^{11}Li. Nuclei with a two-neutron halo represent a three-body system with "Borromean"[2] properties for which no binary subsystem is bound. For these three-body systems also giant halo states of large dimension (for nuclear systems possibly in the order of 100 fm), called Efimov states [6], may occur but under the very restricted condition of single particle s-states with extremely small binding energies of less than 1 keV. Multi-neutron or proton halos are not yet realized in nuclei [2]. For instance, ^8He can be viewed to consist predominantly of an α-particle coupled to a tetraneutron [7, 8]. It does not show halo character and is even stronger bound than ^6He by a factor of two. The same holds for ^{14}Be (possible $^{10}Be + 4n$ structure [9]) reflecting the strong correlation of nucleons in nuclei.

Since for halo nuclei very few degrees of freedom decouple from all others within many-body systems "cluster" formation [10] becomes important and is highly favored especially close to the decay threshold opening up the unique prospect of exploring the quantum many-body problem in a completely new regime.

On the other hand, to incorporate the new features of these exotic nuclei into the basic microscopic theory for nuclear structure, the "shell model", needs a revision. In the shell model approach the nucleons move in a mean field potential $V(r)$ which is created by their mutual interaction. In addition to this potential in the non-relativistic approach a spin-orbit potential $V_{so}(r)$ which is proportional to $-1/r\, dV(r)/dr$ has to be added which energetically favors those nucleons for which orbit and spin are in the same direction. For the exotic nuclei which are more extended and dilute the spin-orbit potential is strongly reduced. In addition there are forces between nucleons called residual interactions like pairing interaction or core polarization. The first favors the coupling of like-nucleons to total angular momentum $J = 0$. Therefore these move in opposite direction but in the same orbit. The latter denotes the case where the mean field potential is dynamically deformed by the interaction with the halo nucleon(s).

For the description of halo nuclei the residual interactions are becoming very important. Since separation energies of the valence nucleon(s) are one order of magnitude lower than for stable nuclei bound and unbound configurations can strongly interact. Moreover, quite in contrast to stable nuclei, considerable low-lying strength of different multipolar-

---

[2]The name comes from the heraldic emblem of the noble Italian family of Borromeo, three rings combined in such a way that if any of the rings is removed the remaining two will fall apart.

ities has been predicted close to the particle threshold [11, 12], and the associated strength distributions are a characteristic signature of the specific halo structure. This large strength can be understood in a single particle approach by the optimal matching of the wave function of the continuum scattering states with the halo wave function, penetrating far into the classical forbidden region [13] or by coherent excitation, i.e. low-frequency oscillations of halo nucleons against the core [14], also called soft giant resonance. Part of this large strength near particle threshold may be shifted into the bound region, i.e. into bound states which do not anymore follow ordinary shell model systematics. The shell structure is now dominated by residual interactions within the valence component rather than by mean-field dynamics. It is thus modified and shell closures may be altered or even vanish. For nuclei near the neutron drip line the magic neutron number, $N = 8$, does not exist anymore [15] but a new magic number, $N = 16$, appears [16].

Neutron and proton halos have attracted much interest and their properties have been studied in recent years using atomic physics [17], β-decay experiments [18], elastic scattering [19], interaction cross section [20], and momentum-distribution measurements [21, 22], and by spectroscopy in the continuum [23, 24]. For two-neutron halo nuclei measurements of momentum and energy correlations between the particles in the exit channel are especially important [25]. Experiments with rare isotope beams to study halo nuclei were performed mainly at DUBNA, GANIL, GSI, ISOLDE, MSU and RIKEN.

In the following mainly results with relativistic beams of energies between 0.2 and 1.4 GeV obtained at GSI will be presented. Emphasis is laid on breakup reactions where the halo nucleus disintegrates into the nuclear core and the valence nucleon(s).

2.   Reaction Cross Sections

Since beams of beta-unstable nuclei have to be produced by nuclear reactions like spallation, fragmentation or fission, their intensities are very low and therefore only reactions with rather large cross sections can be studied. To achieve high luminosities, thick targets have to be used and the reaction products should be measured with almost 4π acceptance and highly efficient detectors. Reaction studies at the intermediate-energy regime have the advantage that reaction cross sections to breakup nuclei are large (up to b), thick targets can be used (up to the order of g/cm^2) and that in the laboratory system all reaction products are emitted in a small forward cone (of about 100 $mrad$). Thus, all particles emitted and correlations between them can be measured covering essentially 4π

solid angle. Despite low statistics the physics information which can be extracted with a 4π measurement of all reaction products is often superior to a conventional experiment where these are measured with high statistics but only in parts of the full solid angle.

An effective means of searching for unusual features in nuclei, such as extended matter distributions, is the measurement of interaction cross sections which represents the total probability that a projectile nucleus will interact with a target nucleus and change the projectile identity. At intermediate energies it is related to the size of the interacting nuclei. These relative simple measurements can be performed with beam intensities as low as 0.01 *particles/s*. For example at GSI the matter distributions of a wide range of light nuclei have been probed by this method [26]. A more accurate technique providing detailed information on nuclear matter distribution is elastic proton scattering at intermediate energies in inverse kinematics [19] but with larger beam intensities.

For all stable nuclei it could be shown that their central densities are almost the same (≈ 0.17 *nucleons/fm^3*) and that the half density radius is proportional to the atomic number to the power of one-third ($A^{1/3}$). The surface thickness (diffuseness) is, roughly independent of mass (≈ 1 *fm*) and the spatial proton and neutron distributions are strongly overlapping. This is not anymore the case for nuclei far from stability where skins (proton and neutron radii are different but not their diffuseness) and halos (proton and neutron radii as well as their diffuseness are strongly different) occur. This is illustrated in Fig. 1. Starting from ^6Li up to ^9Li the radius determined by experiment is nearly constant (≈ 2.4 *fm*). By adding only two neutrons the radius of ^{11}Li increases to 3.5 *fm* [5] which corresponds to a large spatial extend of these neutrons with a rms radius, as sketched in Fig. 1, close to that of ^{208}Pb.

3.   Halo Breakup Reactions

Halo breakup reactions leading to a disintegration into the core and halo nucleons can be induced basically by two different mechanisms:
(i) Projectile-target nucleon-nucleon collisions may remove ("knockout") halo nucleons, eventually leading to unbound nuclear subsystems. It was observed experimentally [27] that single-nucleon knockout of a halo nucleon is the dominant process on light targets. At the intermediate energies accessible, e.g. at GSI, the velocities of the halo nuclei ($\beta \approx 0.6-0.9$) are much larger than the Fermi velocity of their halo nucleons ($\beta < 0.1$). Because of the short interaction time the internal degree of freedom of the halo nuclei may be considered to be "frozen" during

the collision process and thus a "snapshot" of the halo is obtained in momentum space. Nuclear structure effects and reaction dynamics are thus less entangled.

(ii) Excitation into the continuum due to the nuclear (low-Z targets) or electromagnetic (high-Z targets) mean field of the reaction partners. This mechanism allows for studies of the halo continuum response. This information is essential since the weak binding leads to continuum admixtures strongly effecting the ground state.

4.   Momentum Distributions

Other evidence for the halo nature of nuclei comes from narrow momentum distributions of projectile residues. In a knockout reaction of a single halo nucleon the remaining fragment should acquire the internal momentum of the removed nucleon. An example of these types of experiments is shown in Fig. 2 where differential cross sections for fragments detected after halo nucleon knockout from $^{12,17,19}C$-beams are plotted

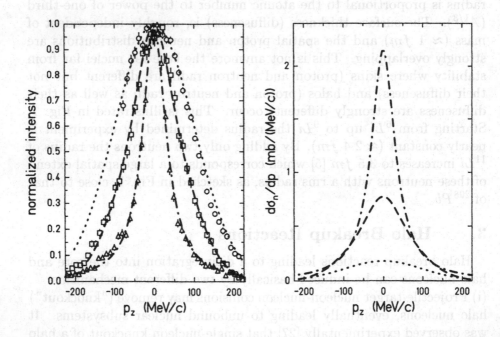

Figure 2 Longitudinal momentum distributions of ^{11}C (dotted line), ^{16}C (dashed line) and ^{18}C (dashed-dotted line) stemming from $^{12,17,19}C$ breakup reactions in a carbon target and normalized to unity on the left [22] and on the right to the one-neutron removal cross sections [28]. The squares, open circles and triangles represent the experimental data.

as a function of their momentum along the beam axis (longitudinal momentum) [22]. The experiment was performed at the Fragment Recoil Separator (FRS) at the GSI. The momentum induced in the breakup reactions was measured by position detection after the FRS, independent of the relatively large momentum spread of the incident 1 GeV/u secondary beam. The distributions are sensitive to the wave function of the removed nucleon. A narrow momentum distribution indicates a large spatial extent (according to the Heisenberg uncertainty principle) and by its shape it is sensitive to the orbital momentum component of the removed nucleon. While for ^{12}C a Gaussian shaped momentum distribution with a width (FWHM) of 220 MeV/c and a small one-neutron removal cross section of 44.7 mb are measured, for ^{19}C a Lorentzian shaped momentum distribution with a small width of 68 MeV/u and a large one-neutron removal cross section of 233 mb is obtained [28]. The Fourier transformation of a Lorentzian results in a Yukawa function (asymptotically). The observation of a Lorentzian shape together with the small momentum width and the large neutron removal cross section thus proves that ^{19}C is a halo nucleus.

For all known two-neutron halo nuclei the residual fragments are particle-unstable and only the momenta of their decay products (core plus one neutron) can be measured, but these are influenced by strong final-state interaction. Therefore it is necessary to perform kinematically complete experiments where the momentum vector of each decay product is measured and the momentum of the particle-unstable residual fragment can be reconstructed. For instance, the momentum distribution of ^{10}Li can be obtained by the measured momenta of ^{9}Li and the neutron [29]. By the width ($\Gamma = 55\ MeV/c$) and the shape of the ^{10}Li momentum distribution it can be shown that in ^{11}Li the halo neutrons occupy the $1s_{1/2}$ and $0p_{1/2}$ orbitals with comparable amplitudes. This is surprising because this nucleus should have a $N = 8$ closed shell and one would not expect such a mixed ground state wave function.

5.   Spectroscopy of Intermediate Unbound Systems

The single halo-nucleon knockout from two-neutron halo nuclei considered here yields unbound intermediate systems. The study of these unbound systems is important to understand the structure of two-neutron halo nuclei since they represent subsystems. An invariant mass analysis of their decay products gives access to spectroscopic information on resonances in the continuum. For the example of ^{10}Li the corresponding

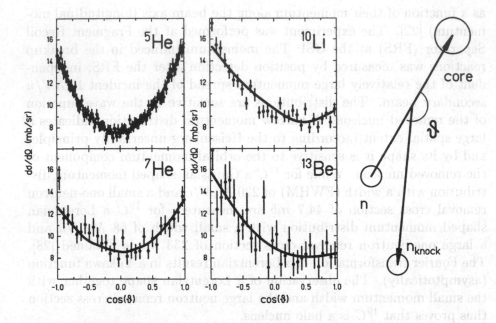

Figure 3 Angular correlation between the direction of relative movement of the core fragments and the neutrons and the direction of momentum of the residual intermediate system $^{5,7}He$, ^{10}Li and ^{13}Be obtained with 240 MeV/u ^6He, 226 MeV/u ^8He, 264 MeV/u ^{11}Li and 287 MeV/u ^{14}Be beams, respectively. A schematic diagram of the reaction is shown on the right side of the figure. On the left side the angular correlations resulting from a single neutron knockout from $^{6,8}He$ are shown [8]. The experimental data e.g. in case of ^5He can be well described by a dominant ^5He ground state configuration plus a 7 % admixture [30] of the first excited $1/2^-$ state (solid line). In case of ^{10}Li and ^{13}Be an asymmetric angular correlation is obtained showing the presence of interference between s and p states of different parity. For ^{10}Li approximately equal strengths of s and p states are existing in its wave function [9, 29].

excitation energy spectrum shows contributions from s- and p-waves [27] in a similar ratio as observed in the momentum distribution [29].

The different angular momentum components could also be determined in a more model-independent way by measuring the angular correlation between the momentum of the recoiling intermediate nucleus and the relative momenta of its decay products. Such correlations are shown in Fig. 3. In $^{5,7}He$ only $l = 1$ neutrons are involved yielding a symmetric but anisotropic distribution [30]. An asymmetric distribution resulting by the interference of states with different parity [31] is observed for ^{10}Li and ^{13}Be reflecting the involvement of $l = 0$ and $l = 1$ components. For instance for ^{10}Li due to the very sensitive interference term, by measuring the angular correlations, the amplitudes and relative phase of the interfering s and p waves could be determined simply by angular momentum and spin coupling. Equal weights for s and p components and a relative phase of $45(10)°$ were obtained [29].

6. Continuum Excitation

The extremely weak binding of halo nuclei gives rise to strong effects associated with coupling to the continuum. Therefore, it is essential to study both resonant and non-resonant continuum transitions and to decompose the multipole strength. Inelastic scattering in the nuclear and the Coulomb fields of light and heavy targets, respectively, gives access to the relevant information. In case of two-neutron halo nuclei the study of inelastic scattering is more complicated because no two-body exit channel exists and the three-body exit channel (core+n + n) has to be measured always in a kinematically complete experiment. This is illustrated for the example of ^6He in Fig. 4. It shows the excitation energy spectra of ^6He as obtained with C and Pb targets, reconstructed by the invariant mass method from the measured $\alpha + n + n$ coincidences. The excitation spectrum in nuclear inelastic scattering which dominates for the C target is presented on the left side of the figure. The peak structure at $E_x = 1.8\ MeV$ belongs to the first excited 2^+ state in the α+n+n continuum. Analysis of the angular correlation gave evidence that the main component of this state is connected with two neutrons being in a singlet state [32]. The $B(E2)$ value could be determined to $B(E2, 0^+ \rightarrow 2^+) = 3.2(0.6)\ e^2 fm^4$ [24].

The smooth continuum cross section at higher energies is composed of different multipolarities (mainly $E2$) which could be disentangled by inelastic scattering angular distributions [32]. On the right side of Fig. 4 the excitation energy spectrum obtained with the Pb target is presented. Most of the inelastic cross section ($\sigma_{inel} = 0.65(11)\ b$) results from electromagnetic excitation ($\sigma_{em} = 0.52(11)\ b$) and the observed strength can be attributed to electric dipole excitation. The cross section for an electric dipole transition is well described by a three-body model [12]. In contrast to stable nuclei a large fraction of the electric dipole strength is located at low excitation energies and 10% of the energy-weighted Thomas-Reiche-Kuhn sum rule is found below $E_x = 5\ MeV$. Up to 10 MeV excitation energy 100 % of the non-energy-weighted cluster sum rule is exhausted. From this a root-mean-square distance $r_{\alpha-2n} = 3.4(4)\ fm$ [24] between the α-particle and the two valence neutrons can be extracted. Together with a measured halo radius of $3.0(3)\ fm$ [19] a distance between the two neutrons of $4(1)\ fm$ is obtained. Similar results were obtained for ^{11}Li and ^{14}Be [9, 27]. The angular distribution

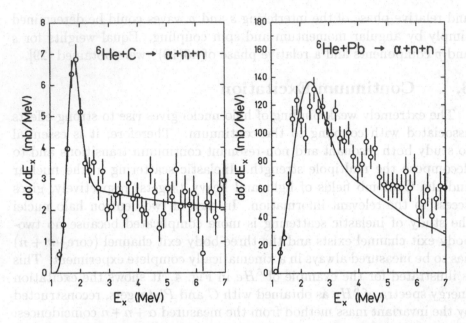

Figure 4 Excitation energy spectra of ^6He deduced from the invariant mass of the $\alpha + n + n$ exit channel obtained with a C target (left) and a Pb target (right) for a ^6He beam energy of 240 MeV/u [24, 32]. The differential cross sections $d\sigma/dE_x$ are corrected for detector acceptances. Left: At the excitation energy of $E_x = 1.80$ MeV the known 2^+ resonance is present. The solid line is a guide for the eye. The continuum strength at high energies appears more or less structureless. Right: The full line represents calculated electromagnetic cross sections using the $dB(E1)/dE_x$ distribution from a three-body model [12] folded with the experimental resolutions.

of inelastic scattering on the lead target shows Fraunhofer oscillations. Both the Coulomb and the nuclear part of the interaction potential play a role. The oscillation clearly indicates a delayed nature of the resonant breakup [32]. Moreover, for electromagnetic excitation, the two-body correlations $\alpha - n$ and $n - n$ could be investigated in the $\alpha + n + n$ channel. Comparison between the data for ^6He and ^8He has shown that the structure of the latter is very complex and supports its description as a five-body system [8]. The same holds for ^{14}Be [9].

7. Astrophysical Aspects

In the past years it was discussed that the post collapse phase in a type-II supernova offers the ideal site for the r-process forming the heavier elements in the outer shells of the collapsing iron core. In the preceeding α-process, elements up to $A \leq 100$ are built. The bottleneck in this nucleosynthesis process is the formation of nuclei with

$A \leq 9$ from nucleons and α-particles. Two-step processes such as $^4He(2n, \gamma)^6He$ and $^6He(2n, \gamma)^8He$ were considered to be potentially relevant to bridge the instability gap at $A = 5$ and $A = 8$ [33, 34]. In the Coulomb breakup on a lead target the exactly inverse process, i.e. absorption of a (virtual) γ quantum followed by two-neutron emission is measured. The model-dependent assumptions on which conclusions are based so far are experimentally checked [24]. However, it could be shown that the $^4He(2n, \gamma)^6He(\alpha, n)^9Be(\alpha, n)^{12}C$ process contributes only with 0.5 % as well as the triple α-process to the formation of ^{12}C with the same percentage. These processes cannot compete with the $^4He(\alpha n, \gamma)^9Be(\alpha, n)^{12}C$ chain in a type-II supernova scenario, but other scenarios such as production of r-process elements in the coalescence of two neutron stars are still under discussion for which the relevance of two-neutron capture processes is to be explored [35].

8. Outlook

Halo nuclei offer excellent experimental opportunities to study marginally bound quantum systems. To bind two-nucleon or multinucleon systems three- or eventually many-body forces become important reflecting the correlations of nucleons in nuclei. First steps have been done in studying extensively e.g. proton-neutron and alpha-neutron systems ($^2H - {}^7H$; $^5He - {}^{10}He$, respectively). With the new generation of rare isotope accelerators and with the new developments of experimental equipment this exploration can be extended to a much wider mass region. The extension can serve for developing new approaches in nuclear studies and provides a step towards a better understanding the many-body problem in quantum mechanics in fairly exotic systems.

Acknowledgments

The author is grateful to all collaborators in the experiments performed at GSI. He is in particular indebted to L. Chulkov, H. Emling, H. Geissel, B. Jonson, H. Lenske, G. Münzenberg, I. Mukha and H. Simon for many stimulating discussions, and to A. Richter for a critical reading of the manuscript.

This work was supported by the German Federal Minister for Education and Research (BMBF) under the contract 06DA915I and by GSI under the contract DARIK.

References

[1] I. Tanihata, J. Phys. G: Nucl. Part. Phys. 22 (1996) 157

[2] P.G. Hansen, A.S. Jensen and B. Jonson, Ann. Rev. Nucl. Part. Sci. 45 (1995) 591

[3] B. Jonson and K. Riisager, Phil. Trans. R. Soc. Lond. A296 (1998) 2063

[4] A.S. Jensen and K. Riisager, Phys. Lett. B480 (2000) 39

[5] J.S. Al-Khalili et al., Phys. Rev. C54 (1996) 1843

[6] U.M. Efimov, Comm. Nucl. Part. Phys. 19 (1990) 271

[7] I. Tanihata et al., Phys. Lett. B289 (1992) 261

[8] K. Markenroth et al., Nucl. Phys. A, in press

[9] H. Simon et al., to be published

[10] J. Wurzer and H.M. Hofmann, Phys. Rev. C55 (1997) 688

[11] S.A. Fayans, Phys. Lett. B267 (1991) 443

[12] B.V. Danilin et al., Nucl. Phys. A632 (1998) 383

[13] C. Dasso, S. Lenzi and A. Vitturi, Nucl. Phys. A611 (1996) 124

[14] P.G. Hansen and B. Jonson, Europhys. Lett. 4 (1987) 409

[15] T.A. Navin et al., Phys. Rev. Lett. 85 (2000) 266

[16] A. Ozawa et al., Phys. Rev. Lett. 84 (2000) 5493

[17] E. Arnold et al., Phys. Lett. B281 (1992) 16

[18] M.J.G. Borge et al., Phys. Rev. C55 (1997) R8

[19] G.D. Alkhazov et al., Phys. Rev. Lett. 78 (1997) 2313

[20] I. Tanihata et al., Phys. Rev. Lett. 55 (1985) 2676

[21] T. Kobayashi et al., Phys. Rev. Lett. 60 (1998) 2599

[22] T. Baumann et al., Phys. Lett. B439 (1998) 256

[23] H.G. Bohlen et al., Nuovo Cimento A111 (1998) 842

[24] T. Aumann et al., Phys. Rev. C59 (1999) 1252

[25] G. Schrieder, Prog. Part. Nucl. Phys. 42 (1999) 27

[26] T. Suzuki et al., Phys. Rev. Letter 75 (1995) 3241

[27] M. Zinser et al., Nucl. Phys. A619 (1997) 151

[28] D. Cortina-Gil et al., Europhys. J., to be published

[29] H. Simon et al., Phys. Rev. Lett. 83 (1999) 496

[30] L.V. Chulkov and G. Schrieder, Z. Phys. A359 (1997) 231

[31] L.C. Biedenharn and M.E. Rose, Rev. Mod. Phys. 25 (1953) 729

[32] D. Aleksandrov et al. Nucl. Phys. A669 (2000) 51

[33] J. Görres et al., Phys. Rev. C52 (1995) 2231

[34] V. Efros et al., Z. Phys. A355 (1996) 101

[35] H. Oberhummer et al., Nucl. Phys. A, to be published

Molecular Structure in Deformed Neutronrich Light Nuclei

W. von Oertzen

Hahn Meitner Institut, D-14109 Berlin, Glienicker Str. 100, Germany and
Fachbereich Physik, Freie Universität Berlin
oertzen@hmi.de

Keywords: Nuclear structure, neutronrich nuclei, alpha particle clustering, molecular orbitals, rotational bands

Abstract   Nuclear clustering and molecular resonances based on α-particles and strongly bound substructures in nuclei with $N = Z$ have been studied since many decades. Excited states close to the decay thresholds show particularly strong clustering, as described by the Ikeda diagram. This diagram can be extended to neutronrich nuclei, in these cases strongly deformed isomeric states consisting of clusters and loosely bound neutrons will appear. Molecular structures in beryllium isotopes and in other neutronrich light nuclei (carbon and neon) are discussed.

1.   Alpha-Clustering and the IKEDA-Diagramm

Clustering has became manifest in heavy-ion reactions as resonances observed in the excitation functions in various combinations of light α-cluster nuclei in the energy regime from the barrier up to regions with excitation energies of $30-50$ MeV. In particular in cases like $^{12}C + {}^{12}C$ [1] and $^{24}Mg + {}^{24}Mg$ [2] the structure of highly excited states can be related to strongly deformed shapes (to super- and hyper-deformation) in the compound nucleus, predicted from cranked α-cluster models [3, 4], and others. Thus in light $N = Z$ nuclei strong deformations and clustering are intrinsically connected, a fact which has been emphasized repeatedly [5, 6]. Ikeda [7, 8] formulated in 1968 a concept, that clustering and deformations in nuclei will become relevant for states in nuclei, which are close to the thresholds for the decomposition into clusters. This concept is depicted as the *Ikeda-diagram* in Fig. 1, where the threshold energies for the decay into substructures are given in MeV. We observe a systematic change in structure related to

55

D. N. Poenaru et al. (eds.), Nuclei Far from Stability and Astrophysics, 55–66.
© 2001 *Kluwer Academic Publishers. Printed in the Netherlands.*

α-clustering and heavier clusters. The concept of alpha clustering shows its power also in describing details of nuclear spectroscopy in light nuclei [10], as well as in heavier nuclei [9]. With the increased interest in the structure of neutronrich light(exotic) nuclei, this concept can be extended by adding neutrons and considering loosely bound threshold states. We expect to observe molecular structures based on clusters and on covalently bound neutrons [11]. The covalent binding of neutrons brings us to a new diagramm *(Covalent Molecular Binding diagramm): covalently bound molecular structures will appear in neutronrich nuclei close to the thresholds for decomposition into neutrons, α-clusters, and other clusters.*

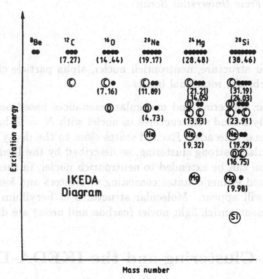

Figure 1 The Ikeda-Diagram for light α-cluster nuclei. The threshold energies for the decomposition into clusters are indicated.

2.   Neutronrich Light Nuclei

The last decade has witnessed a dramatic increase of interest in the structure of neutronrich nuclei [10]. One of the most dramatic aspects of these " exotic nuclei" is the weak binding energy of the last neutron, which for $l = 0$ (orbital angular momentum) gives rise to the "Halo"-phenomenon. More peculiar are the observations of three-body states, also refered to as Borromean nuclei [12]. In this case the total nuclear system is bound, although the constituent 2-body-subsystems are unbound. Most famous examples are ^9Be, for which the two subsystems ^8Be and ^5He are unbound, and ^{11}Li. The binding of the three-body system is obtained by a purely quantal effect, which is the sharing of the valence neutron between two strongly bound centers. This sharing of valence particles is the essence of molecular covalent binding. A second important effect observed in calculations for neutronrich light nuclei by Kanada-En'yo and H. Horiuchi [13, 14] based on the method of Antisymmetrized Molecular Dynamics (AMD), is the fact that neutron excess seems to increase the deformation of the cores, and to favor intrinsic clustering. This is explained as being due to the strength of the neutron-proton interaction, which is optimized for larger neutron excess by offering a larger surface with a strongly deformed core.

3.   Alpha-Nucleus Potentials

The formation of molecular structures relies on special properties of the nuclear potential between the two cores (clusters), namely the occurrence of a "molecular" potential, with attraction at large distances and a repulsion at small distances. It turns out that there are two quite conspicuous cases, where *local potentials* have only a small attractive part and a strong repulsion at small distances.

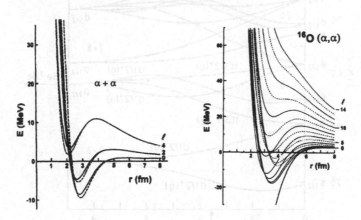

Figure 2 Two examples of shallow local potentials for the interaction of α-particles with α-particles — the case of ^8Be, and with ^{16}O — the case of ^{20}Ne.

These two cases are shown in Fig. 2, which shows the cases of α-particles interacting with α-particles and with ^{16}O nuclei. The curves are actually the result of a reduction of the potential to a phase equivalent shallow potential, which does not exhibit anymore unphysical bound states. For the case of the $\alpha+\alpha$ potential the molecular potential created by Ali and Bodmer [15] for ^8Be (dashed curves) is seen to coincide with the supersymmetric local potential. For a more detailed coverage of this point and a discussion of the numerous literature I refer to Ref. [9]. The repulsion at small distances can be interpreted as the effect of the Pauli blocking, in the case of the $\alpha+\alpha$ potential the nucleons of the second cluster have to move up into the next major shell.

4.   Symmetries of Molecules

Over the last decades the molecular orbital model [17–20] for neutrons has been applied to nuclear collisions, in which single nucleons, mostly neutrons are exchanged between two nuclear cores. This has been done for low energy reactions and for cases, where strongly bound cores (typically alpha-cluster nuclei) and valence nucleons can be defined in a clear way. In a collision process this approach corresponds to a choice of basis states, which takes the two-center structure of the nuclear reaction into account and the couplings appear as radial and Coriolis couplings [18]. It can be seen as an equivalent approach to the coupled reaction channel calculations based on the asymptotic basis states of the separated nuclei.

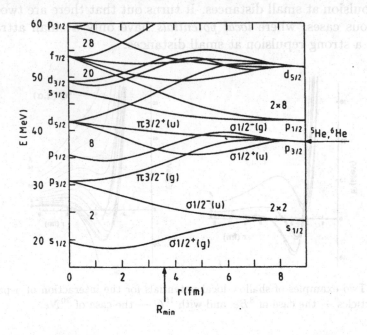

Figure 3 Correlation diagram of molecular orbitals in a two-center shell model picture; the molecular orbitals are labeled by their quantum numbers. R_{min} is the distance (3.7 fm), where the minimum of the $\alpha - \alpha$ potential occurs.

The most extensive study of molecular orbitals has been done for the $^{12}C + ^{13}C$ system and some heavier systems [17, 18]. These studies were limited to scattering states, because the core-core potential between the two heavier nuclei generally becomes strongly attractive and strongly absorptive, once the Coulomb barrier is reached.

In this molecular orbital model and in the two-center shell model (in this case not only valence nucleons, but all nucleons are considered), a correlation diagram has to be drawn. Such diagrams are well known from atomic physics (see also Ref. [16], p.328). The diagram merges at small distances with the Nilsson-diagram of the deformed compound nucleus.

The Molecular Two Center States

$$\Phi_{p,K}^{J,\pi} = 1/[2(1+(-)\Delta_K^p(R))]^{1/2} \{\phi_1 + (-)^P \phi_2\}$$

Quantum numbers :

p=(+), gerade; p=(−), ungerade;

Parity, $\pi = (-) \times p$; l- orbital angular momentum

Projection, K , of : J - total angular momentum

Projection of l: σ- orbits, m = o

π- orbits, m = 1, etc

Figure 4 Construction of molecular orbitals in a two-center shell model picture; the molecular orbitals are labeled by their quantum numbers.

The two-center state is obtained as a linear combination of nuclear orbitals (LCNO). The molecular orbitals are classified according to the well known quantum numbers of molecular valence states, see Figs. 3, 4: the K-quantum number for the projection of the spin, the σ and π orbitals for the $m = 0$, and $m = 1$ projections of the orbital angular momenta l, the parity, and the *gerade* and *ungerade*-symmetry due to the identity of the two molecular cores. With this correlation diagram we are able to discuss the structure of the isotopes ^9Be, ^{10}Be and ^{11}Be, if we consider [11] the population of the molecular orbits in beryllium for a distance corresponding to the potential minimum, or in the Nilsson model for a deformation of $\beta_2 = 0.6-0.7$.

5.   Beryllium and Boron Isotopes

The ^8Be - nucleus can be regarded as the first superdeformed nucleus with an axis ratio of 2:1. The ^8Be is unbound, and it has been known for more than two decades, that the ^9Be nucleus is an example of molecular binding, where two α-particles are bound by the covalent neutron. We note (see Fig. 3), that at the distance, where the α - α potential has its minimum, the $K = 3/2^-$ orbit crosses the $K = 1/2^+$ orbit and becomes the lowest state at the smaller distances. Thus the unusual level sequence of ^9Be is well understood. In a systematic survey of the structure of the beryllium isotopes [11] all(!) known states of the isotopes ^9Be, ^{10}Be and ^{11}Be can be grouped into rotational bands with molecular configurations by using the correlation diagram of Fig. 3. The structure of the states in the Be-isotopes is thus determined by the "driving forces" of the evolution of the π and σ orbitals as function of distance between the two cores. This information on excited states in the isotopes of ^9Be, ^{10}Be and ^{11}Be is compiled in Fig. 5, where the low lying states of these isotopes are grouped together to form rotational bands.

The ground state of ^{10}Be can be interpreted as a $(\pi)^2$ configuration with smaller deformation, whereas the excited 0^+ at 6.179 MeV in ^{10}Be is interpreted as the $(\sigma)^2$ configuration, which attains its lowest energy at much larger distances of the two α-particles with a corresponding much larger moment of inertia. For the $K = 1/2$ bands a Coriolis decoupling pattern is observed in ^9Be and in ^{11}Be with the same decoupling parameter. Note the $K = 3/2^-$ band in ^{11}Be, with excited states [22] established to high excitation energy, and which starts at 3.95 MeV, more than 3 MeV above the particle threshold, see Fig. 5. Inspecting the correlation diagram we recognize that this band must have the covalent neutron configuration $(\pi)x(\sigma)^2$, which results in a particular stability and a large deforma-

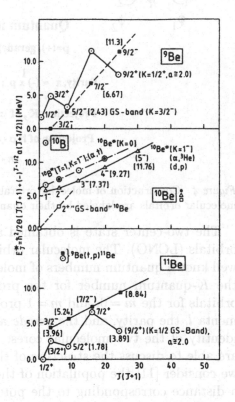

Figure 5 Energy levels of the isotopes of ^9Be, ^{10}Be and ^{11}Be grouped into rotational bands plotted as function of angular momentum $J(J + 1)$.

tion due to the properties of the $(\sigma)^2$-band. The moment of inertia of this band is thus similar to that of the excited $K = 0^+$ band in ^{10}Be.

These molecular states must me considered as true super-deformed shape isomers. This aspect is observed in a dramatic way through the selectivity for the population of states in ^{11}Be. We have studied [22] the population of states in ^{11}Be by, (i) single nucleon transfer on ^{10}Be, and

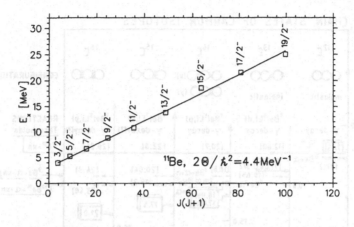

Figure 6 Excitation energies of states of the $K = 3/2^-$ band in ^{11}Be (populated in the 2n-transfer reaction on ^9Be) plotted as function of spin $J(J + 1)$, the spins are presumed to follow the pattern of a rotational band.

in, (ii) two–neutron transfer on ^9Be: whereas for the reaction on ^{10}Be only three low lying single particle states (with the configurations $p_{1/2}$, $p_{3/2}$ and $d_{5/2}$) are observed, the two–neutron transfer populates strongly the higher spin members of the $K = 3/2^-$ rotational band, which have been completely absent in the one neutron transfer spectrum.

The two–neutron transfer spectrum shows resonant structure, which extends up to a projected spin value of 19/2. This value can be obtained in a deformed oscillator-shell model by counting the spins of the individual three valence nucleons: the maximum spin of the valence neutrons would be $[(3/2(p3/2) + 8/2(\text{for two neutrons with } j= 5/2)] = 11/2$; the total spin can than be reached by adding the maximum spin of the ^8Be-system, which is known to be 4, giving the total sum of 19/2!

The molecular properties of the states in Be-isotopes have found striking confirmation from "model independent" calculations using the method of Antisymmetrized Molecular Dynamics (AMD) of Kanada-En'yo and Horiuchi [14, 23]. Particularly impressive are the density distributions obtained for the ground state 0^+ and second 0^+ state of ^{10}Be, which have been obtained by projection on spin and parity [23]. The negative parity band is based on a peculiar mixed (π)x(σ) configuration.

The molecular structure is still visible in heavier isotopes of Be, like ^{12}Be, where the persistence of the strong clustering of the α-particles prevents the formation of a closed shell with $N = 8$. At higher excitation energies experiments suggest the formation of a rotating Dimer decaying into two ^6He fragments [24]. Using the concept of isospin the properties of these strongly deformed configurations can be also traced to excited states of isotopes of boron [11].

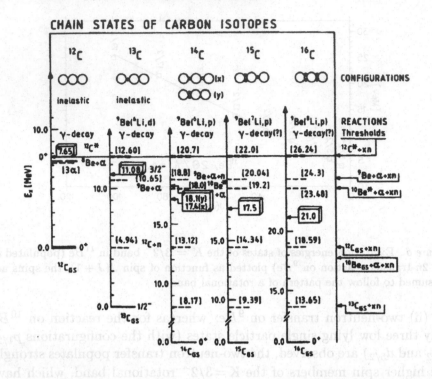

Figure 7 Chain states of carbon isotopes with covalent valence neutrons. The thresholds for the decay into α-particles and neutrons are aligned to the same level.

6.   Polymers

Based on the covalently bound structures of the system of two α-particles we can continue the discussion of extreme deformations with three α-particles by showing the schematic diagrams for the shape-isomers of the carbon isotopes in Fig. 7. The three-α-particle chain state, corresponding to a 3:1 deformation is taken to be the second 0^+ in ^{12}C (a more detailed discussion is given in Ref. [11]). The Trimers are build with the covalent binding of two times two (or three) neutrons by using the well established knowledge of the covalent bonds in ^{10}Be and ^{11}Be. Actually, we may expect that covalently bound shape isomers in ^{16}C are indeed rather stable configurations (as it is the case in atomic molecules). Two bonds in the covalent configuration between the α-particles will influence each other so as to increase the total binding energy of the chain, which is expected to be more than 5 MeV (see also [25]).

There are many other nuclear molecules and polymers, which can be build on repulsive α-nucleus potentials and additional covalent neutrons. A particular conspicuous example is discussed in the next section [26].

7.   Asymmetric Shapes, ^{21}Ne

The structure of α-cluster nuclei, in particular of ^{20}Ne, has been discussed in the $(\alpha + {}^{16}O)$ cluster model over the last decades [5]. The particular feature of the two bands with $K = 0^+$ and $K = 0^-$ is well described by a $(\alpha + {}^{16}O)$ local potential; this point has been summarized recently by Ohkubo [9]. A peculiar feature in nuclear structure arises for an intrinsic asymmetric shape, which has no good parity and K-quantum number zero, $K = 0$. We have a band with states of alternating parity, $J^\pi = (0^+, 1^-, 2^+, 3^-,$ etc). The fact that this band is split in ^{20}Ne is due to a large probability to change between the two shapes by some tunneling process. This case is discussed for example in Vol. II of the book of Bohr and Mottelson [27] and in the book of Herzberg [16] (pp. 129). A particular case occurs for $K \neq 0$, where two independent bands will occur, a phenomenon also known as *signature splitting* [16, 27].

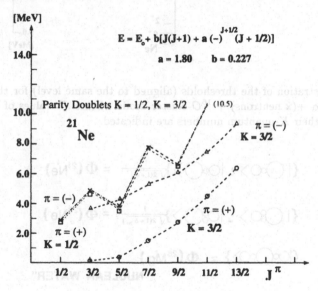

$$E = E_o + b[J(J+1) + a\,(-)^{J+1/2}\,(J + 1/2)]$$

$$a = 1.80 \qquad b = 0.227$$

Figure 8 Plot of the excitation energies of known states in ^{21}Ne rotational bands as rotational bands. The compilation is made in order to illustrate the parity-doublets expected from signature inversion of the structures shown in Fig. 10

In order to study the structure of ^{21}Ne in a model which consists of one neutron bound in a *two-center system with two unequal cores*, namely α and ^{16}O, we have to look into the structure of an asymmetric top with $K \neq 0$, with an intrinsic violation of parity, as illustrated in Fig. 10. Important features for the formation of this covalent molecule can be deduced from an energy diagram, as shown in Fig. 9, which shows the positions of the single particle orbits of the neutron at the two centers, namely for α and ^{16}O. For this system a *remarkable coincidence* occurs:

the $p_{3/2}$ resonance of ^5He at 890 keV, is almost degenerate with the $d_{3/2}$-state also a resonance, at 941 keV in ^{17}O. We note that this particular situation is known as the *quasi-resonance condition* for the sharing of valence particles in molecular science.

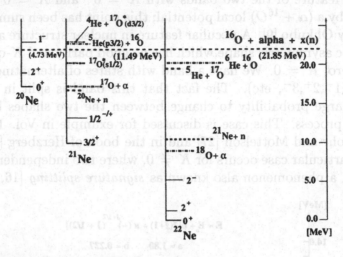

Figure 9 Illustration of the thresholds (aligned to the same level) for the neon isotopes for the α +(x neutrons) + ^{16}O-cluster model. Some J^π values of band heads in ^{21}Ne with their K-quantum numbers are indicated.

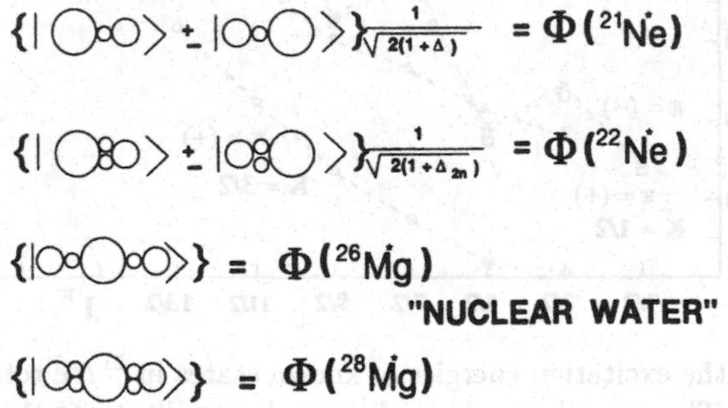

$$\{|\bigcirc\!\!\infty\rangle \pm |\infty\!\!\bigcirc\rangle\}\tfrac{1}{\sqrt{2(1+\Delta)}} = \Phi(^{21}\dot{N}e)$$

$$\{|\bigcirc\!\!8\!\!\circ\rangle \pm |\circ\!\!8\!\!\bigcirc\rangle\}\tfrac{1}{\sqrt{2(1+\Delta_{2n})}} = \Phi(^{22}\dot{N}e)$$

$$\{|\infty\!\!\bigcirc\!\!\infty\rangle\} = \Phi(^{26}\dot{M}g)$$

"NUCLEAR WATER"

$$\{|\circ\!\!8\!\!\bigcirc\!\!8\!\!\circ\rangle\} = \Phi(^{28}\dot{M}g)$$

Figure 10 Schematic illustration of the structure of molecular shape isomers in the neon and magnesium isotopes based on the (α + ^{16}O-cluster) model plus some neutrons. For each K-quantum number $\neq 0$ a parity doublet will occur. The splitting of the bands will be proportional to the overlap Δ.

For the two-center states in ^{21}Ne we take these two $j = 3/2$ orbits and construct the $K = 1/2$ and $K = 3/2$ states. There is no reflection symmetry in these intrinsic states, therefore states with definite parity are constructed by using superpositions, which produce a definite signature. These configurations are schematically shown in Fig. 10. I use the concept of parity doublets [16, 27] to establish rotational bands as shown in Fig. 8. The outcome of these considerations is that there must be (in the cluster model, with $\alpha + {}^{16}O$ + neutron) two bands with $K = 1/2$ and $3/2$ each with the parities ($+$ and $-$). This situation with a finite K-value is well known in asymmetric atomic molecules as cited by Herzberg [16]. The interesting question of the signature splitting and of the purity of these configurations must be answered by experiments.

Gamma-spectroscopy of the neon isotopes has been reported mostly more than 10 – 20 years ago [28, 29]. In Fig. 8 states in ${}^{21}Ne$ are plotted as rotational bands in such a manner that the parity doublets, which may occur due to the signature splitting become visible [26]. The experimental systematics suggests the existence of two $K = 1/2$ bands with opposite parity and *very small splitting*. This result could be interpreted as due to a rather stable molecular structure based on the σ-bond between the α-particle and the ${}^{16}O$ clusters. This molecular interpretation is supported by an interesting finding [29] about the extreme retardation by *four orders of magnitude* of the $E1$-transition from the excited $1/2^-$- state to the $3/2^+$- ground-state, when compared to the expectations deduced from conventional configuration assignments. The states of the $K = 3/2$ bands, which most likely will exhibit strong mixing with other configurations have large splitting.

Applying the observed results and extending the cluster model, we can construct (also in accordance with the Ikeda-diagramm) more extended molecules, whose band heads can be estimated by using the information from the neon-isotopes. The existence of intrinsically symmetric stable molecules of "Nuclear Water", $(He)_2O$, can be predicted; their wave functions are schematically shown in Fig 10. All the shapes discussed here are true shape isomers, with large deformations, which can *not* be easily described by a few terms in the Legendre-function expansions for nuclear shapes. Radioactive beam facilities will give access to these nuclei and thus to new regions of deformations in light nuclei and clustering will play the dominant role in this domain.

References

[1] K.A. Erb and D.A. Bromley, Treatise in Heavy Ion Science, Vol.3, p. 201, ed. D.A. Bromley

[2] R.W. Zurmuehle, Proceedings of 5th Conf. Clustering Aspects in Nucl. Physics (1988), J. Phys. Soc. Japan 58 (1989) 37

[3] S. Marsh and W.D. Rae, Phys. Lett. B180 (1986) 185

[4] J. Zhang, et al., Nucl. Phys. A575 (1994) 61

[5] M. Freer, A.C. Merchant, J. Phys. G23 (1997) 261 and refs. therein

[6] B.R. Fulton and W.D.M. Rae, J. Phys. G16 (1990) 333; B.R. Fulton, Z. Phys. A349 (1994) 227 and references therein

[7] K. Ikeda, et al., Prog. Theor. Phys. (Jap), Suppl. (1968) 464

[8] H. Horiuchi, et al., Prog. Theor. Phys. (Jap.), Suppl. 52 (1972) Chapt. 3

[9] S. Ohkubo et al., Prog. Theor. Phys. (Jap.) Suppl. 132 (1998) 1

[10] Proceedings of the Fourth Int. Conf. on Radioactive Beams, OMIYA(Japan), June 1996, Nucl. Phys. A616 (1997) 1c

[11] W. von Oertzen, Z. Phys. A354 (1996) 37 and Z. Phys. A357 (1997) 355; Il Nuovo Cimento 110A (1997) 895

[12] M.V. Zhukov, et al., Phys. Reports 231 (1993) 151

[13] H. Horiuchi, Y.K. Kanada-En'yo, Nucl. Phys. A616 (1997) 394c

[14] Y.K. Kanada-En'yo, H. Horiuchi, A. Ono, Phys. Rev. C52 (1995); Phys. Rev. C56 (1997) 1844

[15] S. Ali and A.R. Bodmer, Nucl. Phys. 80 (1966) 99

[16] G. Herzberg, Molecular Spectra and Molecular Structure Vol. I, Spectra of Diatomic Molecules (1950), van Nostrand Comp. Inc., Princeton

[17] W. von Oertzen, Nucl. Phys. A148 (1970) 529; W. von Oertzen and H.G. Bohlen, Phys. Rep. C19 (1975) 1

[18] B. Imanishi and W. von Oertzen, Phys. Rep. 155 (1987) 29

[19] J.Y. Park, W. Greiner and W. Scheid, Phys. Rev. C21 (1980) 958; A. Thiel J. Phys. G16 (1990) 867

[20] M. Seya, M. Kohno, S. Nagata, Prog. Theor. Phys. 65 (1981) 204

[21] H. Horiuchi, et al., Z. Phys. A349 (1994) 142

[22] H.G. Bohlen, et al., Il Nuovo Cimento 111A (1998) 841

[23] Y.K. Kanada-En'yo, H. Horiuchi and A. Dote, J. Phys. G24 (1998) 1499; Phys. Rev. C60 (1999) 064304

[24] M. Freer, et al., Phys. Rev. Lett. 82 (1999) 1383

[25] N. Itagaki et al., contribution to these proceedings, 2000

[26] W. von Oertzen, to be published (2000)

[27] A. Bohr and B. Mottelson, Nuclear Structure, Vol. II (1975), Benjamin, Inc., Reading Mass.

[28] P.M. Endt, Nucl. Phys. A521 (1990) 1 and references therein

[29] A.A. Pilt et al., Canad. J. of Physics 50 (1972) 1286

Gamma-Ray Spectroscopy of Exotic Nuclei: Physics and Tools

G. de France

GANIL, BP 55027, F-14076 Caen Cedex 5, France

defrance@ganil.fr

Keywords: Gamma-ray, spectroscopy, magic numbers, exotic nuclei, EXOGAM

Abstract Radioactive beams obtained via fragmentation of the projectile on a primary target have shown to be a powerfull tool to produce exotic nuclei. To get beams of exotic nuclei, new facilities have been developped recently. At GANIL the SPIRAL facility is ready and will deliver the first radioactive beam before the end of 2000. However, when looking into details, severe limitations might reduce the hopes if R&D efforts are not done. The low intensities, wide range of experimental conditions, and the radioactive nature of the beam itself are some of these constraints. The design of the future (and already under construction) spectrometers intended to be used with radioactive beams integrates these new difficulties in their technical design. Based on simple realistic experiments, the physics will be described and the expected performance of new generation detectors will be discussed. New ideas for large efficiency spectrometers will be briefly described, based on the experience which has been acquired through the R&D performed in some large collaborations.

1. Introduction

Atomic and nuclear physics have always been looking to "regularities" among the elements to get more insight in the understanding of nature. The more famous example is of course the periodic table of the elements first established by D. Mendeleyev in 1869. In grouping the elements by their similarities in terms of radii, boiling points, ionization potential, electronegativity, electron affinity, etc. Mendeleyev realized what he called the periodic law: "the properties of the elements are a periodic function of their atomic masses" which was a major step forward in the understanding of elements properties. There were however some inconsistencies in this arrangement and it was not until 1914 that H.G.J. Moseley was able to determine experimentally the atomic number of the elements.

67

D. N. Poenaru et al. (eds.), Nuclei Far from Stability and Astrophysics, 67–78.
© 2001 *Kluwer Academic Publishers. Printed in the Netherlands.*

He then rearranged the classification according to increasing Z and this cleared up the observed inconsistencies. It is indeed fascinating to observe that all the chemical or atomic quantities mentioned above are increasing when going from right to left and top to bottom of the periodic table. In the nuclear world, one of the first observed "regularity" was the two neutron separation energy (S2n) revealing in a spectacular way for example the well known magic gaps and already pointing towards pairing with the odd-even staggering in Sn versus S2n. What we call "regularities" translate directly in "symmetry" in nuclear physics and is obviously associated to quantum physics which strengthen the interest in searching for new "regularities". The level schemes of even-even isotopes are also a striking example. Spins and parities of all the ground states are 0^+ and the first excited states of this class of isotopes is always 2^+ with the exception of ^{16}O, ^{40}Ca, ^{90}Zr and ^{208}Pb which all are magic. Along the same line the energy of the first 2^+ states illustrates very clearly the occurrence of magic numbers among the elements.

Many observation of these regular behaviors are well explained in terms of single particle features. As an example, one can mention the clusterization of long-lived isomers in odd-A nuclei around closed shells, a phenomenon well explained by shell model. In the vicinity of magic numbers, single particle orbitals with large difference in angular momentum may lie quite close to each other. Around $Z = 50$, for instance, the $\pi p_{1/2}$ and $\pi g_{9/2}$ levels are responsible for the well known $M4$ isomerism. A similar feature is observed for the $E2$ isomers around $N = 50$.

Another interesting quantity reveals the influence of closed shells: the deformation. Not only the static ground state quadrupole deformation which is beautifully related to magicity, but also the deformation deduced from excitation spectra related to rotational motion. In this latter case, this is a highly collective behavior which generates spectra with excitation energies proportional to $J(J + 1)$ like those observed for non spherical molecules rotating around an axis perpendicular to the symmetry axis. Deformation affects in a dramatic manner single particle spectra as illustrated for example in single particle energies obtained from axially symmetric deformed harmonic oscillator. This causes some specific orbitals (the intruders) diving from a major shell N into lower ones. The coexistence of these levels of various major shells occasionally generates "new" magic number at large deformation like at 2 : 1 axis ratio. And indeed in the vicinity of these gaps, the nuclei are sufficiently stabilized to allow superdeformed bands to be observed. The interesting point is that some orbitals have a dominant influence over the whole system but the behavior of superdeformed nuclei is primarily of collective nature. This is probably the most spectacular example of a delicate interplay between single particle and collective behavior.

Today as in the past, physics in the vicinity of magic numbers, their robustness as a function of mass, isospin and the search for new symmetry along the chart of nuclei is a major input for nuclear structure. This is true in a number of domains which I will describe in the first part of this lecture. In the second part I will detail the new tools which are now mandatory in gamma-ray spectroscopy to go beyond what is feasible today. More precisely new arrays which are designed to work with radioactive beams are under construction. EXOGAM is one of these and I will use it as a typical example of these new tools.

2. Physics at the Closed Shells

Identification of New Magic Elements.   The discovery of new magic numbers is first of all mandatory to test and develop nuclear models. This is a basic milestone to understand the evolution of shell structure with exoticity. To get more detailed informations in terms of structure of these nuclei, it is necessary to study the neighboring nuclei of the magic elements. This is the case if one wants to learn about residual interactions.

The latest doubly magic elements discovered are ^{100}Sn [5, 6], ^{78}Ni [8] and ^{48}Ni [7]. ^{100}Sn is the heaviest $N = Z$ doubly magic nuclei stable against p-emission from the ground state. The specific interest in the vicinity of this isotope is the study of p-rich nuclei and the stringent test of mass models. Mapping the p-drip line is also of prime importance to learn about rp-process and the associated waiting points. Just at the opposite, ^{78}Ni is located in the r-process path region and it allows test of mass models in the neutron rich part of the chart. It is also the doubly magic nucleus which exhibits the largest A/Z ratio: $A/Z = 2.79$. Finally, the lately discovered ^{48}Ni has several interesting properties. It first of all has the largest isospin projection T_z value ($T_z = -4$) ever observed; it is also the unique element whose mirror nucleus (^{48}Ca) is stable against particle emission, allowing therefore symmetry studies. The Ni isotope chain comprises 3 doubly magic nuclei ($A = 48, 56, 78$), the two extremes having a difference of 30 neutrons. Finally, it is a good candidate together with ^{45}Fe for direct $2p$ emission from the ground state. This makes many good reasons to have a special look towards these magic elements.

The production method for these extremely rare nuclei consists in the high energy fragmentation of a beam onto a target. For ^{100}Sn at GANIL, a beam of ^{112}Sn at 63 MeV per nucleon was fragmented on a natural Ni target located between the two superconducting solenoids of the SISSI device (see [5]). The fragments are then selected first by a standard alpha shape spectrometers, separated and identified in flight in the LISE spectrometers after a flight path of roughly 118 meters. The identification is made on the basis of redundant informations giving time of flight and energy loss in several detectors. After a drastic rejection of the events which do not match all the conditions, we get the usual energy loss versus time of flight plot also called identification plot. The one obtained in the ^{48}Ni experiment is shown in Fig. 1.

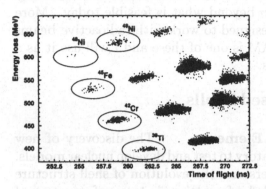

Figure 1 Two dimensional identification plot for energy loss as a function of time of flight in the ^{48}Ni experiment.

It is interesting to see how far we can go with "only" 4 ± 2 good events. ^{48}Ni is calculated to be particle unstable by all common models. Even Z nuclei have a $1p$ emission energetically forbidden as compared to $2p$ emission in the very p-rich elements. Finally the Q-values for $2p$ emission $(Q2p)$ are predicted to be between less than $1\ MeV$ and more than $3\ MeV$ corresponding to lifetimes between 10^{-14} and 1 second! The plot of the expected half life as a function of $Q2p$ is given in Fig. 2 for a large number of usual models. Knowing the transmissions for $^{48,49}Ni$ and the cross section decrease of $\sigma(^{49}Ni)/\sigma(^{48}Ni) \sim 26$, we can deduce the number of counts expected for ^{48}Ni: 5.7 ± 2.2. With a flight path of 118 m, (time of flight 1.32 μs) we are then able to estimate the lifetime for ^{48}Ni: $T_{1/2}(^{48}Ni) \sim 2.6^{+\infty}_{-2.1}\ \mu s$.

Thus the lower limit at a 1σ level is $T_{1/2}(^{48}Ni) \sim 0.5\ \mu s$ which translates to $Q2p \sim 1.5\ MeV$ if we assume a spectroscopic factor of 1. These values lead to the experimental limits given in Fig. 2 which is a real test for the models. Indeed, models which have masses calculated explicitly for this mass region gives Q values closer to the present result than global models which lead to Q values larger than 1.5 MeV and therefore to much shorter half lives. In the case of microscopic calculations, the results depend strongly on the parametrization of the Skyrme interaction.

The basic physics motivation in the search for new magic numbers and the associated test of the models is also at the origin of the quest for superheavy elements. Where ^{100}Sn, ^{48}Ni, and ^{78}Ni test the isospin degree of freedom, the first ones in the p-rich direction, ^{78}Ni in the n-rich one, experiments on superheavy elements are aiming at exploring the next stable shell gaps. These elements are stabilized entirely by shell effects since their macroscopic energy becomes unstable with respect to shape defor-

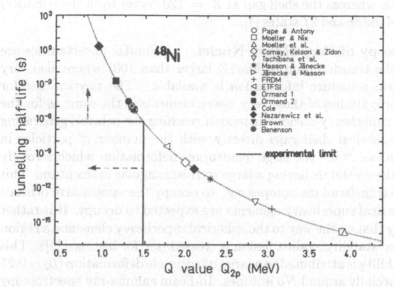

Figure 2 Barrier penetration half-life as a function of the two-proton Q-value.

mation due to the strong Coulomb repulsion. The discovery of elements with $Z = 100$ to $Z = 112$ and the today's controversy regarding the elements 114 [1] 116 [2, 3] and 118 [3] indicates the tremendous activity in this quest and demonstrates the importance of this issue. Indeed, nuclei submitted to extreme conditions of excitation energy, spin, or isospin are of major importance to verify and improve the validity of nuclear models. Testing these models with nuclei having high charge and mass is equally fundamental.

Experimentally, very little is known about the structure of superheavy elements, except the α-decay chains and estimates of the lifetimes. This lack of knowledge is a severe limitation when comparing to model predictions which, in turn, limits the predictive power of the theoretical models. For example, basic features such as the position of the stabilizing spherical shell gap remains completely uncertain. The results from Nilsson-Strutinsky, Hartree-Fock or Relativistic Mean Field theories vary strongly. While most of the calculations predict $N = 184$ as the next spherical neutron gap, the different treatment of the large Coulomb potential and of the spin-orbit interaction leads to uncertainties in the position of proton shell gap, which could be located either at $Z = 114$ (macroscopic-microscopic models) or at $Z = 126$ (self-consistent calculations). It should also be noted that relativistic mean-field calculations predict $Z = 120$ and $N = 172$. The different predictions are due to the fact that the gap at $Z = 114$ originates from the spin-orbit splitting of the 2f state, whereas the shell gap at $Z = 120$ comes from the (smaller) splitting of the $3p$ and $2f$ states [4].

Spectroscopy of Very Heavy Nuclei.   Similar uncertainties are found in the transfermium region (Z larger than 100) where also very little nuclear structure information is available. The physics case for spectroscopic studies of these very heavy elements is the same as for the search for superheavy nuclei. Instead of reaching the orbitals generating the next spherical shell gaps directly with the number of particles in the system, we "use" the large quadrupole deformation which strongly influence these orbitals having a large intrinsic angular momentum. This causes the transfermium isotopes e.g. to occupy the same active orbitals as the spherical superheavy elements are expected to occupy. It is rather clear today that on the way to the spherical superheavy elements, a region of enhanced stability against fission is crossed (see for instance [9]). This gain in stability is attributed to a large quadrupole deformation ($\beta_2 \sim 0.25$ – 0.30) especially around No isotopes. In-beam gamma-ray spectroscopy of ^{254}No has been performed with the help of the recoil-decay-tagging technique [10, 11]. Because of the fourfold magic character of the reaction used ($^{48}Ca + ^{208}Pb$), leading to an exceptionally high cross section (3 μb), indicating again the influence and importance of symmetry in the production mechanism, these studies have led to significant progress in the knowledge of the structure of these heavy nuclei. In the case of ^{254}No, spins up to 20^+ have been established. In this region, and in particular for odd-A nuclei, the high atomic number and low transition energies expected for the excited states lead to very high electronic conversion coefficients.

Recent macro-microscopic calculations with a Woods-Saxxon potential [12] reproduce very nicely the extrapolated energy for the 2^+ state ($E2^+ = 44\ keV$ (exp.), 43 keV (calc.) and the quadrupole deformation ($\beta_2 = 0.27$ (exp.), 0.25 (calc.))

From this example we do see how spectroscopic studies of nuclei, which are produced with a rather large cross section as compared to those measured for spherical superheavy elements, contribute to the fundamental knowledge and understanding of the structure of nuclei that we cannot easily reach.

μ-second Isomers.   Another tool to study magicity are the μ-seconds isomers. As already said, isomers are well understood in terms of shell model and strongly associated to the difference in the angular momenta of the final and initial states connected by γ-rays (or electrons), and thus to the spin-orbit term of the interaction. Around closed shells, the transition are pure or have very small mixing. Measuring transitions which decay in isomeric states in the vicinity of the closed shells, is therefore a particularly appropriate tool to have inputs relevant for the residual interactions and the nucleon effective charge.

These studies have been particularly prolific at GANIL using the LISE spectrometer as shown for example in [13]. In these experiments the recoiling ions are fully stripped which blocks the internal conversion channel and allow the flight of short lived isomers over quite a long distance and small losses. The exotic ions are implanted in a silicon detector and the γ-rays decaying the isomeric level are detected by germanium (Ge) detectors surrounding the implantation detector.

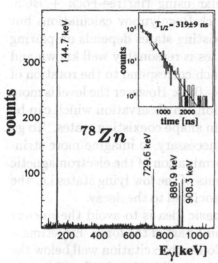

Figure 3 Gamma-ray spectrum measured for the 8^+ isomer decay in ^{78}Zn. The associated lifetime is shown in the insert.

As an example, we can mention the results obtained from the fragmentation of ^{86}Kr at 60.5 MeV/u on a natural Ni target [14]. This experiment was aiming to look at the isomers around the $Z = 28$ shell gap. Several isomers have been identified and studied in some details in ^{78}Zn (see Fig. 3), ^{80}Ge, ^{82}Se, ^{84}Kr, and ^{86}Sr. In this region, the isomers are based on a 2 neutron hole configuration in $g_{9/2}$ ($\nu g_{9/2}^{(-2)}$). Shell model calculations using a realistic interaction [15] reproduce the spectra with a polarization of 0.5 e for both protons and neutrons (the effective charge is a measure of the polarization of the core induced by the valence particle undergoing an electromagnetic transition). The 8^+ state is de-excited by an $E2$ transition ($B(E2 : 8^+ \rightarrow 6^+) = 1.21(5) Wu$ (1.06 Wu calc.)).

From what we know in the ^{100}Sn region and using symmetry arguments it is possible to deduce properties around the $Z = 28$ shell gap [16, 17]. Indeed, p-holes at N=50 and neutrons at $Z = 28$ play a very similar role. In particular the relevant nuclei have an identical single particle structure around the Fermi surface: they are "valence mirror nuclei" or more precisely they have the same structure at one major shell apart. The differences are twofold: 1) they have very different A/Z ratio (isospin) and 2) there is no overlap between neutrons and protons wave functions at $Z = 28$. Which also means no $n - p$ interaction. There is an 8^+ isomer in ^{98}Cd which is associated to the existence of the $N = 50$ shell gap. ^{76}Ni is a 2 proton hole in ^{78}Zn observed in this experiment. Therefore, the 8^+ isomer observed in ^{78}Zn (maximum alignment in the $\nu g_{9/2}^{(-2)}$ configuration), must be also present in ^{76}Ni which is, in turn, a 2 neutron hole in ^{78}Ni. From these symmetry based arguments it is possible to deduce that the existence of the observed isomer in ^{78}Zn is evidence for the persistence of the $Z = 28$ shell gap towards very neutron rich elements.

Shape Coexistence.   It is predicted since already quite a long time [18] that around $A = 70 - 80$, shape coexistence should occur. A large variety of nuclear shapes is anticipated from the availability of several shell gaps. In particular the Kr isotopes are the best candidates. The equilibrium deformation is rather similar using Hartree-Fock + BCS, Nilsson- Strutinsky+BCS or Hartree-Fock-Bogolyubov calculations but the excitation energy for the shape coexisting states depends on pairing interaction. The energy of low lying states is reasonably well known and enhanced $B(E2)$ values are observed which correspond to the rotation of an elongated charge distribution with $\beta \sim 0.35$. However the level is more complicated than a "simple" rigid rotation, an observation which can be interpreted in terms of perturbation from shape coexisting states. To go beyond this assumption, it is therefore necessary to imagine more stringent tests. This can be done by the determination of the electromagnetic transition probabilities and static moments of the low lying states i.e. the determination of the matrix elements associated to the decay.

To measure these probabilities, the basic idea is to avoid the *nuclear* interaction and deal only with the *Coulomb* part of the force. Experimentally, this can be done by performing a Coulomb excitation well below the Coulomb barrier. This reaction mechanism will excite the first 2^+ state which de-excites via gamma to the ground state. The matrix elements related to the decay of the first 2^+ state is related to the cross section of the excitation probability of this state (σ_{2+}). However the Q_0 moment perturbs the measurement. Other low lying collective states or change in the reduced $B(E2)$ values can have similar effects. To eliminate these perturbation it is necessary to perform Coulomb excitation using several targets and to measure precisely the scattering angle dependence of σ_{2+}. Indeed, these measurements are significantly sensitive to Q_0. By giving the precise matrix elements, this sort of data is a very severe test for the current models and no doubt that sub-barrier Coulomb excitation experiment will be extremely interesting with the availability of radioactive beams.

Island of Inversion.   The last example which illustrates the importance and the relevance still today of the study of magic nuclei is related to what is called the "island of inversion". Around $N = 20$, it is already known that there is a sudden change in the slope of the S2n plotted as function of mass and that the energy of the first 2^+ state is decreasing in Mg isotopes when the number of neutrons increases between 14 and 20. Full shell model [20] calculations including both sd and fp shells have demonstrated that the $N = 20$ gap is quenched between the $1f_{7/2}$ and the $1d_{3/2}$ orbitals which crosses each other around ^{32}Mg. This is why it is called island of inversion. This quenching is due to an increasing mixing of the fp shell which is deformation driving. In ^{32}Mg, only the first 2^+ state is known at 885 keV. An experiment has been performed at GANIL [21] to study more in detail ^{32}Mg (among others) and in particular to try to get information on the deformation. Gamma-ray spectroscopy has been made with a few number of Ge detectors. The challenge of this experiment is in the use of the in-beam spectroscopy, at the target point, of exotic nuclei produced in the high energy fragmentation of the beam. The fragments of interest had a velocity of about 30 % of light velocity. In this experiment a second gamma ray line has been detected at 1430 keV. Unfortunately it was not possible to deduce the spin of the levels de-excited by this transition but it was shown that the 2 known lines are in coincidence. If it is a 4^+, then the ratio $E(4^+?)/E(2^+) = 2.6$ which is in between the value expected for a perfect rotor and perfect vibrator. This could be due to the role of the deformation induced by the fp shell mixing.

Several examples that I have described here deal with spectroscopy and the level of precision and efficiency which is needed today requires high performance detectors. This is even more important for future studies related to the advent of radioactive beams such as those expected from the SPIRAL facility at GANIL. In the following section, I will describe in some details EXOGAM, one of these gamma-ray arrays of the new generation which is specifically designed to work with radioactive beams.

3. The EXOGAM Array

Design Specifications. Radioactive beams impose new design considerations to a γ-ray spectrometer. The beam intensity is expected to be much lower than with stable beams, factor of 10 or even 1000 lower. EXOGAM must therefore be designed to maximize the total photopeak efficiency. In maximizing efficiency the spectrum quality must be maintained. The spectrum quality is determined by the peak to total ratio and energy and time resolution. The total efficiency measures the ability of the array to collect statistics. The spectrum quality measures the effectiveness of the array in isolating a single sequence or sequences of gamma-rays from a complex spectrum.

There will be a large variety of nuclear reactions using radioactive beams on which the design of a detection system must be based. The experimental conditions will be very different from one experiment to another in terms of γ-ray energy (from x-rays of tens of keV to γ-ray energies up to $5 - 6\ MeV$), of multiplicity (from one to ~ 15 coincident photons); of recoil velocity (from zero to $\sim 10\ \%$ of light velocity); and of kinematics of the reaction mechanism (from recoiling fusion products emitted at $\sim 0°$ or scattered particles between $0°$ and $180°$). This variety means that the setup of the array must be adapted for each experiment. The radioactive nature of the beam is also a concern and shielding of the detectors becomes an important design criterion.

It is also clear that in addition to the detection of gamma radiation it will be vital to have ancillary detectors available to detect both light and heavy charged particles and neutrons.

Segmented CLOVER Ge Detector.   In order to meet all the design criteria the EXOGAM spectrometer will consist of an array of high resolution germanium detectors each surrounded by an escape suppression shield.

One composite and segmented detector is the segmented CLOVER detector (see [22] for a detailed description). It is made of four individual crystals each electronically segmented into four regions. The EXOGAM array will consist of segmented CLOVER detectors. This segmentation is particularly useful when the emitting nuclei recoil with a large velocity since it allows a better determination of the interaction point of the γ ray in the detector. The detectors can be easily arranged in different configurations as will be shown later. This high degree of versatility is a very important design criterion as in the future the full gamut of nuclear reactions with stable and radioactive beams will be used for nuclear spectroscopy.

GEANT simulation calculations have been carried out to optimize the performances of CLOVER detectors for EXOGAM. The EXOGAM CLOVER will be based on the use of large Ge crystals, 60 mm in diameter and 90 mm long, before shaping. The photopeak efficiency of each CLOVER will be $\epsilon_p\omega \sim 12 \times 10^{-3}$ at 11 cm for a 1.3 MeV γ-ray. The segmentation of the crystals leads to a reduction by a factor of two of the Doppler broadening of the peak as compared to a non-segmented crystal.

Suppression Shield for a Segmented CLOVER Ge Detector.
Each segmented CLOVER Ge detector is surrounded by an escape suppression shield. The shield designed for the EXOGAM CLOVER is based on a new concept in which the shield comprises several distinct elements, a backcatcher, a rear side element and a side shield (see [23] for details). De-

signing suppression shields in this way, from individual elements, creates greater flexibility for different configurations. The shields will be operated in two configurations. The first is with the back catcher and rear-side element, configuration A, and the second with the additional side elements, configuration B.

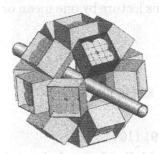

Figure 4 The EXOGAM spectrometer with 16 segmented CLOVERs.

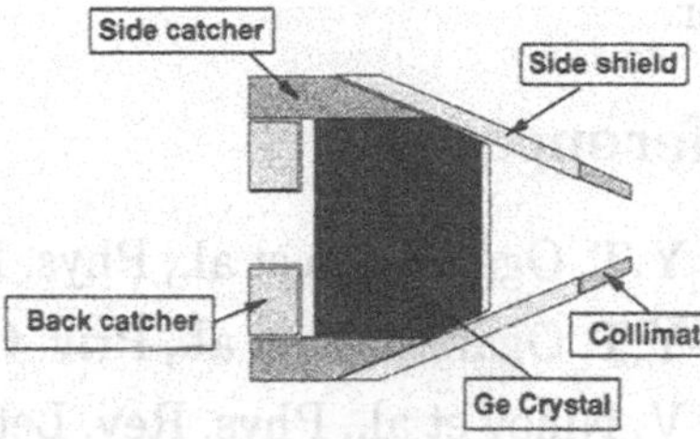

Figure 5 The different elements of the BGO suppression shield for the segmented CLOVER *Ge* detectors (not to scale).

Segmented CLOVERS Arrays.

The EXOGAM segmented CLOVERs can be arranged in different geometries (see Fig. 5). In all the geometries the suppression elements can be used in configurations A and B. Configuration A is the close packed geometry where the *Ge* detectors can essentially touch at the front. Configuration B is the pulled back geometry in which the detectors are farer from the target to allow for the inclusion of the additional side suppression elements. An array geometry for the CLOVERS to be as close as possible to the target is with the detectors on the faces of a cube. An array of 16 CLOVER detectors can be arranged. An isometric projection of the array is shown in Fig. 4. In configuration A the signals from adjacent *Ge* crystals can be summed to increase the efficiency. The calculated increase in efficiency is 6%. The configurations, distances from the target, and performances for arrays of segmented CLOVERs are summarized in Tab. 1.

Table 1 Summary of array geometries for segmented CLOVERs.

Geom.	Shield conf.	d Ge-target (mm)	Phot. eff. 662 keV	Phot. eff. 1.3 MeV	Peak/Total 662 keV	Peak/Total 1.3 MeV
Cube	B	68.3	0.15	0.10	0.72	0.60
16 det.	A	114.1	0.28	0.20	0.57	0.47
16 det.	B	147.4	0.17	0.12	0.72	0.60

4.   Conclusions

With the development of radioactive beams in several places all over the world and the associated new spectrometers, the limits of our current knowledge in nuclear structure will be overtaken. More than a conclusion, I hope that this new step will bring a harvest of surprise. I would also like to thank all the people who contributed to this lecture by one mean or the other.

References

[1] Y.T. Oganessian et al., Phys. Rev. Lett. 83 (1999) 3154

[2] Y.T. Oganessian et al., Priv. Comm.

[3] V. Ninov et al., Phys. Rev. Lett. 83 (1999) 1104

[4] K. Rutz et al., GSI Ann. Rep. (1999) 38; M. Bender et al., in preparation.

[5] M. Lewitowicz et al., Phys. Lett. B332 (1994) 20

[6] R. Schneider et al, Z. Phys. A348 (1994) 241

[7] B. Blank et al., Phys. Rev. Lett. 84 (2000) 1116

[8] Ch. Engelmann et al., Z. Phys. A352 (1995) 351

[9] G. Muenzenberg et al., Z. Phys. A322 (1985) 227

[10] M. Leino et al., Eur. Phys. J. A6 (1999) 63

[11] P. Reiter et al., Phys. Rev. Lett. 82 (1999) 509

[12] I. Muntian, Z. Patyk, A. Sobiczewski et al., GSI Ann. Rep. (1999) 39

[13] R. Grzywacz et al., Phys. Rev. C55 (1997) 1

[14] J.M. Daugas et al., Phys. Lett. B476 (2000) 213

[15] Sinatkas et al., J. Phys. G: Nucl. Part. Phys. 18 (1992) 1377, 1401

[16] R. Grzywacz et al., Phys. Rev. Lett. 81 (1998) 766

[17] M. Gorska et al., Phys. Rev. Lett. 79 (1997) 2415

[18] W. Nazarewicz et al., Nucl. Phys. A435 (1985) 397

[19] R. Bengtsson, Nuclear Structure of the Zirconium Region, Springer (1985) 17

[20] N. Fukunishi et al., Phys. Lett. B296 (2000) 279

[21] M.J. Lopez, PhD thesis, GANIL T 00 01

[22] G. Duchêne et al., NIM A432 (1999) 90

[23] J. Simpson et al., Heavy Ion Physics 11 (2000) 159-188

Experimental and Theoretical Aspects of Proton Radioactivity: Proton Decay of Spherical and Deformed Nuclei*

Cary N. Davids

Physics Division, Argonne National Laboratory

Argonne, IL 60439, USA

davids@anl.gov

Keywords: Proton radioactivity, decay rates, spherical nuclei, deformed nuclei

Abstract   Proton radioactivity is a decay mode found only in nuclei beyond the
proton drip line. It competes with alpha decay, positron decay, and
electron capture. Proton decay is a quantum tunneling phenomenon,
and the decay rate is governed by the properties of the potential barrier set up by the Coulomb, centrifugal, and nuclear potentials. This
presents the opportunity to extract spectroscopic information on a nuclide beyond the proton drip line. Recent experimental developments
will be presented, including gamma spectroscopy of proton emitters,
and fine structure in proton decay. Calculations of proton decay rates
for spherical and deformed proton emitters will be discussed.

1.   Introduction

Nuclei beyond the proton drip line ($S_p < 0$) can decay by proton
emission, in competition with alpha decay, positron decay, and electron
capture [1]. As in alpha decay, proton decay is a quantum tunneling
phenomenon whose rate is governed by the properties of the barrier created by the nuclear, Coulomb, and centrifugal potentials, which act to
confine the proton to the core nucleus. However, the centrifugal potential plays a much more important role in proton decay than it does in
alpha decay. Fig. 1 shows the proton-nucleus potential calculated for

*Supported by the U.S. Department of Energy, Nuclear Physics Division, under Contract
W-31-109-ENG-38.

D. N. Poenaru et al. (eds.), Nuclei Far from Stability and Astrophysics, 79–89.

the decay of the proton emitter ^{167}Ir. From the table of calculated and measured proton half-lives in this figure it can be seen that the half-life is extremely sensitive to both the proton energy and the orbital angular momentum ℓ carried off by the proton. This sensitivity allows spectroscopic information on the parent state to be obtained.

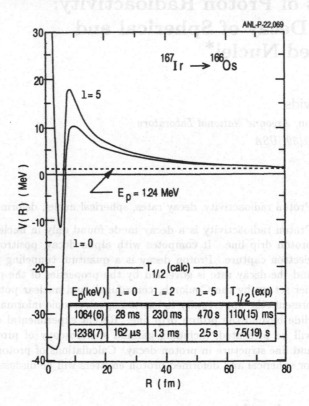

Figure 1 Proton-nucleus potential for the proton emitter ^{167}Ir, calculated for the ℓ-values 0 and 5. It is the sum of Coulomb, nuclear, and centrifugal terms, using the real part of the Becchetti-Greenlees potential [2].

2.   Experimental Techniques

Space does not permit a complete review of experimental techniques used in proton radioactivity research. For a more thorough discussion of this topic, see the review by Woods and Davids [1]. Recent progress in proton emitter research has been propelled by a number of technical advances. The application of the double-sided silicon strip detector (DSSD) [3] to the observation of proton emitters by the Edinburgh group, [4-8] was responsible for a major advance in the field. More recently, work at

Argonne [9-15] and Oak Ridge [16-19] has increased our knowledge of both spherical and deformed proton emitters.

All of these advances were made possible by the background reduction afforded by coupling the DSSD to a second-generation recoil mass spectrometer [20-22]. The reaction products pass through the focal plane of the separator, where their mass/charge ratios are determined. They then proceed on and are implanted into the DSSD. This detector has independently instrumented orthogonal strips on front and back, allowing the X-Y position (pixel) of the implant to be recorded. When a subsequent proton decay occurs in the same pixel, it can be correlated with the implant, and the time interval between implantation and decay is used to calculate the decay half-life.

When the daughter nucleus itself can decay by alpha emission, even more background reduction can be achieved by requiring the sequence of events: a) implantation, b) proton decay, and c) alpha decay in the same pixel. Fig. 2 shows an example, from the decay of ^{167}Ir [10].

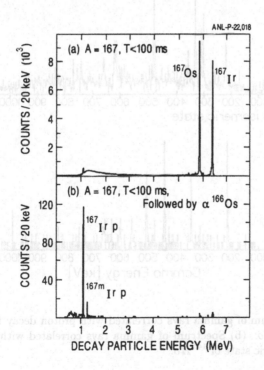

Figure 2 (a) Energy spectrum of decay events in the DSSD from 357 MeV ^{78}Kr + ^{92}Mo, after requiring that the decay occurred in the same pixel within 100 ms following a mass 167 implant. (b) Same as (a) with the additional requirements that a second decay event occurred in the same pixel within 100 ms, having an energy of 6000 keV, the known alpha-decay energy of ^{166}Os.

In addition to studying the energy spectra of emitted protons, one can use a large gamma-ray detector array around the target to observe gamma rays emitted from excited states of the proton emitter itself. This allows us for the first time to obtain spectroscopic information on a nucleus situated well beyond the proton drip line. The technique of Recoil-Decay Tagging (RDT) [23] has been used to study energy levels in the proton emitters ^{147}Tm [24], ^{151}Lu [25], ^{109}I [26], ^{113}Cs [27], ^{167}Ir [28], ^{131}Eu [29], and ^{141}Ho [29]. Fig. 3 shows gamma ray spectra associated with the decay of levels in ^{141}Ho to the $7/2^-$ ground state and to the $1/2^+$ isomeric state [29]. Rotational bands have been identified in this nucleus, providing independent information on the deformation [29].

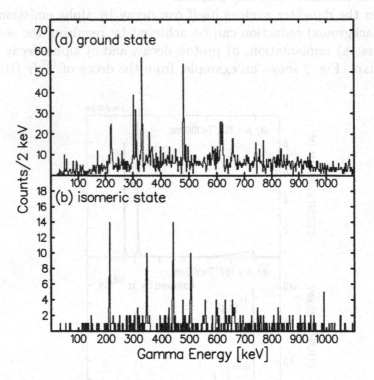

Figure 3   (a) Spectrum of gamma rays correlated with proton decay from the $7/2^-$ ground state of ^{141}Ho. (b) Spectrum of gamma rays correlated with proton decay from the $1/2^+$ isomeric state of ^{141}Ho.

The recent observation of fine structure in the decay of the deformed proton emitter ^{131}Eu [15] has provided an opportunity to test our models of proton decay. The large deformation predicted for the daughter nucleus ^{130}Sm ($\beta_2 \sim 0.33$) [30] means that its first 2^+ state should lie at an excitation energy near 130 keV. This is sufficiently low that proton decay from the $J^\pi = \frac{3}{2}^+$ ^{131}Eu ground state should be possible to both the 0^+ ground state and the first 2^+ state of ^{130}Sm. Fig. 4 shows that this is indeed the case, with a small proton group appearing 122(3) keV below the main ground state group. The lower panel shows a simulation of the decay [15], which takes into account the fact that the 2^+ state, once populated, can decay by emitting either a gamma ray or a conversion electron. The decay to the ground state requires that the emitted proton have only $\ell = 2$, while decay to the 2^+ state can proceed via $\ell = 0, 2,$ or 4. Thus each branch probes different parts of the initial wavefunction. The branching ratio determined from this measurement is 0.24(5) [15].

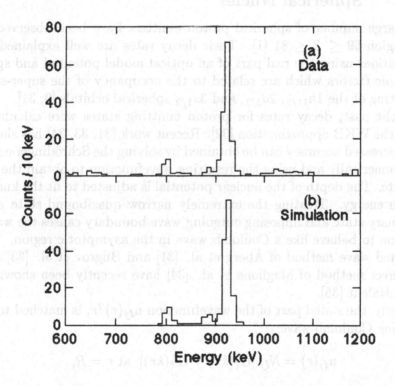

Figure 4 (a) Energy spectrum of decay events in the DSSD from 402 MeV ^{78}Kr + ^{58}Ni, occurring within 100 ms following a mass 131 implant. (b) Simulated energy spectrum of protons from the decay of ^{131}Eu, assuming an implantation depth of 19.3 μm, a 2^+ excitation energy in ^{130}Sm of 122 keV, and a 2^+ decay branching ratio of 0.24(5).

3.   Decay Rates of Proton Emitters

The decay rates of proton emitters are calculated as a quantum-mechanical barrier penetration, supplemented by a spectroscopic factor. The spectroscopic factor represents the overlap between the parent state, and the daughter state plus a free proton. This is not necessarily equal to 1, because the proton orbital involved may be partially occupied in the daughter state. The majority of the known proton emitters have been treated as spherical nuclei. However, the rare earth proton emitters ^{131}Eu [14, 15], ^{140}Ho [19], and ^{141}Ho [14] have been recently shown to be strongly deformed. Decay rate calculations for spherical and deformed nuclei will be discussed in Subsections 3.1 and 3.2, respectively.

3.1.   Spherical Nuclei

A large number of spherical proton emitters have been observed in the region $69 \leq Z \leq 81$ [1]. Their decay rates are well explained by calculations using the real part of an optical model potential and spectroscopic factors which are related to the occupancy of the super-shell consisting of the $1h_{11/2}$, $2d_{3/2}$, and $3s_{1/2}$ spherical orbitals [9, 31].

In the past, decay rates for proton emitting states were calculated using the WKB approximation [32]. Recent work [31, 33, 34] has shown that increased accuracy can be obtained by solving the Schrödinger equation numerically and using the resulting wavefunction to obtain the decay rate. The depth of the nuclear potential is adjusted to fit the known proton energy. Treating the extremely narrow quasibound state as a stationary state and imposing outgoing wave boundary causes the wavefunction to behave like a Coulomb wave in the asymptotic region. The distorted wave method of Åberg et al. [31] and Bugrov et al. [33] and the direct method of Maglione et al. [34] have recently been shown to be equivalent [35].

Briefly, the radial part of the wavefunction $u_{\ell j}(r)/r$, is matched to an outgoing Coulomb wave:

$$u_{\ell j}(r) = N_{\ell j} \left[G_\ell(kr) + iF_\ell(kr) \right] \text{ at } r = R, \tag{1}$$

where R is a large distance outside the range of the nuclear field. Here $F_\ell(kr)$ and $G_\ell(kr)$ are the regular and irregular Coulomb wavefunctions, respectively, calculated for the asymptotic relative kinetic energy, $(\hbar k)^2/(2\mu)$, of the proton and the daughter nucleus, μ being the reduced mass. The function $u_{\ell j}(r)/r$ is normalized by requiring that

$$\int_0^R [u_{\ell j}(r)]^2 dr = 1,$$

where $R < R_o$, the classical external turning point. In practice, integrating out to $R = 25\ fm$ is adequate for the known spherical proton emitters. Once $u_{\ell j}(r)$ is normalized, then $N_{\ell j}$ can be determined from (1):

$$N_{\ell j} = \frac{u_{\ell j}(R)}{G_\ell(kR) + iF_\ell(kR)}. \tag{2}$$

By calculating the radial probability flux through a sphere, using the radial wavefunction on the right hand side of equation (1), one can then express the decay rate $\Gamma_{\ell j}$ (or the mean lifetime τ) in terms of the matching amplitude $N_{\ell j}$ as in Ref. [34],

$$\Gamma_{\ell j} = \frac{\hbar}{\tau} = \frac{\hbar^2 k}{\mu}|N_{\ell j}|^2. \tag{3}$$

We note that, strictly speaking, the matching condition (1) requires a complex energy and a complex wavefunction [34]. However, since the imaginary part of the resonance energy (i.e. the decay width) is usually extremely small, one could just as well use energies and wavefunctions that are real [31]. In that case, one would replace equation (2) by

$$N_{\ell j} = \frac{u_{\ell j}(R)}{G_\ell(kR)}. \tag{4}$$

This condition will be quite close to (2) because it is usually possible to find a value for R such that $G_\ell(kR) \gg F_\ell(kR)$.

3.2.   Deformed Nuclei

As mentioned above, several deformed rare-earth proton emitters have been discovered. The picture we consider here is that of a deformed odd-A nucleus, consisting of a proton strongly coupled to an axially-symmetric even-even core. Using the spherical formalism to calculate the decay rates results in large discrepancies with the experimental rates. This is not surprising, since the potential barrier seen by the outgoing proton is orientation-dependent. As a result, the total angular momentum j of the proton is no longer a good quantum number, and only the parity and K, the projection of j on the symmetry axis, are good quantum numbers. A number of authors have considered this problem [34,36-43]. We proceed by using deformed Coulomb and nuclear potentials, and expanding the wavefunction in spherical components in the intrinsic system:

$$\psi_K(\mathbf{r}) = \sum_{\ell j} \frac{\phi_{\ell j}(r)}{r}|\ell j K\rangle, \tag{5}$$

where the sum is over $j \geq |K|$. Here the spin-angular wavefunction $|\ell j K\rangle$ is shorthand for

$$|\ell j K\rangle = \sum_{m_\ell m_s} \langle \ell m_\ell \tfrac{1}{2} m_s | j K \rangle Y_\ell^{m_\ell}(\hat{\mathbf{r}}) \chi(m_s).$$

We then insert (5) into the Schrödinger equation $H\psi_K(\mathbf{r}) = E\psi_K(\mathbf{r})$. When we project this equation with the spin-angular part of the wavefunction $|\ell j K\rangle$ we obtain a set of coupled differential equations for the radial wavefunctions $\phi_{\ell j}(r)$:

$$\left[-\frac{\hbar^2}{2\mu}\frac{d^2}{dr^2} + \frac{\hbar^2 \ell(\ell+1)}{2\mu r^2} - E \right] \phi_{\ell j}(r) = -\sum_{\ell' j'} \langle \ell j K | V_{def}(\mathbf{r}) | \ell' j' K \rangle \phi_{\ell' j'}(r).$$

$$(6)$$

Next we make a multipole expansion of the total deformed interaction, $V_{def}(\mathbf{r})$, consisting of Coulomb, nuclear, and spin-orbit potentials:

$$V_{def}(\mathbf{r}) = \sum_{\lambda=0} V_\lambda(r) P_\lambda(\cos(\theta')), \qquad (7)$$

where θ' is the angle with respect to the symmetry axis. Only even values of λ appear in the summation because of the axial symmetry. The spin-angular matrix elements on the right hand side of the coupled equations are then [44]

$$\langle \ell j K | P_\lambda(\cos(\theta')) | \ell' j' K \rangle = \langle j K \lambda 0 | j' K \rangle \langle j' \tfrac{1}{2} \lambda 0 | j \tfrac{1}{2} \rangle, \quad \lambda \text{ even.} \qquad (8)$$

The (real) resonance solutions to the coupled equations (6) are obtained by matching each $\phi_{\ell j}(r)$ to the irregular Coulomb function $G_\ell(kR)$. In Ref. [36] this is done at a relatively small distance from the nucleus, 15 fm. This significantly reduces the amount of computation required. The total wavefunction $\psi_K(\mathbf{r})$ is then normalized by requiring

$$\sum_{\ell j} \int_0^R [\phi_{\ell j}(r)]^2 dr = 1,$$

where $R \approx 100\ fm$. The distorted wave Green's function method [35] is used to calculated the decay width, taking into account the differences between intrinsic and laboratory frames. For a given angular momentum ℓ_p carried off by the proton, the decay width of an odd-A proton emitter with spin I and projection K ($I = K$), where the daughter nucleus with atomic number Z is left in a state with spin R_d, is given by

$$\Gamma_{\ell_p j_p}^{R_d I K} = \frac{4\mu}{\hbar^2 k} \frac{2(2R_d+1)}{(2I+1)} \langle j_p K R_d 0 | I K \rangle^2 \left| \mathcal{M}_{\ell_p j_p}^K \right|^2,$$

where

$$\mathcal{M}^K_{\ell_p j_p} = \sum_{\lambda \ell j} \langle \ell_p j_p K | P_\lambda(cos(\theta')) | \ell j K \rangle \int_0^R dr F_{\ell_p}(kr) \left[V_\lambda(r) - \delta_{\lambda 0} \frac{Ze^2}{r} \right] \phi_{\ell j}(r).$$

The wavenumber k of the proton is determined from the decay Q-value Q_p, atomic screening correction E_{sc}, and excitation energy E_x of the daughter state:

$$\frac{(\hbar k)^2}{2\mu} = Q_p - E_{sc} - E_x.$$

By carrying out the integration to a radius $R = 100$ fm, the effect of the long-range Coulomb field is taken into account. Pairing has not been considered here. It would tend to reduce the decay width by $30-50$ %.

Table 1 Proton decay widths (in units of 10^{-20} MeV) for $^{131}Eu(\frac{3}{2}^+)$ decay to the 0^+ and 2^+ states of ^{130}Sm, the branching ratio to the 2^+ state, and similarly for the decays of $^{141g,m}Ho$ to the 0^+ and 2^+ states of ^{140}Dy. The 2^+ widths Γ_2 have been calculated with $E_{2+} = 122$ keV for ^{130}Sm [15] and $E_{2+} = 160$ keV for ^{140}Dy.

Nucleus	Γ_0	Γ_2	$\frac{\Gamma_2}{\Gamma_0}$
$^{131}Eu(\frac{3}{2}^+)$	2.88	0.929	0.244
Experiment [15]	1.71(24)	0.54(13)	0.24(5)
$^{141g}Ho(\frac{7}{2}^-)$	16.5	0.45	0.023
Experiment [14, 29]	10.9(10)	—	≤ 0.01
$^{141m}Ho(\frac{1}{2}^+)$	21700	330	0.015
Experiment [19]	5700(2140)	—	—
Experiment [29]	7020(1080)	—	≤ 0.01

Using this formalism, we have calculated the decay widths for the deformed proton emitters $^{131}Eu(\frac{3}{2}^+)$, $^{141g}Ho(\frac{7}{2}^-)$, and $^{141m}Ho(\frac{1}{2}^+)$, as well as the branching ratios for the decays to the 2^+ states of the daughter nuclei ^{130}Sm and ^{140}Dy, respectively. The results are shown in Tab. 1. Good agreement with experiment is obtained. These calculations can be improved by the addition of pairing.

Acknowledgments

The author wishes to acknowledge the contribution of his many colleagues to this work, especially H. Esbensen, D. Seweryniak, A. A. Sonzogni, and P. J. Woods.

References

[1] P.J. Woods and C.N. Davids, Annu. Rev. Nucl. Part. Sci. 47 (1997) 541

[2] F.D. Becchetti and G.W. Greenlees, Phys. Rev. 182 (1969) 1190

[3] P.J. Sellin et al., Nucl. Instr. and Meth. A311 (1992) 217

[4] R.D. Page et al., Phys. Rev. Lett. 72 (1994) 1798

[5] K. Livingston et al., Phys. Lett. B312 (1993) 46

[6] R.D. Page et al., Phys. Rev. Lett. 68 (1992) 1287

[7] P.J. Sellin et al., Phys. Rev. C47 (1993) 1933

[8] K. Livingston et al., Phys. Rev. C48 (1993) R2151

[9] C.N. Davids et al., Phys. Rev. Lett. 76 (1996) 592

[10] C.N. Davids et al., Phys. Rev. C55 (1997) 2255

[11] R.J. Irvine et al., Phys. Rev. C55 (1997) R1621

[12] J. Uusitalo et al., Phys. Rev. C59 (1999) R2975

[13] G.L. Poli et al., Phys. Rev. C59 (1999) R2979

[14] C.N. Davids et al., Phys. Rev. Lett. 80 (1998) 1849

[15] A.A. Sonzogni et al., Phys. Rev. Lett. 83 (1999) 1116

[16] J.C. Batchelder et al., Phys. Rev. C57 (1998) R1042

[17] C.R. Bingham et al., Phys. Rev. C59 (1999) R2984

[18] T.N. Ginter et al., Phys. Rev. C61 (1999) 014308.

[19] K. Rykaczewski et al., Phys. Rev. C60 (1999) 011301

[20] A.N. James et al., Nucl. Instr. and Meth. A267 (1988) 144

[21] C.N. Davids et al., Nucl. Instr. and Meth. B70 (1992) 358

[22] J.D. Cole et al., Nucl. Instr. and Meth. B70 (1992) 343

[23] E.S. Paul et al., Phys. Rev. C55 (1997) R2137

[24] D. Seweryniak et al., Phys. Rev. C56 (1997) R723

[25] C.-H. Yu et al., Phys. Rev. C58 (1998) R3042

[26] C.-H. Yu et al., Phys. Rev. C59 (1999) R1834

[27] C.J. Gross et al., in *Proceedings of the 2nd International Conference on Exotic Nuclei and Atomic Masses*, Shanty Creek, USA, 1998, ed. B. M. Sherrill, D. J. Morrissey, and C. N. Davids (AIP, New York, 1998) p. 444.

[28] M.P. Carpenter et al., Acta Physica Polonica 30 (1999) 581

[29] D. Seweryniak et al., submitted for publication

[30] P. Möller et al., At. Data Nucl. Data Tables 59 (1995) 185

[31] S. Åberg, P.B. Semmes, and W. Nazarewicz, Phys. Rev. C56 (1997) 1762; Phys. Rev. C58 (1998) 3011

[32] S. Hofmann, in *Nuclear Decay Modes*, ed. D.N. Poenaru (IOP Publishing, 1996), p. 143

[33] V.P. Bugrov et al., Sov. J. Nucl. Phys. 41 (1985) 717

[34] E. Maglione, L.S. Ferreira, and R.J. Liotta, Phys. Rev. Lett. 81 (1998) 538

[35] C.N. Davids and H. Esbensen, Phys. Rev. C61 (2000) 054302

[36] H. Esbensen and C.N. Davids, submitted for publication

[37] V.P. Bugrov and S.G. Kadmensky, Sov. J. Nucl. Phys. 49 (1989) 967

[38] S.G. Kadmensky and V.P. Bugrov, Phys. of Atomic Nuclei 59 (1996) 399

[39] L.S. Ferreira, E. Maglione, and R.J. Liotta, Phys. Rev. Lett. 78 (1997) 1640

[40] E. Maglione, L.S. Ferreira, and R. J. Liotta, Phys. Rev. C59 (1999) R589.

[41] L.S. Ferreira and E. Maglione, Phys. Rev. C61 (2000) 021304(R)

[42] E. Maglione and L. Ferreira, Phys. Rev. C61 (2000) 047307

[43] A.T. Kruppa et al., Phys. Rev. Lett. 84 (2000) 4549

[44] A. Bohr and B. R. Mottelson, "Nuclear Structure", Vol. I (W. A. Benjamin, New York, 1969), Equation 3A-14

Present and Future Nuclear Shell Model

Takaharu Otsuka

Department of Physics, University of Tokyo
Hongo, Bunkyo-ku, Tokyo, 113-0033 Japan
otsuka@phys.s.u-tokyo.ac.jp

Keywords: Shell model, Monte Carlo shell model

Abstract   The present status and future perspectives of the nuclear shell model are discussed. The development and limitation of the conventional shell model calculations are shown, and stochastic approaches are introduced. As one of such approaches, the Quantum Monte Carlo Diagonalization (QMCD) method has been proposed. The formulation of the QMCD method is presented with an illustrative example. While the QMCD method is a general method for solving the quantum many-body interacting systems, its application to the nuclear shell model is referred to as the Monte Carlo Shell Model (MCSM). A test of the MCSM is presented confirming the feasibility of the MCSM. The MCSM provides us with a breakthrough in the shell model calculation: the level structure of low-lying states can be studied with realistic interactions for a wide, probably basically unlimited, variety of nuclei.

1.   Introduction

The nuclear shell model has been successful in the description of various aspects of nuclear structure, partly because it is based upon a minimum number of natural assumptions, and partly because all dynamical correlations in the model space can be incorporated appropriately. Although the direct diagonalization of the Hamiltonian matrix in the full valence-nucleon Hilbert space is desired, the dimension of such a space is too large in many cases, preventing us from performing the full calculations. In order to overcome this difficulty, quantum Monte Carlo approaches have been introduced. As one of them, the Shell Model Monte Carlo (SMMC) method has been proposed successfully [1]. However, the SMMC is basically restricted to ground-state and thermal properties, and is not a proper tool for studying excited states, i.e., level scheme

D. N. Poenaru et al. (eds.), Nuclei Far from Stability and Astrophysics, 91–102.
© 2001 *Kluwer Academic Publishers. Printed in the Netherlands.*

and $E2$ properties. Moreover, the SMMC suffers from the so-called minus sign problem for realistic interactions. As a completely different approach, the Quantum Monte Carlo Diagonalization (QMCD) method has been proposed for solving quantum many-body systems with a two-body interaction [2-5]. The QMCD can describe not only the ground state but also excited states, including their energies, wave functions and hence transition matrix elements. The sign problem is irrelevant to the QMCD. Thus, based upon the QMCD method, we introduce the Monte Carlo Shell Model as its application to the nuclear shell model calculation. The MCSM has become a new tool for clarifying the structure of the ground and low-lying states of nuclei.

2.   Development and Limitations of Conventional Shell Model

The nuclear shell model has been started by Mayer and Jensen in 1949 as a single-particle model. Afterwards, many valence particles are treated in the shell model, which then became a many-body theory or calculational method.

In the shell model calculation, one introduces single-particle state first. This state can be given as an eigenstate of a spherical single-particle potential, for instance, Harmonic Oscillator or Woods-Saxon potential, and can be written as a product of the radial wave function, a Spherical Harmonics, a spin wave function and an isospin wave function. The spherical Harmonics (*i.e.*, orbital-angular-momentum wave function) and spin wave function can be coupled to definite total angular momentum and its z-projection.

Single-particle states relevant to a given nucleus are grouped into the core part and the valence part. The core part is completely occupied, and is frozen like a vacuum. The valence part is usually one major shell on the top of the core part, and called valence shell. The valence shell is partly occupied. In the shell model calculation, one generates all possible Slater determinants in the valence shell. The Slater determinants can be classified according to the z-component of angular momentum, denoted by M. The number of Slater determinants for a given value of M is called (M-scheme) shell model dimension.

In order to obtain the eigenstate of the Hamiltonian for the given valence shell, one calculates matrix elements of the Hamiltonian for Slater determinants. This is a straightforward calculation. However, this calculation should be done for all combinations of Slater determinants as bra and ket vectors of the matrix elements. Once all matrix elements are calculated, one should diagonalize the matrix. So, if the shell model dimension is large, the number of matrix elements is certainly much larger (about half of the square of the dimension), and the actual calculation becomes very difficult. By recent (conventional) shell-model codes like ANTOINE by Caurier, VECSSE by Sebe or MSHELL by Mizusaki, one can handle up to shell model dimension $\sim$ 100 million at technical edge, while practical calculations up to a few tens million dimension can be done. An example can be found in [6], where the levels of ^{52}Fe are calculated by the code ANTOINE.

Although the conventional shell model calculation has thus been developed significantly, the dimension can be much larger in many real nuclei and is indeed much beyond the reach of the future development. For instance, certain unstable nuclei to be discussed in this lecture (though not in this lecture note) require calculations with more than 10 billion dimensions. This is already very far beyond the limit of the existing shell model codes.

3.   Formulation of QMCD

We have seen the difficulties the shell model calculations are facing. In order to overcome those difficulties, one has to introduce an alternative approach. That is stochastic methods to many-body problems. We now turn to this subject.

3.1.   Illustrative Explanation of the Quantum Monte Carlo Method

We first present an illustrative explanation of the quantum Monte Carlo method, taking a toy Hamiltonian. Although the discussions are quite incomplete, one may find a simple and intuitive picture of the quantum Monte Carlo calculations. We begin with the imaginary-time evolution operator

$$e^{-\beta H}, \tag{1}$$

for a given Hamiltonian, H. Here, β is a real parameter. If this operator in Equ. 1 is acting on a state $|\Psi^{(0)}\rangle$, one obtains

$$e^{-\beta H} |\Psi^{(0)}\rangle = \sum_i e^{-\beta E_i} c_i |\phi_i\rangle, \tag{2}$$

where E_i is the i-th eigenvalue of H, $|\phi_i\rangle$ is its eigenfunction, c_i stands for amplitude: $|\Psi^{(0)}\rangle = \sum_i c_i |\phi_i\rangle$. For β large enough, amplitudes of the ground and low-lying states become larger in Equ. 2 than others. This means that $e^{-\beta H}$ behaves as a projection operator for those states with lower energies. At $\beta \to \infty$, only the ground state survives. Thus, the operator $e^{-\beta H}$ has a nice feature, but its actual handling is very complicated for H containing a two-body (or many-body) interaction.

The Hubbard-Stratonovich transformation [7, 8] can be used to ease this difficulty. We shall explain this transformation taking a simple illustrative example. A toy Hamiltonian is taken as,

$$H = \frac{1}{2}VO^2, \tag{3}$$

where V is a coupling constant and O is a one-body operator. For simplicity, we assume $V < 0$. The imaginary-time evolution operator is then written as,

$$e^{-\frac{1}{2}\beta VO^2}. \tag{4}$$

At this point, we note a well-known formula,

$$\int_{-\infty}^{\infty} d\sigma\, e^{-a\,(\sigma+c)^2} = \sqrt{\pi/a} \quad (a > 0), \tag{5}$$

or

$$e^{-a\,c^2} = \sqrt{a/\pi} \int_{-\infty}^{\infty} d\sigma\, e^{-a\,\sigma^2 - 2a\sigma c} \quad (a > 0), \tag{6}$$

where σ means the variable for integration, and a and c are parameters. By substituting $-\frac{1}{2}\beta V^2 \to a$ and $O \to c$, the operator in Equ. 4 can be written as,

$$e^{-\frac{1}{2}\beta VO^2} = \int_{-\infty}^{\infty} d\sigma \sqrt{\frac{\beta|V|}{2\pi}} \cdot e^{-\frac{\beta}{2}|V|\sigma^2} \cdot e^{-\beta|V|\sigma O}. \tag{7}$$

This is an exact expression, but is not practical at all. The idea of the quantum Monte Carlo calculation comes in now: the integration in Equ. 7 is approximated by a Monte Carlo (MC) sampling as,

$$e^{-\frac{1}{2}\beta VO^2} \approx \sum_{MC:\sigma} \sqrt{\frac{\beta|V|}{2\pi}} \cdot e^{-\beta|V|\sigma O}, \tag{8}$$

where MC stands for Monte Carlo sampling. Here, we introduce a Gaussian,

$$G(\sigma) = e^{-\frac{\beta}{2}|V|\sigma^2}, \tag{9}$$

and the σ variable, called usually *auxiliary field*, is sampled by using the probability weight of Equ. 9. Equ. 8 becomes exact if the MC sampling is complete.

By combining Equs. 2, 3 and 8, one can see that, apart from the normalization, the ground state can be obtained for $\beta \to \infty$ by

$$|\Phi_g\rangle \sim \sum_{MC:\sigma} e^{-\beta h(\sigma)} |\Phi^{(0)}\rangle, \tag{10}$$

where $h(\sigma)$ is called the *one-body Hamiltonian* and is defined as

$$h(\sigma) = V \sigma O. \tag{11}$$

So far, we assumed $V < 0$. In the case of $V > 0$, the above argument should be modified, and the general definition of $h(\sigma)$ is given by

$$h(\sigma) = s V \sigma O, \tag{12}$$

where s takes the following values,

$$s = 1 \ (i) \ for \ V < (>) \ 0. \tag{13}$$

The ground state energy is given by

$$E_g = \frac{\langle \Phi_g | H | \Phi_g \rangle}{\langle \Phi_g | \Phi_g \rangle}, \tag{14}$$

where Φ_g stands for an unnormalized wave function of the ground state. By combining Equ. 10 with this, one obtains

$$E_g \sim \frac{\langle \Phi^{(0)} | H \sum_{MC:\sigma} e^{-\beta h(\sigma)} | \Phi^{(0)} \rangle}{\langle \Phi^{(0)} | \sum_{MC:\sigma} e^{-\beta h(\sigma)} | \Phi^{(0)} \rangle}. \tag{15}$$

The Shell Model Monte Carlo (SMMC) method [1] is based on this equation. Although this expression seems to be appropriate, the calculation does not converge usually, because of variance due to the sampling. Therefore, various techniques have been introduced as described for instance in [1]. We also note that, in the SMMC, the MC sampling is not made for wave functions but for the expectation values in Equ. 15. The minus-sign problem occurs generally in most of quantum Monte Carlo calculations. We shall explain briefly what the minus-sign problem is. The denominator of the right-hand side of Equ. 15 should be positive definite in principle. This quantity fluctuates in the MC sampling, however. The variance can be much larger than the mean value, producing vanishing or negative values for the denominator. The calculation then becomes unstable. This is called the minus-sign problem in general, and is indeed a common and serious difficulty in quantum Monte Carlo calculations, although it may not occur for the present toy Hamiltonian. We note that there are special cases without the minus-sign problem [1].

3.2. Basic Idea of QMCD

We now move on to the QMCD method which is the major subject of this lecture. We then return to Equ. 10, which was written as

$$|\Phi_g\rangle \sim \sum_{MC:\sigma} e^{-\beta h(\sigma)} |\Phi^{(0)}\rangle. \tag{16}$$

For this equation, we introduce a new interpretation that the process on the right-hand side should produce all basis vectors needed for describing the ground state. We can therefore introduce a basis vector

$$|\Phi(\sigma)\rangle \propto e^{-\beta h(\sigma)} |\Psi^{(0)}\rangle. \tag{17}$$

By having different values of the random number, σ, we can generate different state vectors, $|\Phi(\sigma)\rangle$'s by Equ. 17. We diagonalize the Hamiltonian in a subspace spanned by those basis states. To be more precise, let us first suppose that some basis states are already obtained. We then create another basis state $|\Phi(\sigma)\rangle$ from a value of the random number σ. We diagonalize the Hamiltonian in a subspace spanned by the previously obtained basis states and this newly created state. One can then check contribution of this state for lowering the energy eigenvalue being calculated. If the contribution from this new basis vector is not sufficient, one forgets this sate, and should look for another basis vector generated from another value of the random number σ. We repeat this iterative process and obtain more basis states, until we come to reasonable convergence of the energy eigenvalue. One can thus select only important basis vectors. By having such selected basis vectors, the ground state energy and wave function are obtained as a result of the diagonalization of the Hamiltonian in a subspace spanned by these basis vectors. This is expected to be a good approximation provided that all important bases are included. We have thus proposed a new method, called QMCD method [2]. In the following subsections, we explain practical processes for realistic calculations.

3.3.   Monte Carlo Method for General Cases

A general shell model Hamiltonian should consists of single particle energies and a two-body interaction:

$$H = \sum_{i,j=1}^{N_{sp}} \epsilon_{ij} c_i^\dagger c_j + \frac{1}{4} \sum_{i,j,k,l=1}^{N_{sp}} v_{ijkl} c_i^\dagger c_j^\dagger c_l c_k, \tag{18}$$

where $c_i^\dagger$ (c_i) denotes the creation (annihilation) operator of a nucleon in a single particle state i. The dimension of the single particle states is denoted as N_{sp}; $N_{sp} = 24$ for the sd shell and 40 for the pf shell. This Hamiltonian can be rewritten in the quadratic form of one-body operators O_α:

$$H = \sum_{\alpha=1}^{N_f} (E_\alpha O_\alpha + \frac{1}{2} V_\alpha O_\alpha^2), \tag{19}$$

where the number of the O_α's, called N_f, can be at most N_{sp}^2 and usually appears to be much smaller. For the Hamiltonian in Equ. 19, one can generalize $h(\sigma)$ in Equ. 12 to a one-body Hamiltonian $h(\vec{\sigma}_n)$ as

$$h(\vec{\sigma}_n) = \sum_\alpha (E_\alpha + s_\alpha V_\alpha \sigma_{\alpha n}) O_\alpha, \tag{20}$$

where $s_\alpha = \pm 1$ $(= \pm i)$ if $V_\alpha < 0$ (> 0).

The imaginary time evolution operator $e^{-\beta H}$ becomes in this case

$$e^{-\beta H} \approx \int_{-\infty}^{\infty} \prod_\alpha d\sigma_\alpha \left(\frac{\beta |V_\alpha|}{2\pi}\right)^{1/2} \cdot G(\sigma) \cdot e^{-\beta h(\vec{\sigma})}, \tag{21}$$

where $\vec{\sigma}$ means a set of random numbers (auxiliary fields). This is an approximate relation because H contains many terms in general, but becomes exact at the limit of small β. The Gaussian weight factor $G(\sigma)$ in the general case is defined by

$$G(\sigma) = e^{-\sum_\alpha \frac{\beta}{2}|V_\alpha|\sigma_\alpha^2}. \tag{22}$$

Equ. 21 can be applied to smaller imaginary time intervals by dividing β into N_t time slices:

$$e^{-\beta H} = \prod_{n=1}^{N_t} e^{-\Delta \beta H}, \tag{23}$$

where $\Delta\beta = \beta/N_t$. In this case, Equ. 21 is applied for individual small interval $\Delta\beta$ instead of β.

3.4.   Basis Generation in General Cases

The basis generation process for general cases is outlined as follows (See [2] also).

1 We take an initial intrinsic state $|\Psi^{(0)}\rangle$ which can be determined by a mean-field method, for instance, Hartree-Fock. The initial energy is calculated as $E^{(0)} = \langle \Psi^{(0)} | H | \Psi^{(0)} \rangle$.

2 A set, σ, of random numbers (auxiliary fields) is given.

3 We calculate a wave function $|\Phi(\sigma)\rangle$ for the present set σ:

$$|\Phi(\sigma)\rangle \propto \prod_{n=1}^{N_t} e^{-\Delta\beta h(\vec{\sigma}_n)} |\Psi^{(0)}\rangle. \tag{24}$$

where all time slices in Equ. 23 are included, and the collective symbol σ means $\{\vec{\sigma}_1, \vec{\sigma}_2, \cdots, \vec{\sigma}_{N_t}\}$ with $\vec{\sigma}_i$ corresponding to the i-th time slice.

4 The state $|\Phi(\sigma)\rangle$ is a candidate for the new basis vector. It is ortho-normalized, by means of the Gram-Schmidt method, with respect to all other basis states obtained previously.

5 By including this state, we diagonalize the Hamiltonian H, and obtain an improved ground state energy E.

6 The energy eigenvalue E is compared to the eigenvalue obtained in the previous iteration. If the present eigenvalue is lowered sufficiently compared to the eigenvalue in the previous iteration, $|\Phi(\sigma)\rangle$ is adopted as the new basis state. Otherwise, $|\Phi(\sigma)\rangle$ is discarded. Steps from (ii) to (v) are iterated.

The iteration is continued until the the energy E converges reasonably well.

The adopted basis states are called **QMCD bases**. The number of QMCD bases is referred to as the **QMCD basis dimension**, which is increased as the above steps are repeated. We emphasize that energies and wave functions are determined by the diagonalization, and that we can obtain excited states as well as the ground state, for instance, by monitoring energy eigenvalues of excited states.

In practical shell model calculations, it is convenient to adopt basis states in the form of Slater determinants:

$$|\Phi\rangle = \prod_{\alpha=1}^{N} a_\alpha^\dagger |-\rangle, \qquad (25)$$

where N denotes the number of valence nucleons, $|-\rangle$ is an inert spherical core, and $a_\alpha^\dagger$ represents the nucleon creation operator in a canonical single-particle state α, which is a linear combination of the spherical bases.

$$a_\alpha^\dagger = \sum_{i=1}^{N_{sp}} c_i^\dagger D_{i\alpha}. \qquad (26)$$

We can specify the basis state $|\Phi\rangle$ in terms of an $N_{sp} \times N$ complex matrix D. Note that the operation of an exponential of any one-body operator $T = \sum_{ij} T_{ij} c_i^\dagger c_j$ on a state gives a new matrix $D'_{i\alpha} = \sum_j T_{ij} D_{j\alpha}$, while the form of a Slater determinant remains.

The above procedure is the first version of QMCD. Although it works quite well for simple systems [2], it turned out that, in order to carry out realistic large-scale shell-model calculations, we have to improve the method as discussed in the next section.

4.   Improvement of QMCD

4.1.   Basis Generation around the Mean Field Solution

Since the number of manageable bases is finite in practice, we should select bases of higher importance. We thus choose $|\Psi^{(0)}\rangle$ from a Hartree-Fock local minima or some states of similar nature, although $|\Psi^{(0)}\rangle$ is not specified in Equ. 24. In fact, the Hartree-Fock ground state is the "best" single Slater determinant by definition, and fits well to the present scheme.

At the same time, $h(\vec{\sigma}_n)$ in Equ. 20 is rewritten as

$$h(\vec{\sigma}_n) = h_{\mathrm{MF}} + \sum_\alpha V_\alpha s_\alpha \sigma_\alpha O_\alpha, \qquad (27)$$

where $h_{\mathrm{MF}} = \sum_\alpha (E_\alpha + V_\alpha c_\alpha) O_\alpha$, with c_α's denoting c-numbers (see Equs. (4-5) of [4]). By choosing appropriate values for the c_α's, h_{MF} can be set a mean-field potential, for instance, Hartree-Fock potential of the given shell-model Hamiltonian. Since h_{MF} is independent of the auxiliary fields σ_α and the sampling is made around $\sigma_\alpha = 0$, the QMCD bases $|\Phi(\sigma)\rangle$ are generated around local minima of the relevant mean fields. This process is useful for yrast states by combining with the angular momentum projection discussed later. On the other hand, this process is not very relevant to non-yrast states, unless states in a pronounced local minima are studied.

4.2.   Restoration of Angular Momentum

Since a nucleus has rotational symmetry, the restoration of the total angular momentum is quite crucial. The basis state in Equ. 24 is not an eigenstate of the angular momentum, however. This is because the initial state $|\Psi^{(0)}\rangle$ breaks the rotational symmetry in general, and moreover the operator $e^{-\Delta\beta h(\vec{\sigma}_n)}$ has nothing to do with any symmetry in general.

In the QMCD generation of the basis vectors, the restoration of the rotational symmetry should be fulfilled because the eigenstate has this symmetry. However, this is a very slow process. We therefore restore the rotational symmetry by means of angular momentum projection of matrix elements [5]. More details of this treatment can be found in [5].

4.3.   Compression of QMCD Basis Space

In the first version of QMCD, the QMCD bases are generated according to Equ. 24 [2]. This version has been shown to be good for simple systems [2]. It was realized, however, that this version is not efficient enough for handling realistic shell model systems [4], even after implementing the angular momentum projection.

If one generates basis states by a stochastic method, each basis should contain unnecessary fluctuations which are nothing but noise components. This deficiency is inherent to the stochastic process. We therefore revise the basis-generation method so that each basis contains more relevant components lowering the energy and less irrelevant components to be canceled by other bases. The Hilbert space used for the Hamiltonian diagonalization can then be much compressed. In fact, this *basis compression* enables us to carry out some QMCD calculations which are otherwise practically infeasible. This compression process is one of the characteristic differences of the QMCD method from other quantum Monte Carlo approaches.

5.   Test of QMCD for ^{48}Cr

By combining all the above improvements, the current version of QMCD has been constructed [5]. The validity of QMCD has been confirmed by comparing to the result of the exact diagonalization of the same Hamiltonian. Here, the nucleus ^{48}Cr is taken, and the exact result is obtained from Ref. [9]. Fig. 1 shows energies of yrast states of ^{48}Cr.

As shown in Fig. 1, the ground-state energy has been reproduced within 130 keV with the QMCD dimension 40. This result is already ~ 100 keV below the lower edge of the error bar of the SMMC result with the temperature $T = 0.5$ MeV [10]. Since the angular momentum projection is made for each basis, the addition of a new basis (i.e., increase of the dimension) implies inclusion of more dynamical degrees of freedom.

6.   Summary

In summary, we have presented a new method for solving quantum many-body problems where particles (fermions or bosons) are interacting through a two-body interaction. We have presented from the first version of the QMCD formulation to the current version including the compression of the basis space and the precise treatment of the angular momentum. In the QMCD calculation, favorable bases are selected based upon their contribution to the energy eigenvalue of the state being

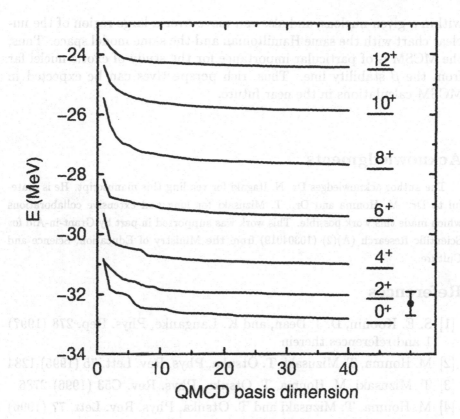

Figure 1 Energies of yrast states of ^{48}Cr obtained by QMCD calculation compared with the energies obtained by the exact diagonalization [9]. The QMCD energy eigenvalues are shown as functions of the basis dimension. The point with error bar at far right is the ground state energy of the SMMC calculation with finite temperature $T = 0.5\ MeV$ [10].

investigated. In other words, the basis vectors are taken according to their "importance". This nice feature can be expressed by (stochastic) **importance truncation** to the exact calculation [11]. There is a natural relation to the mean-field solutions which provide us with the best single Slater determinant, *etc.*, by definition. Note that the minus-sign problems of other quantum Monte Carlo methods are irrelevant to the QMCD method.

The application of QMCD to the nuclear shell model is called **Monte Carlo Shell Model (MCSM)**. There have several applications, which are discussed in my lecture but cannot be included in this note. I just mention published references [12, 13].

Because one can handle many valence orbits and many valence particles in the MCSM calculations, one can describe a wide variety of states within a given nucleus, and also can move over a large region of the nuclear chart with the same Hamiltonian and the same model space. Thus, the MCSM is of particular importance for the study of exotic nuclei far from the β stability line. Thus, rich perspectives can be expected in MCSM calculations in the near future.

Acknowledgments

The author acknowledges Dr. N. Itagaki for reading this manuscript. He is grateful to Dr. M. Honma and Dr. T. Mizusaki for long and extensive collaborations which made this work possible. This work was supported in part by Grant-in-Aid for Scientific Research (A)(2) (10304019) from the Ministry of Education, Science and Culture.

References

[1] S. E. Koonin, D. J. Dean, and K. Langanke, Phys. Rep. 278 (1997) 1 and references therein

[2] M. Honma, T. Mizusaki, T. Otsuka, Phys. Rev. Lett. 75 (1995) 1284

[3] T. Mizusaki, M. Honma, T. Otsuka, Phys. Rev. C53 (1996) 2786

[4] M. Honma, T. Mizusaki and T. Otsuka, Phys. Rev. Lett. 77 (1996) 3315

[5] T. Otsuka, M. Honma and T. Mizusaki, Phys. Rev. Lett. 81 (1998) 1588

[6] C.A. Ur, et al., Phys. Rev. C53 (1996) 2786

[7] J. Hubbard, Phys. Rev. Lett. 3 (1959) 77

[8] R.L. Stratonovich, Dokl. Akad. Nauk. SSSR 115 (1957) 1097 [transl: Soviet Phys. Dokl. 2 (1957) 416]

[9] E. Caurier, et al., Phys. Rev. C50 (1994) 225

[10] K. Langanke, et al., Phys. Rev. C52 (1995) 718

[11] T. Otsuka, T. Mizusaki, and M. Honma, J. Phys. G25 (1999) 699

[12] T. Mizusaki et al., Phys. Rev. C59 (1999) R1846

[13] Y. Utsuno et al., Phys. Rev. C60 (1999) 054315

Variational Approach to Coexistence Phenomena in N=Z Nuclei

A. Petrovici

National Institute for Physics and Nuclear Engineering,
R-76900 Bucharest, Romania

spetro@ifin.nipne.ro

Keywords: Nuclear structure, shape coexistence, proton-rich nuclei

Abstract   Results are presented concerning shape coexistence and neutron-proton correlations at low and high spins in the $N = Z$ nuclei ^{68}Se and ^{56}Ni obtained within the *complex* version of the Excited Vampir variational approach using a rather large model space and realistic effective interactions. Oblate-prolate coexistence at low and intermediate spins dominates the structure of the nucleus ^{68}Se. The coexistence of small and moderately deformed oblate and prolate configurations, strongly mixed in some cases, may explain the irregularities observed for the yrast line, as well as the presence of other nonyrast states in the nucleus ^{56}Ni. The first deformed band is well reproduced and predictions on the structure of the next deformed bands are presented.

1.   Introduction

Many nuclear structure problems require the use of single particle basis-systems which are far too large to allow for a complete diagonalization of a suitable chosen effective many-nucleon Hamiltonian as it is done in the Shell-model Configuration-Mixing (SCM) approach [1]. The microscopic description of the shape coexistence and shape transition phenomena dominating the structure of the $N \simeq Z$ medium mass nuclei is an obvious example. For such problems one has therefore to truncate the complete SCM expansion of the nuclear wave functions to a numerically feasible number of A-nucleon configurations without loosing the essential degrees of freedom relevant for the particular states under consideration. In order to avoid the difficulties associated with the truncation prescriptions we did follow in the last years another avenue which starts from

103

the ideas of mean-field theory and selects the relevant degrees of freedom directly from the nuclear Hamiltonian via variational procedures. The *complex* Vampir approaches are completely microscopic procedures that employ chains of variational calculations based on symmetry-projected and essentially *complex* Hartree-Fock-Bogoliubov (HFB) vacua which include neutron-proton as well as general unnatural-parity two nucleon correlations. These models can account for the delicate interplay between collective and single particle degrees of freedom, treat like-nucleon and neutron-proton pairing correlations on the same footing and are numerically feasible for rather large model spaces and general two-body forces. The shape coexistence phenomena dominating the structure of even mass nuclei in the $A \simeq 70$ region at low and high spins have been described rather well by a variable mixing of more or less deformed prolate and oblate configurations in the frame of the *complex* Excited Vampir and Excited Fed Vampir models [2-9].

Recently, two distinct rotational bands were found in the $N = Z$ nucleus ^{68}Se, the ground state band having properties consistent with collective oblate rotation, and the excited band having characteristics consistent with prolate deformation [10, 11]. These observations require selfconsistent microscopic calculations which can afford a unified description of the properties of low as well as high spin states. On the other hand the doubly magic nucleus ^{56}Ni and the neighboring nuclei are expected to manifest delicate interplay between single particle and collective degrees of freedom. Recently, the observation of two rotational bands beside the irregular sequence of yrast states, as well as some other medium spin nonyrast states in ^{56}Ni have been reported [12]. The simultaneous description of the properties of the yrast and nonyrast states including the identified deformed bands in medium mass $N \simeq Z$ nuclei is a challenge for the theoretical nuclear models. Furthermore, since these nuclei are accessible for the most sophisticated microscopical theoretical models presently developed, they may also help to improve our knowledge on the effective interactions in rather large model spaces. The structure of the nucleus ^{56}Ni has been studied by extensive large-scale shell-model and cranked Hartree-Fock and Hartree-Fock-Bogoliubov calculations [12] and also by the Monte Carlo shell model based on Quantum Monte Carlo diagonalization method (QMCD) [13, 14].

In the following section we shall briefly describe the *complex* Excited Vampir variational procedure used for the presently reported results and define the effective Hamiltonian. In section 3 we shall discuss the results on shape coexistence in ^{68}Se and ^{56}Ni. Finally we shall present some conclusions in section 4.

Z. A. Hosaini et al. (eds.), Nuclei Far from Stability and Astrophysics, 103–114.
© 2001 Kluwer Academic Publishers. Printed in the Netherlands.

2.   Theoretical Framework

In the variational procedure described below the relevant degrees of freedom are determined dynamically by the effective Hamiltonian of the considered system solving the general HFB problem with symmetry projection before variation. In order to investigate the lowest few states of a given spin and parity in a particular nucleus first the Vampir solutions obtained by starting from intrinsically prolate and oblate deformed trial configurations are constructed. The most bound ones are selected as the optimal mean-field description of the yrast states by single symmetry-projected HFB determinants. Then the Excited Vampir approach is used to construct additional excited states by independent variational calculations. Finally, for each considered spin and parity the residual interaction between the lowest orthogonal configurations is diagonalized.

We define the model space as in our earlier calculations for nuclei in the $A \simeq 70$ mass region [8]: a ^{40}Ca core is used and the valence space consists out of the $1p_{1/2}, 1p_{3/2}, 0f_{5/2}, 0f_{7/2}, 1d_{5/2}$ and $0g_{9/2}$ oscillator orbits for both protons and neutrons. For the corresponding single-particle energies we take (in units of the oscillator energy $\hbar\omega = 41.2A^{-1/3}$) $0.040, -0.270, 0.300, -0.560, 0.157$ and 0.029 for the proton, and $-0.070, -0.332, 0.130, -0.690, 0.079$ and -0.043 for the neutron levels, respectively.

As effective two-body interaction a renormalized nuclear matter G-matrix based on the Bonn One-Boson-Exchange potential (Bonn A) is used. It is worthwhile to mention a few particular aspects of the renormalization procedure. The G-matrix is modified by two short range ($0.707\ fm$) Gaussians for the isospin $T = 1$ proton-proton and neutron-neutron matrix elements with strengths of $-40\ MeV$ and $-30\ MeV$, respectively. The isoscalar spin 0 and 1 particle-particle matrix elements are enhanced by an additional Gaussian with the same range. Assuming that for $N \simeq Z$ nuclei the interaction should be almost charge symmetric, it was considered well justified to introduce an additional short-range (same range as above) Gaussian for the isospin $T = 1$ neutron-proton matrix elements of strength $-35\ MeV$. All these modifications enhance the various pairing correlations and thus have the tendency to lessen the deformation of the considered nucleus. Indeed the onset of deformation has to be influenced by additional modifications. The interaction contains monopole shifts for some of the diagonal isospin $T = 0$ and $T = 1$ matrix elements. Some of these shifts had been introduced in our earlier calculations in order to influence the onset of deformation in the $A \simeq 70$ mass region [8]. However, new monopole shifts have been introduced for nuclei belonging to the $A \simeq 60$ mass region [9].

3.   Results and Discussion

3.1.   The Nucleus ^{68}Se

In the nucleus ^{68}Se we calculated the lowest few positive-parity states up to spin 16^+. The first 0^+ minimum is oblate deformed ($\beta_2 = -0.32$) in the intrinsic system, the orthogonal spherical minimum ($\beta_2 = 0.05$) is situated 52 keV above it and the prolate deformed ($\beta_2 = 0.34$) configuration is 300 keV higher in energy with respect to the latter. We should mention that changing the $T = 0$ monopole shifts involving neutrons and protons occupying the spherical orbitals $0f_{7/2}(0f_{5/2})$ and $0g_{9/2}$ by 40 keV the relative position of the deformed oblate and prolate minima is reversed. For spin 2^+ the lowest minimum is also oblate deformed and separated from the first prolate one by 195 keV. Already at spin 4^+ the first minimum is prolate deformed and is situated 58 keV below the oblate one. The Vampir solution is prolate deformed for the higher spins. The first excited state is oblate deformed up to spin 10^+, but at higher angular momenta other prolate deformed states become energetically favored with respect to the oblate one. The final wave functions indicate strong oblate-prolate mixing not only for low spins, but also for some intermediate spin states, like 10^+ and 12^+.

At intermediate spins the Excited Vampir states are decaying by many $E2$ branches, as it is illustrated by the level scheme presented in Fig. 1. The significant $B(E2; I \rightarrow I - 2)$ strengths are given in Tab. 1. The strengths for the secondary branches are indicated in parentheses. In order to illustrate the collectivity of the investigated states we give also the $B(E2)$ strengths (in the square brackets) for branches which are not shown in Fig. 1. As effective charges we used here $e_p = 1.5$ and $e_n = 0.5$ for protons and neutrons, respectively. We organized the states in bands according to the $B(E2)$ values connecting them. The strong mixing of states results not only in a strong fragmentation of the total $B(E2)$ decay of a given state, but is responsible for the fact that some of the calculated states can not be attached to any particular band structure. Also the decay path indicates that some of the higher bands prefer at spins 12^+ and 14^+ to decay to some other bands. The bands labeled $o(p)_1$ and $p(o)_3$ manifest a particular structure. The first (second) 4^+ state is based on a prolate (oblate) configuration, but the main decaying branch of this state takes place to yrast (first excited) 2^+ state based on an oblate (prolate) configuration. Thus, the quadrupole deformation of the sates building the two discussed bands is changing

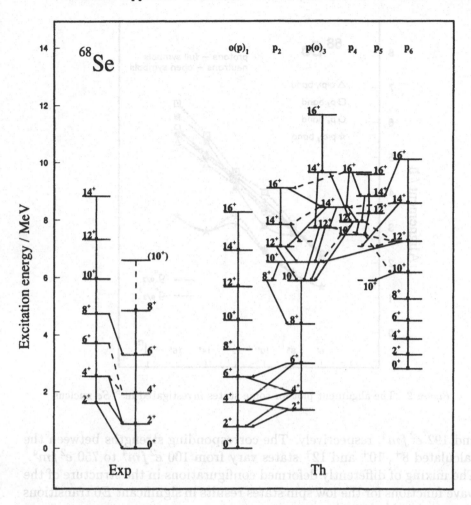

Figure 1 The theoretical spectrum of ^{68}Se for even spin positive parity states calculated within the *complex* Excited Vampir approximation is compared to the experimental results [12]. The labels o and p are for states based on intrinsically oblate and prolate deformed configurations, respectively. The $M1$, $\Delta I = 0$ transitions are indicated by dashed lines.

sign with increasing spin. A particular situation was found at spin 16^+. The yrast 16^+ state shows a main $B(E2)$ decaying branch to the second 14^+ state and a secondary one to the third 14^+ state, but energetically it is situated 57 keV below the second 14^+. Since we did not calculate the negative-parity states we do not have results concerning the possible $E1$ decays of the yrast 16^+ state.

It is worthwhile to mention that we found strong $E2$, $\Delta I = 0$ transitions. The $B(E2)$'s for transitions between the first excited and the yrast state for spins 2^+, 4^+, 6^+ amounts to 1309 $e^2 fm^4$, 1237 $e^2 fm^4$

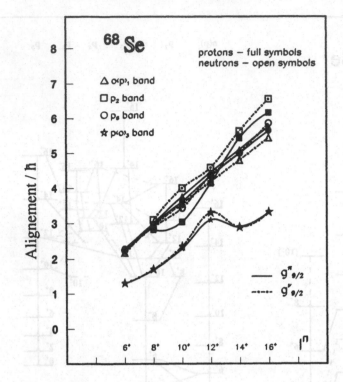

Figure 2 The alignment plot for some states investigated in ^{68}Se nucleus.

and 192 e^2fm^4, respectively. The corresponding strengths between the calculated 8^+, 10^+ and 12^+ states vary from 100 e^2fm^4 to 750 e^2fm^4. The mixing of differently deformed configurations in the structure of the wave functions for the low spin states results in significant $E0$ transitions for spin 0^+, 2^+ and 4^+ states.

At intermediate spins the states are also connected by $M1$, $\Delta I = 0$ transitions. In Fig. 2 we present the alignment plot which represents the contribution of the $g_{9/2}$ protons and neutrons to the total angular momentum. As it is expected, in all the illustrated bands protons and neutrons contribute about equal amounts: we obtained simultaneous alignment. This can also be inferred from the calculated g-factors. For all the investigated states theory yields values around 0.5. We should mention here that the alignment is smaller for the band $p(o)_3$ dominated by oblate deformed configurations at higher spins.

To summarize, the *complex* Excited Vampir results indicate a strong influence of particular $T = 0$ matrix elements of the effective Hamiltonian involving neutrons and protons occupying the $0f_{5/2}$ ($0f_{7/2}$) and $0g_{9/2}$ spherical orbits on the structure of the yrast and first excited band at low and intermediate spins. The oblate-prolate coexistence and mix-

Table 1 $B(E2; I \to I - 2)$ values (in $e^2 fm^4$) for some states of the nucleus ^{68}Se presented in Fig. 1. The strengths for the secondary branches are given in parentheses and the labels indicate the end point of the transition. In Fig. 1 the transitions corresponding to the strengths given in brackets are not shown. As effective charges $e_p = 1.5$ and $e_n = 0.5$ have been used.

$I[\hbar]$	$o(p)_1$	p_2	$p(o)_3$	p_4	p_5	p_6
2^+	966					572
4^+	1381 $(156)p(o)_3$		1381 $(144)o(p)_1$			1190
6^+	1402 $(330)p(o)_3$		1340 $(315)o(p)_1$			1628
8^+	1710 $[38]p(o)_3$		1690 $[46]o(p)_1$			1843 $[48]o(p)_1$
10^+	1656	1312	1612			1820 $[71]o(p)_1$ $[49]p(o)_3$
12^+	1537	882 $(686)p(o)_3$ $[262]p_4$	686 $(708)p_2$	718 $(317)p(o)_3$	$437p_4$ $[76]p_2$	1756 $(169)p_2$ $[74]o(p)_1$
14^+	1435	1026 $(148)p_6$	997 $(497)p_4$ $[73]p_2$	448 $(610)p(o)_3$ $(258)p_2$	973 $(352)p_4$ $[54]p_6$	1753 $(88)p_2$
16^+	1311	590 $(436)p_4$ $[285]p_5$	1603	664 $(227)p_5$ $[259]$	728 $[436]$	1688 $(116)p_5$

ing are very sensitive to small changes of the strengths of these matrix elements around the value used for this mass region.

3.2.   The Nucleus ^{56}Ni

In the nucleus ^{56}Ni recent experimental results clearly indicate the coexistence of an irregular sequence of yrast states (identified up to a tentative angular momentum 14^+) and two deformed bands (extended up to tentative angular momenta 12^+ and 17, respectively) [12].

We investigated the nature of the lowest few even spin positive parity states up to $I^\pi = 16^+$ within the *complex* Excited Vampir approach using the above mentioned large model space. These are the first microscopic calculations for ^{56}Ni including in the model space the $0g_{9/2}$ and $1d_{5/2}$ spherical orbitals besides the full pf-shell. In order to study the possible mixing in the structure of the lowest states for a given spin up to eight Excited Vampir configurations have been taken into account. Strong mixing with the dominant component representing less than 70% in the final wave function was obtained for particular states of spin 0^+, 2^+, 4^+, 8^+, 10^+, 12^+, 16^+. For all investigated states the dominant projected determinant for the yrast solution has a small oblate deformation in the intrinsic system (slowly varying with the spin), except for spins 0^+ and 6^+ characterized by a small prolate deformation. The second minimum for all investigated states is a strongly prolate deformed one except for the spin 10^+. At spin 10^+ three almost spherical minima are energetically favored with respect to the strongly prolate deformed one. It is worthwhile to mention that the first eight orthogonal minima at spin 10^+ are bunched in less than 2.1 MeV and the third and the fifth minimum are separated by only 130 keV. This fact may explain the strong mixing characterizing the structure of the final Excited Vampir wave functions for spin 10^+. It should be mentioned that even after the diagonalization of the residual interaction the lowest five 10^+ states are still bunched in 1.450 MeV. A similar situation is found for the 8^+ states. At spin 8^+ the lowest five orthogonal minima are separated by 1.6 MeV before and also after the diagonalization of the residual interaction in the eight-dimensional A-nucleon basis. The highest density of states is found for spin 12^+. The separation energy between the second and the fifth orthogonal minimum at spin 12^+ is 325 keV and 400 keV before and after the diagonalization of the residual interaction, respectively.

Except for the strongly deformed prolate minima, the calculated Excited Vampir configurations building the states presented in Fig. 3 are characterized in the intrinsic system by moderate or small quadrupole deformations. The first well deformed prolate minimum is characterized in the intrinsic system by $\beta_2 \simeq 0.3$ up to spin 12^+. At spin 14^+ the corresponding deformation is decreasing to $\beta_2 = 0.26$, and to $\beta_2 = 0.20$ at spin 16^+. The second, even more deformed minimum ($\beta_2 \simeq 0.32$) appears as the fourth minimum at spin 14^+, as a third one at 16^+ and becomes yrast at spin 18^+. These latest deformed configurations are characterized by the promotion of one proton and one neutron in the

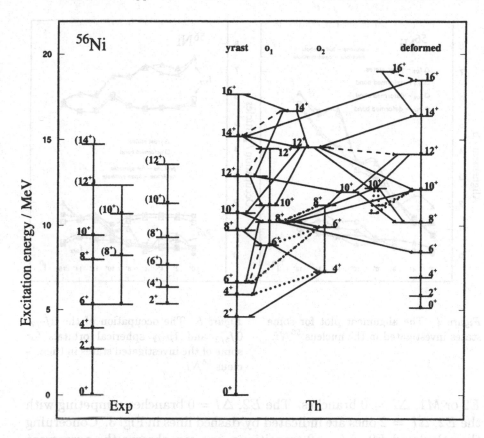

Figure 3 The theoretical spectrum of ^{56}Ni for even spin positive parity states calculated within the *complex* Excited Vampir approximation is compared to the experimental results [12]. The labels *o* and *p* are for states based on intrinsically oblate and prolate deformed configurations, respectively. Significant $E2$ and $M1$, $\Delta I = 0$ transitions are indicated by dashed and dotted lines, respectively.

$0g_{9/2}$ orbital. The structure of the wave functions at intermediate and high spins suggests that the dimension of the Excited Vampir basis has to be increased in order to improve the accuracy of the final solutions.

The high density of states for a given spin and the strong mixing of states revealed by the structure of the wave functions result in a very complicated decay pattern as illustrated in Fig. 3 and Tab. 2. Beside the yrast line we organized the states in a deformed band and short links solely on the basis of the $B(E2)$ values connecting them. Then we tried to find the fastest ways to enter the yrast line and so to identify the most probable candidates for the experimental linkage of the gamma rays in bands. Due to the very strong mixing of some intermediate spin states, either the $B(E2)$ strength for the decay of the corresponding states is strongly fragmented, or competitive decay paths are represented by some

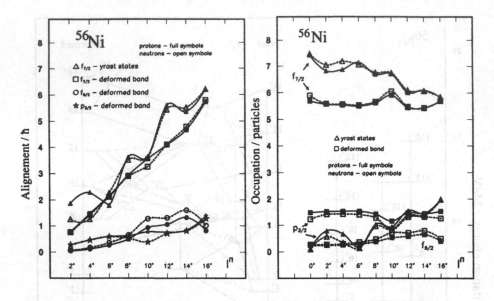

Figure 4 The alignment plot for some states investigated in the nucleus ^{56}Ni.

Figure 5 The occupation of the $0f_{7/2}$, $0f_{5/2}$ and $1p_{3/2}$ spherical orbitals for some of the investigated states in the nucleus ^{56}Ni.

$E2$ or $M1$, $\Delta I = 0$ branches. The $E2$, $\Delta I = 0$ branches competing with the $E2$, $\Delta I = 2$ ones are indicated by dashed lines in Fig. 3. Concerning the calculated $E2$, $\Delta I = 2$ transitions, one can observe the agreement between the measured $B(E2; 2_1^+ \rightarrow 0_1^+) = 120_{-24}^{+24}\ e^2 fm^4$ [15] and the theoretical value $115\ e^2 fm^4$ obtained for the corresponding transition.

The alignment plot giving the angular momentum contribution of the $0f_{7/2}$, $0f_{5/2}$ and $1p_{3/2}$ neutrons and protons in the direction of the total angular momentum is presented in Fig. 4. The changes in structure with increasing spin for the yrast line suggested by the irregularities observed in the level scheme appear also in the alignment plot for the particles occupying the $0f_{7/2}$ spherical orbital. This fact is corroborated by the trends in the evolution of the occupation of some single particle orbitals building the model space presented in Fig. 5.

In summary, we obtained a complex structure for the even spin positive parity states up to spin $I^\pi = 16^+$ in ^{56}Ni. Small deformed oblate and prolate configurations (almost spherical for particular spins), strongly mixed in some cases, may explain the irregularities observed for the yrast line, as well as the presence of other non-yrast states connected with them by $E2$ or $M1$ transitions. The first deformed band is well reproduced and predictions concerning the next deformed even spin band indicates the occupation of the $0g_{9/2}$ spherical orbital by one proton and

Table 2 $B(E2; I \rightarrow I - 2)$ values (in $e^2 fm^4$) for some states of the nucleus ^{56}Ni presented in Fig. 3. The strengths for the secondary branches are given in parentheses and the labels indicate the end point of the transition. In Fig. 3 the transitions corresponding to the strengths given in brackets are not shown. As effective charges $e_p = 1.5$ and $e_n = 0.5$ have been used.

$I[\hbar]$	yrast	o_1	o_2	deformed
2^+	115			348
	[10][34][20][14]			[81]
4^+	94		56	615
	[18]		[50]	[11]
6^+	44	86	57	673
	[30]	$(33)o_2[17][22][79]$	[32]	[24]
8^+	3	163	59	563
	$(39)o_1$	$(38)yrast$		
		$(134)deformed(41)o_2$	[12][28]	[10][30][13][29]
10^+	2	29		315
	$(36)o_1[11]$	$(25)yrast[10]$		$(89)o_1$
12^+	3	19		360
	$(12)o_1(76)$	[10][19]		$(46)(209)$
14^+	68			520
	$(7)o_1(8)$			(27)
16^+	33			222
	(24)			$(22)yrast$

one neutron. The Excited Vampir decaying paths for the other nonyrast sequences of states offer a microscopic description for the experimental states identified up to now and could guide the future experiments.

4.   Summary

We presented new microscopic results concerning the shape coexistence in ^{68}Se and ^{56}Ni nuclei at low and high spins. A variable mixing of prolate and oblate more or less deformed configurations could explain the irregularities observed at low spins. The complex band structures observed at high spins can be explained by a strong mixing of coexisting states based on configurations having different quadrupole, hexadecapole and octupole deformations in the intrinsic system as well as different pairing properties. The whole picture for the low spin states is strongly dependent on particular neutron-proton correlations.

Acknowledgments

The results reported here have been obtained in collaboration with Prof. K.W. Schmid and Prof. Amand Faessler from the University of Tübingen, Germany. The work was partly supported by the Internationales Büro des Bundesministerium für Bildung und Forschung under the WTZ contract RUM-040-97.

References

[1] P.J. Brussard and P.W.M. Glaudemans, *Shell-Model Applications in Nuclear Spectroscopy*, North-Holland, Amsterdam, 1977; E. Caurier *et al.*, Phys. Rev. C50 (1994) 225

[2] A. Petrovici, K.W. Schmid, A. Faessler, Nucl. Phys. A571 (1994) 77

[3] A. Petrovici, K.W. Schmid, A. Faessler, Nucl. Phys. A605 (1996) 290

[4] A. Petrovici et al., Phys. Rev. C53 (1996) 2134

[5] A. Petrovici, K.W. Schmid, A. Faessler, Z. Phys. A359 (1997) 19

[6] A. Petrovici, K.W. Schmid, A. Faessler, Nucl. Phys. A647 (1999) 197

[7] A. Petrovici et al., Progr. Part. Nucl. Phys. 43 (1999) 485

[8] A. Petrovici, K.W. Schmid, A. Faessler, Nucl. Phys. A665 (2000) 333

[9] A. Petrovici, K.W. Schmid, A. Faessler, will appear in Nucl. Phys. A

[10] C. Baktash and S.D. Paul, private communication

[11] S.M. Fischer et al., Phys. Rev. Lett. 84 (2000) 4064

[12] D. Rudolph et al., Phys. Rev. Lett. 82 (1999) 3763

[13] T. Otsuka, M. Honma, T. Mizusaki, Phys. Rev. Lett. 81 (1998) 1588

[14] T. Mizusaki et al., Phys. Rev. C59 (1999) R1846

[15] G. Kraus et al., Phys. Rev. Lett. 73 (1994) 1773.

Relativistic Theory of Nuclear Structure far from the Valley of β-Stability

P. Ring

Physics Department of the Technical University Munich
*D-85748 Garching, Germany**
ring@ph.tum.de

Keywords: Exotic nuclei, relativistic mean field theory

Abstract   Exotic nuclei far from the line of beta-stability have gained considerable
interest in recent years both on the experimental as on the theoretical
side. New phenomena are expected such as dramatic changes in the
nuclear shell structure. A reliable description of the spin-orbit term
and its isospin dependence is therefore necessary. On the other hand
the Fermi surface comes very close to the continuum limit in this region.
A careful treatment of the continuum and its coupling to the states in
the Fermi see is essential. Therefore we report on investigations in
the framework of relativistic Hartree-Bogoliubov (RHB) theory in the
continuum for nuclei close to the drip lines. Depending on the shell
structure at the continuum limit we find the formation of neutron skins
and neutron halos.

1.   Introduction

New accelerators with radioactive beams allow the experimental study
of nuclei far from stability. This allows us to extend our understanding of
nuclear structure considerably because of the large isospin values in this
region. Theoretical investigations predict considerable changes in the
shell structure, which provide the basic understanding of many effects
in nuclear structure.

In recent years Relativistic Mean-Field (RMF) models have been successfully applied in calculations of the properties of nuclear matter and

*Supported in part by the Bundesministerium für Forschung und Bildung under the contract
06 TM 979.

115

D. N. Poenaru et al. (eds.), Nuclei Far from Stability and Astrophysics, 115–126.
© 2001 *Kluwer Academic Publishers. Printed in the Netherlands.*

of finite nuclei throughout the periodic table [1]. Using only six or seven parameters they allow a fully self-consistent description of many nuclear properties with high accuracy. These methods form an effective field theory for the relevant degrees of freedom of QCD at very low energies.

As compared to non-relativistic mean field methods such as density dependent Hartree-Fock (DDHF) calculations these models have the advantage to provide a consistent description of the spin-orbit term and its isospin dependence.

An essential problem in the theoretical description of drip-line nuclei arises from the closeness of the Fermi level to the particle continuum: particle-hole and particle-particle excitations reach the continuum. The coupling between bound states and the particle continuum has to be taken into account explicitly. The Relativistic Hartree Bogoliubov (RHB) theory [2, 3], which is a relativistic extension of the Hartree Fock Bogoliubov theory, provides a unified description of mean-field and pairing correlations. A fully self-consistent RHB theory in coordinate space [4, 5] correctly describes the coupling between bound and continuum states. The theory provides a framework for describing the nuclear many-body problem as a relativistic system of baryons and mesons not only in the valley of β-stability but also in regions with large neutron or proton excess even close to the drip-lines.

2.   The RHB Model

In comparison with conventional non-relativistic approaches, relativistic models explicitly include mesonic degrees of freedom and describe the nucleons as Dirac particles. Nucleons interact in a relativistic covariant manner through the exchange of virtual mesons: the isoscalar scalar σ-meson, the isoscalar vector ω-meson and the isovector vector ρ-meson. The model is based on the one boson exchange description of the nucleon-nucleon interaction. We start from the effective Lagrangian density [6]. To obtain a realistic description of finite nuclei one has to take into account the density dependence of the parameters. This is conventionally done in a phenomenological way in the model proposed by Boguta and Bodmer [7] by non-linear interactions in the meson sector.

The lowest order of the quantum field theory is the *mean-field* approximation. The Dirac equation reads

$$\left\{ -i\boldsymbol{\alpha} \cdot \boldsymbol{\nabla} + \beta(m + g_\sigma\sigma) + g_\omega\omega^0 + g_\rho\tau_3\rho_3^0 + e\frac{(1 - \tau_3)}{2}A^0 \right\}\psi_i = \varepsilon_i\psi_i$$

(1)

The effective mass $m^*(\mathbf{r})$ is defined as

$$m^*(\mathbf{r}) = m + g_\sigma\,\sigma(\mathbf{r}),$$

(2)

and the potential $V(\mathbf{r})$ is

$$V(\mathbf{r}) = g_\omega\, \omega^0(\mathbf{r}) + g_\rho\, \tau_3\, \rho_3^0(\mathbf{r}) + e\frac{(1 - \tau_3)}{2} A^0(\mathbf{r}). \tag{3}$$

In order to describe ground-state properties of spherical open-shell nuclei, pairing correlations have to be taken into account. For nuclei close to the β-stability line, pairing has been included in the relativistic mean-field model in the form of a simple BCS approximation [8]. However, for nuclei far from stability the BCS model presents only a poor approximation. In particular, in drip-line nuclei the Fermi level is found close to the particle continuum. The lowest particle-hole or particle-particle modes are often embedded in the continuum, and the coupling between bound and continuum states has to be taken into account explicitly. The BCS model does not provide a correct description of the scattering of nucleonic pairs from bound states to the positive energy continuum. It leads to an unbound system, because levels in the continuum are partially occupied. Including the system in a box of finite size leads to unreliable predictions for nuclear radii depending on the size of this box. In the non-relativistic case, a unified description of mean-field and pairing correlations is obtained in the framework of the Hartree-Fock-Bogoliubov (HFB) theory in coordinate space [9]. It has be extended to relativistic Hartree Bogoliubov (RHB) theory [2, 3], with the following RHB-equations:

$$\begin{pmatrix} \hat{h} - \lambda & \hat{\Delta} \\ -\hat{\Delta}^* & -\hat{h} + \lambda \end{pmatrix} \begin{pmatrix} U_k \\ V_k \end{pmatrix} = E_k \begin{pmatrix} U_k \\ V_k \end{pmatrix}. \tag{4}$$

where $\hat{h}$ is the Dirac operator given in Equ. 1 and $\hat{\Delta}$ is the pairing field.

HFB-theory, being a variational approximation, results in a violation of basic symmetries of the nuclear system, among which the most important is the non conservation of the number of particles. In order that the expectation value of the particle number operator in the ground state equals the number of nucleons, Equ. 4 contains a chemical potential λ which has to be determined by the particle number subsidiary condition. The column vectors denote the quasi-particle wave functions, and E_k are the quasi-particle energies.

In the coordinate space representation of the pairing field $\hat{\Delta}$ is

$$\Delta_{ab}(\mathbf{r}, \mathbf{r}') = \frac{1}{2} \sum_{c,d} V_{abcd}(\mathbf{r}, \mathbf{r}')\kappa_{cd}(\mathbf{r}, \mathbf{r}'). \tag{5}$$

where a, b, c, d denote all quantum numbers, apart from the coordinate $\mathbf{r}$, that specify the single-nucleon states. $V_{abcd}(\mathbf{r}, \mathbf{r}')$ are matrix elements of a general two-body pairing interaction, and the pairing tensor is defined as

$$\kappa_{cd}(\mathbf{r}, \mathbf{r}') = \sum_{E_k > 0} U_{ck}^*(\mathbf{r}) V_{dk}(\mathbf{r}').$$

(6)

3. Spin-Orbit Reduction

The spin-orbit interaction plays a central role in the physics of nuclear structure. It is rooted in the basis of the nuclear shell model, where its inclusion is essential in order to reproduce the experimentally established magic numbers. In non-relativistic models based on the mean field approximation, the spin-orbit potential is included in a phenomenological way. Of course such an ansatz introduces an additional parameter, the strength of the spin-orbit interaction. The value of this parameter is usually adjusted to the experimental spin-orbit splittings in spherical nuclei, for example ^{16}O. Its isospin dependence would require an additional parameter, which is neglected in most of the DDHF calculations. On the other hand, in the relativistic framework the nucleons are described as Dirac spinors. This means that in the relativistic description of the nuclear many-body problem, the spin-orbit interaction and its isospin-dependence arises naturally from the Dirac-Lorenz structure of the effective Lagrangian. No additional strength parameter is necessary, and relativistic models reproduce the empirical spin-orbit splittings.

Many properties of nuclei along the line of beta stability have been successfully described in the framework of models based on the mean-field approximation. Conventional non-relativistic models that include density dependent interactions with finite range (Gogny) or zero-range (Skyrme) forces, have been extensively used to describe the structure of stable nuclei. More recently, it has been shown that models based on the relativistic mean-field theory [6] provide an elegant and economical framework, in which properties of nuclear matter and finite nuclei, as well as the dynamics of heavy-ion collisions, can be calculated (for a recent review see [1]). In comparison with conventional non-relativistic approaches, relativistic models explicitly include mesonic degrees of freedom and describe the nucleons as Dirac particles. Non-relativistic models and the relativistic mean-field theory predict very similar results for many properties of beta stable nuclei. However, cases have been found where the non-relativistic description of nuclear structure fails. An example is the anomalous kink in the isotope shifts of Pb nuclei [10]. This phenomenon could not be explained neither by the Skyrme model, nor by the Gogny approach. Nevertheless, it is reproduced very naturally in relativistic mean-field calculations. A more careful analysis [11] has shown that the origin of this discrepancy is the isospin dependence of the spin-orbit term.

In the following we present results for the chain of Sn and Ni isotopes. We find that in the framework of relativistic mean field theory, the magnitude of the spin-orbit potential is considerably reduced in light drip line nuclei. With the increase of the neutron number, the effective one-body spin-orbit interaction becomes weaker. This result in a reduction of the energy splittings between spin-orbit partners. The reduction of the spin-orbit potential is especially pronounced in the surface region, and does not depend on a particular parameter set used for the effective Lagrangian. These results are at variance with those calculated with the non-relativistic Skyrme model. It has been shown that the differences have their origin in the isospin dependence of the spin-orbit terms in the two models. If the spin-orbit term of the Skyrme model is modified in such a way that it does not depend so strongly on the isospin, the reduction of the spin-orbit potential is comparable to that observed in relativistic mean-field calculations.

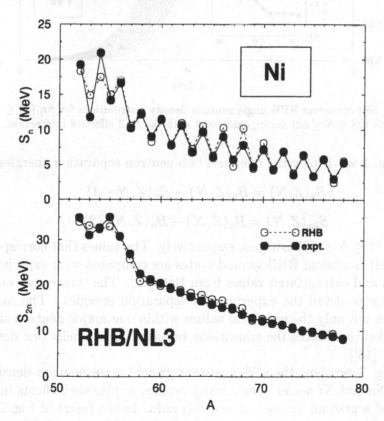

Figure 1 One-neutron and two-neutron separation energies in Ni-isotopes calculated with the parameter set NL3.

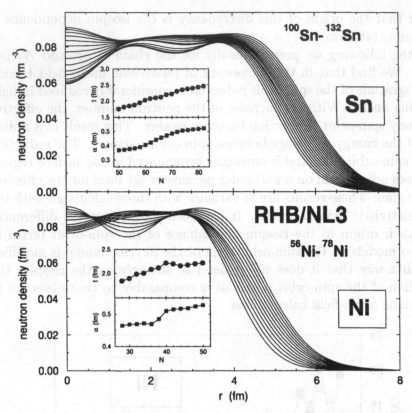

Figure 2 Self-consistent RHB single-neutron density distributions for Sn ($50 \leq N \leq 82$) and Ni ($28 \leq N \leq 50$) nuclei, calculated with the NL3 effective interaction.

In Fig. 1 we display the one- and two-neutron separation energies

$$S_n(Z, N) = B_n(Z, N) - B_n(Z, N-1) \qquad (7)$$

$$S_{2n}(Z, N) = B_n(Z, N) - B_n(Z, N-2) \qquad (8)$$

for Ni ($24 \leq N \leq 50$) isotopes, respectively. The values that correspond to the self-consistent RHB ground-states are compared with experimental data and extrapolated values from Ref. [12]. The theoretical values reproduce in detail the experimental separation energies. The model describes not only the empirical values within one major neutron shell, but it also reproduces the transitions between major shells (for details see Ref. [13]).

In Fig. 2 we show the self-consistent ground-state neutron densities for the Sn and Ni nuclei. The density profiles display shell effects in the bulk and a gradual increase of neutron radii. In the insert of Fig. 2 we include the corresponding values for the surface thickness and diffuseness parameter. The surface thickness t is defined to be the change in radius required to reduce $\rho(r)/\rho_0$ from 0.9 to 0.1 (ρ_0 is the maximal value of the neutron density; because of shell effects we could not use for ρ_0 the density in the center of the nucleus). The diffuseness parameter α is determined by fitting the neutron density profiles to the Fermi distribution

$$\rho(r) = \rho_0 \left(1 + exp(\frac{r - R_0}{\alpha})\right)^{-1}, \qquad (9)$$

where R_0 is the half-density radius. By adding more units of isospin the value of the neutron surface thickness increases and the surface becomes more diffuse. The increase in t and α is more uniform in Sn, and both parameters increase approximately forty percent from ^{100}Sn to ^{132}Sn. A somewhat smaller increase in the surface thickness is observed for Ni isotopes. The diffuseness parameter for Ni is essentially a step function: $\alpha \approx 0.45 \; fm$ for $N < 40$ and $\alpha \approx 0.50 \; fm$ for neutrons in the $1g_{9/2}$ orbital. We will show that the observed changes in surface properties result from the reduction of the spin-orbit term in the effective single-nucleon potential.

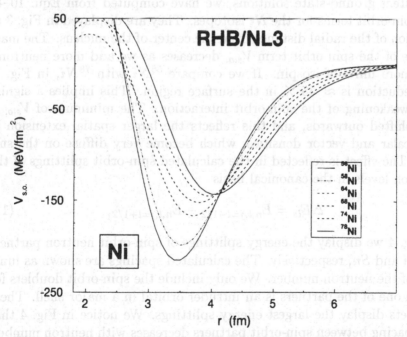

Figure 3 Radial dependence of the spin-orbit term of the potential in self-consistent solutions for the ground-states of Sn ($50 \leq N \leq 82$) nuclei.

In Ref. [14] we have shown that in the framework of the relativistic mean-field model the magnitude of the spin-orbit term in the effective single nucleon potential is greatly reduced for light neutron rich nuclei.

With the increase of the number of neutrons the effective spin-orbit interaction becomes weaker and this results in a reduction of the energy splittings for spin-orbit partners. The reduction in the surface region was found to be as large as $\approx 40\%$ for Ne isotopes at the drip-line. The spin-orbit potential originates from the addition of two large fields: the field of the vector mesons (short range repulsion), and the scalar field of the sigma meson (intermediate attraction). In the first order approximation, and assuming spherical symmetry, the spin orbit term can be written as

$$V_{s.o.} = \frac{1}{r}\frac{\partial}{\partial r}V_{ls}(r), \tag{10}$$

where V_{ls} is the spin-orbit potential [15]

$$V_{ls} = \frac{m}{m_{eff}}(V - S) \qquad \text{with} \qquad m_{eff} = m - \frac{1}{2}(V - S). \tag{11}$$

V and S denote the repulsive vector and the attractive scalar potentials, respectively. Using the vector and scalar potentials from the NL3 self-consistent ground-state solutions, we have computed from Equ. 10–11 the spin-orbit terms for the Ni isotopes. They are displayed in Fig. 3 as function of the radial distance from the center of the nucleus. The magnitude of the spin-orbit term $V_{s.o.}$ decreases as we add more neutrons, i.e. more units of isospin. If we compare ^{56}Ni with ^{78}Ni, in Fig. 3, the reduction is $\approx 35\%$ in the surface region. This implies a significant weakening of the spin-orbit interaction. The minimum of $V_{s.o.}$ is also shifted outwards, and this reflects the larger spatial extension of the scalar and vector densities, which become very diffuse on the surface. The effect is reflected in the calculated spin-orbit splittings of the neutron levels in the canonical basis

$$\Delta E_{ls} = E_{n,l,j=l-1/2} - E_{n,l,j=l+1/2}, \tag{12}$$

In Fig. 4 we display the energy splittings of spin-orbit neutron partners for Ni and Sn, respectively. The calculated spacings are shown as function of the neutron number. We only include the spin-orbit doublets for which one of the partners is an intruder orbital in a major shell. These doublets display the largest energy splittings. We notice in Fig. 4 that the spacing between spin-orbit partners decreases with neutron number. The effect is stronger in Ni than in Sn.

4.   Halo-Phenomena

In Fig. 5 we show the calculated density distribution for the neutrons in the isotopes ^9Li and ^{11}Li. It is clearly seen that the increase of the

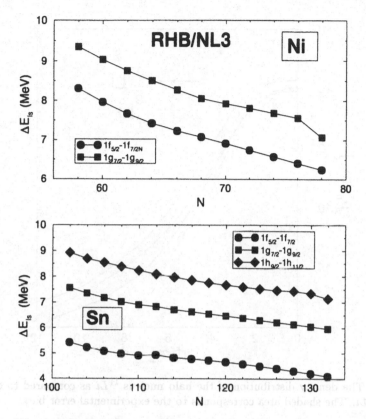

Figure 4 Energy splittings between spin-orbit partners for neutron levels in Ni and Sn isotopes, as functions of neutron number.

matter radius is caused by a large neutron halo in the nucleus ^{11}Li. Its density distribution is in very good agreement with the experimental density of this isotope shown with its error bars by the shaded area.

As shown in Fig. 6, the microscopic structure of this halo can be understood by the fact that the Fermi level for the neutrons is very close to the continuum limit in close vicinity to the $\nu 1p_{1/2}$ below the continuum and to the $\nu 2s_{1/2}$ level in the continuum. Pairing correlations cause a partial occupation of both the $\nu 1p_{1/2}$ and the $\nu 2s_{1/2}$ level, i.e. a scattering of Cooper pairs into the continuum. This is in contrast to earlier calculations using Skyrme forces and relativistic mean field without pairing, where the last occupied $\nu 1p_{1/2}$ level had to be shifted artificially very close to the continuum limit by an adjustment of the potential. In contrast to these investigations the halo is not formed by two neutrons occupying the $1p_{1/2}$ level very close to the continuum limit, but is is formed by Cooper-pairs scattered mainly in the two levels $1p_{1/2}$ and $2s_{1/2}$. This is made possible by the fact that the $2s_{1/2}$ comes down

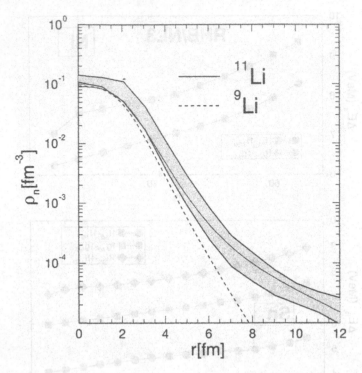

Figure 5 The density distribution of the halo nucleus ^{11}Li as compared to that of the core ^{9}Li. The shaded area corresponds to the experimental error bars.

close to the Fermi level in this nucleus and by the density dependent pairing interaction coupling the levels below the Fermi surface to the continuum. In contrast to earlier explanations which use the accidental coincidence that one single particle level is so close to continuum threshold that the tail of its wave function forms a halo, this is a much more general mechanism, which could possibly be observed in other halo nuclei also. One needs only several single particle levels with small orbital angular momenta and correspondingly small centrifugal barrier close, but not directly at, the continuum limit.

In fact going along the neutron drip line there are several such regions, in particular the region where the $2p$ and the $3p$ levels come close to the continuum limit. In the first case a multi-particle halo in the region of heavy Ne-isotopes has been predicted in Ref. [5].

5.   Summary

Summarizing we can conclude that we have to go beyond the simple relativistic mean field model in order to describe halo nuclei properly. We have to take into account pairing correlations and the coupling to the

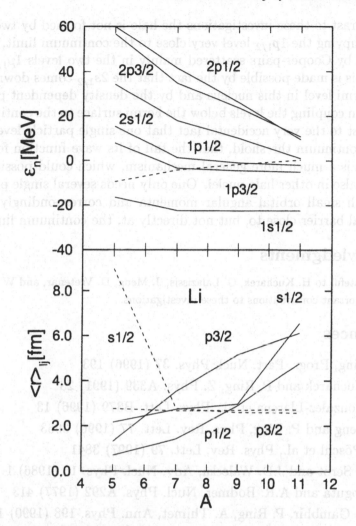

Figure 6 *Li*-isotopes: single particle energies ϵ_n for neutrons in the canonical basis (upper part) and contributions to the matter radius for various spin-parity channels (lower part) as a function of the mass number. In the upper part the dashed line corresponds to the chemical potential, and in the lower part dashed lines correspond to proton contributions.

continuum in the framework of relativistic Hartree Bogoliubov theory. A density dependent force of zero range has been used in the pairing channel, whose strength is adjusted for the isotope ^7Li to a similar calculation with Gogny's force D1S. Good agreement with experimental values is found for the total binding energies and the radii of the isotope chain from ^6Li to ^{11}Li. In excellent agreement with the experiment we obtain a neutron halo for ^{11}Li without any artificial adjustment of the potential, as it was necessary in earlier calculations.

In contrast to these investigations the halo is not formed by two neutrons occupying the $1p_{1/2}$ level very close to the continuum limit, but is is formed by Cooper-pairs scattered mainly in the two levels $1p_{1/2}$ and $2s_{1/2}$. This is made possible by the fact that the $2s_{1/2}$ comes down close to the Fermi level in this nucleus and by the density dependent pairing interaction coupling the levels below the Fermi surface to the continuum. In contrast to the very accidental fact that one single particle level is so close to continuum threshold, that the tail of its wave function forms a halo, this is a much more general mechanism, which could possibly be observed also in other halo nuclei. One only needs several single particle levels with small orbital angular momenta and correspondingly small centrifugal barrier close to, but not directly at, the continuum limit.

Acknowledgments

I am grateful to H. Kucharek, G. Lalazissis, J. Meng, D. Vretenar, and W. Pöschl for the important contributions to these investigations.

References

[1] P. Ring, Progr. Part. Nucl. Phys. 37 (1996) 193

[2] H. Kucharek and P. Ring, Z. Phys. A339 (1991) 23

[3] T. Gonzalez-Llarena et al., Phys. Lett. B379 (1996) 13

[4] J. Meng and P. Ring, Phys. Rev. Lett. 77 (1996) 3963

[5] W. Pöschl et al., Phys. Rev. Lett. 79 (1997) 3841

[6] B.D. Serot and J.D. Walecka, Adv. Nucl. Phys. 16 (1986) 1

[7] J. Boguta and A.R. Bodmer, Nucl. Phys. A292 (1977) 413

[8] Y.K. Gambhir, P. Ring, A. Thimet, Ann. Phys. 198 (1990) 132

[9] J.Dobaczewski, H. Flocard and J.Treiner, Nucl. Phys. A422 (1984) 103

[10] M.M. Sharma, G.A. Lalazissis, and P. Ring, Phys. Lett. B317 (1993) 9; Phys. Rev. Lett. 74 (1995) 3744

[11] M.M. Sharma et al., Phys. Rev. Lett. 74 (1995) 3744

[12] G. Audi and A. H. Wapstra, Nucl. Phys. A595 (1995) 409

[13] G.A. Lalazissis, D. Vretenar, and P. Ring, Phys. Rev. C57 (1998) 2294

[14] G.A. Lalazissis et al., Phys. Lett. B418 (1998) 7

[15] W. Koepf and P. Ring, Z. Phys. A339 (1991) 81

Treatment of Pairing Correlations in Nuclei Close to Drip Lines

Nguyen Van Giai[a], M. Grasso[b], R.J. Liotta[c] and N. Sandulescu[d]

[a] *Institut de Physique Nucléaire, IN2P3-CNRS,*
Université Paris-Sud, 91406 Orsay Cedex, France

[b] *Università di Catania, 95129 Catania, Italy*

[c] *Royal Institute of Technology, Frescativ. 24, 10405 Stockholm, Sweden*

[d] *Institute of Atomic Physics, P.O.Box MG-6, Bucharest, Romania*

nguyen@ipno.in2p3.fr

Keywords: Pairing correlations, HFB, continuum

Abstract   We discuss the HFB equations in coordinate representation, a suitable method for handling the full effects of the continuous quasiparticle spectrum. We show how the continuum HFB equations can be solved with the correct asymptotic conditions instead of the discretization conditions which are commonly used in the literature. The continuum HFB method is illustrated with a model where the mean field and pairing field have simple forms. The relationship with the continuum Hartree-Fock-BCS (HF-BCS) approximation is also discussed. Realistic HFB and HF-BCS calculations based on Skyrme interactions are compared for the case of a neutron-rich nucleus.

1. Introduction

In many nuclear systems pairing correlations have an important influence on physical properties. This situation occurs when the energy difference between an occupied orbital and a neighboring unoccupied one is relatively small, thus enabling a nucleon pair to be promoted to the unoccupied level by the interaction among the pair. Well known theoretical approaches such as the Hartree-Fock-Bogoliubov (HFB) or the Hartree-Fock-BCS (HF-BCS) approximations have been developed [1] to treat the pairing effects in nuclei.

In usual situations, i.e., when the nuclear system is stable and far away from the drip lines the active orbitals are well bound and it may

D. N. Poenaru et al. (eds.), Nuclei Far from Stability and Astrophysics, 127–138.
© *2001 Kluwer Academic Publishers. Printed in the Netherlands.*

be sufficient to solve the HFB or HF-BCS equations within the discrete subspace of those active orbitals. When one approaches a drip line this is no longer true since the active orbitals must also include states belonging to the continuous single-particle spectrum. One of the most convenient tools for the treatment of pairing correlations in the presence of continuum coupling is given by the HFB approach. The standard procedure is to discretize the quasiparticle continuum by solving the HFB equations in a finite basis of orthonormal functions, or by imposing to the solutions of HFB equations, written in coordinate space, a box boundary condition at some distance R_b [2, 3, 4]. There may be circumstances where a discretized solution is unsatisfactory unless an extremely large box radius R_b is used. Such should be the case for halo nuclei, for instance. In these lectures, we show how the HFB equations can be solved with correct boundary conditions. Based on these solutions we examine how much the continuum, especially the resonant continuum, could affect the pairing properties of nuclei close to the drip lines.

The structure of the lectures is the following: In the next section we introduce the HFB equations in coordinate space [5, 2] and in section 3 we examine the solutions which satisfy the correct boundary conditions, i.e., of scattering wave type. In section 4 these considerations are illustrated for the particular case of a square well potential and a constant pairing field of a finite range. Then we discuss how the resonant continuum can be taken into account in HF-BCS approach. Finally we show how the continuum, calculated in different approximations, can affect the pairing properties of a nucleus close to the neutron drip line.

2.   Hartree-Fock-Bogoliubov Equations in Coordinate Space

Let us denote by H the Hamiltonian of the nuclear system. When treated in the HFB approximation, H will give rise to the mean field and the pairing field in which the quasiparticles are moving. We assume that the two-body effective interactions associated with the Hartree-Fock (HF) mean field and pairing field are zero-range forces so that the total energy E is a functional of the *local* particle density $\rho(r)$ and pairing density $\kappa(r)$:

$$\rho(r) \equiv \langle HFB|\Psi^+(r)\Psi(r)|HFB\rangle ,$$
$$\kappa(r) \equiv \langle HFB|\Psi^+(r)\Psi^+(r)|HFB\rangle , \tag{1}$$

where $\Psi^+(r)$ is a nucleon creation operator and $|HFB\rangle$ is the HFB ground state. Typically, the self-consistent mean field is generated by a Skyrme-type interaction whereas the pairing field is produced by a zero-range, possibly density-dependent interaction.

The HFB equations in coordinate space can be expressed in the following form:

$$\begin{pmatrix} h - \lambda & \Delta \\ \Delta & -(h - \lambda) \end{pmatrix} \begin{pmatrix} U_\alpha(\mathbf{r}) \\ V_\alpha(\mathbf{r}) \end{pmatrix} = E_\alpha \begin{pmatrix} U_\alpha(\mathbf{r}) \\ V_\alpha(\mathbf{r}) \end{pmatrix} , \qquad (2)$$

where λ is the chemical potential, h is the HF Hamiltonian and Δ is the pairing field. The local nucleon density and pairing density can be written in terms of the solutions (U_α, V_α) in the form:

$$\rho(\mathbf{r}) = \sum_\alpha |V_\alpha(\mathbf{r})|^2 , \qquad (3)$$

$$\kappa(\mathbf{r}) = \sum_\alpha U_\alpha(\mathbf{r}) V_\alpha^*(\mathbf{r}) . \qquad (4)$$

For Skyrme forces, the HF and pairing fields may also depend on derivatives of ρ and κ and on some local currents because of the velocity dependence of the interaction. Since all densities appearing in the energy functional are local, Equ. 2 is a set of coupled differential equations which is highly non-linear because the fields h and Δ depend themselves on the solutions U_α, V_α (self-consistency problem). Nevertheless, using Skyrme-type forces brings a major simplification. If one starts instead with a finite range effective force, for instance a Gogny's force, Equ. 2 would be a set of coupled integro-differential equations which is usually solved in a harmonic oscillator basis. The general properties of Skyrme-HF Hamiltonians are well-known and detailed expressions of h and Δ for the case of Skyrme forces can be found for instance in Ref. [2].

The Skyrme-HF Hamiltonian has the general form:

$$h = -\nabla . \frac{\hbar^2}{2m^*(\mathbf{r})} \nabla + V_{HF}(\mathbf{r}), \qquad (5)$$

where the effective mass m^* and HF potential V_{HF} depend on nucleon densities and currents. An important property is that m^* tends to the nucleon mass m and V_{HF} tends to zero at infinity, with the same rate as the densities go to zero. We shall use this property in the next section to establish the asymptotic forms of the solutions of Equ. 2.

For illustration, we can look at the pairing field calculated with a pairing interaction often used in the literature [3, 4]:

$$V(1,2) = V_0 \left[1 - \left(\frac{\rho(\mathbf{r}_1)}{\rho_c} \right)^\gamma \right] \delta(\mathbf{r}_1 - \mathbf{r}_2) . \qquad (6)$$

To be meaningful, such an interaction must be used with an energy cut-off in the quasiparticle spectrum, i.e., the summations in Equs. 3,4 must be limited to $E_\alpha \leq E_{cutoff}$. In this case, the pairing field is:

$$\Delta(\mathbf{r}) = \frac{1}{2}V_0 \left[1 - \left(\frac{\rho(\mathbf{r})}{\rho_c}\right)^\gamma\right]\kappa(\mathbf{r}) . \tag{7}$$

On this example we can see that the local pairing field behaves asymptotically like the local pairing density.

3.   Asymptotic Behavior of HFB Solutions

From the symmetries of Equ. 2 it can be seen that, if $(E_\alpha, U_\alpha, V_\alpha)$ is a solution then $(-E_\alpha, V_\alpha^*, U_\alpha^*)$ is also a solution. Thus, we need to consider only one class of solutions and we choose those with $E_\alpha \geq 0$.

The asymptotic behavior of the solutions of Equ. 2 has been discussed in detail by Bulgac [5]. Let us consider for simplicity the case of spherical symmetry and write:

$$\begin{pmatrix} U_\alpha(\mathbf{r}) \\ V_\alpha(\mathbf{r}) \end{pmatrix} = \frac{1}{r}\begin{pmatrix} u_\alpha(r) \\ v_\alpha(r) \end{pmatrix} \mathbf{Y}_\alpha(\hat{r}, \sigma) . \tag{8}$$

Equ. 2 now becomes a one-dimensional equation in the radial coordinate r for each $\alpha = (l, j)$. At very large distances the HF Hamiltonian h tends to $-(\hbar^2/2m)(\frac{1}{r}\frac{d^2}{dr^2}r - \frac{l(l+1)}{r^2})$ (plus a Coulomb potential for protons) while the pairing field $\Delta(r)$ has vanished. The equations for $u_\alpha(r)$ and $v_\alpha(r)$ are decoupled and one can easily see how the physical solutions must behave at infinity. Thus, for a negative chemical potential λ, i.e., for a bound system, there are two well separated regions in the quasiparticle spectrum:

— between 0 and $-\lambda$ the quasiparticle spectrum is discrete and both upper and lower components $(u_\alpha(r), v_\alpha(r))$ of the HFB wave function decay exponentially at infinity;

— above $-\lambda$ the quasiparticle spectrum is continuous and the physical solutions are such that at infinity the upper component of the HFB wave function has a scattering wave form (see the next section) while the lower component is exponentially decaying. In what follows the continuous HFB wave functions are normalized to a delta function of energy.

Thus, the summations in Equs. 3, 4 should in fact include integrations over the continuum of the quasiparticle spectrum:

$$\rho(\mathbf{r}) = \sum_{0 \leq E_\alpha \leq -\lambda} |V_\alpha(\mathbf{r})|^2 + \int_{-\lambda}^{E_{cutoff}} dE_\alpha |V_{E_\alpha}(\mathbf{r})|^2 ,$$

$$\kappa(\mathbf{r}) = \sum_{0 \le E_\alpha \le -\lambda} U_\alpha(\mathbf{r})V_\alpha^*(\mathbf{r}) + \int_{-\lambda}^{E_{cutoff}} dE_\alpha U_{E_\alpha}(\mathbf{r})V_{E_\alpha}^*(\mathbf{r}) . \quad (9)$$

As we have already mentioned in the introduction, most of the practical calculations are done by solving Equ. 1 with a box boundary condition [2, 3, 4], i.e., by requiring that the solutions $U_\alpha(\mathbf{r}), V_\alpha(\mathbf{r})$ have a node at $r = R$ (they are taken to be identically zero beyond R). This condition makes the spectrum of E_α entirely discrete and allows the use of Equs. 3, 4 instead of Equ. 9. However, it is not clear how accurately one can mock up continuum effects like single-particle resonance contributions by using this discretization procedure. In order to avoid such ambiguities one needs solutions with proper boundary conditions. These solutions are illustrated in the next section for the case of a simple model [5, 6].

4.   HFB Solutions: A Schematic Model

In what follows we discuss the solutions of HFB equations in coordinate space for a mean field given by a square well potential, of depth V_0 and radius a, and a pairing field, Δ, constant inside the same radius a and zero outside. We suppose also that the Fermi level, λ, is given. For such a system the radial HFB equations inside the potential well, i.e. for $r \le a$, are given by:

$$(\frac{1}{r}\frac{d^2}{dr^2}r - \frac{l(l+1)}{r^2} + \alpha^2)u_{lj} - \delta^2 v_{lj} = 0 ,$$

$$(\frac{1}{r}\frac{d^2}{dr^2}r - \frac{l(l+1)}{r^2} + \beta^2)v_{lj} - \delta^2 u_{lj} = 0 . \quad (10)$$

where $\alpha^2 = \frac{2m}{\hbar^2}(\lambda + E + U_0)$, $\beta^2 = \frac{2m}{\hbar^2}(\lambda - E + U_0)$, $\delta^2 = \frac{2m}{\hbar^2}\Delta$ and $U_0 = -(V_0 + V_{sol}\vec{l}.\vec{s})$. The above equations have the following physical solution for any value of the quasiparticle energy :

$$u_{lj} = A_+ j_l(k_+ r) + A_- j_l(k_- r) ,$$

$$v_{lj} = A_+ g_+ j_l(k_+ r) + A_- g_- j_l(k_- r) , \quad (11)$$

where j_l are spherical Bessel functions, $k_\pm = \frac{2m}{\hbar^2}(U_0 + \lambda \pm \sqrt{(E^2 - \Delta^2)})$ and $g_\pm = (E \pm \sqrt{(E^2 - \Delta^2)})/\Delta$.

Outside the potential well both U_0 and Δ are zero and the HFB equations are decoupled. In this case the type of solutions depends on the quasiparticle energy. Thus, for $E < -\lambda$ the solutions have the form:

$$u_{lj} = Ah_l^{(+)}(\alpha_1 r) ,$$

$$v_{lj} = Bh_l^{(+)}(\beta_1 r) , \quad (12)$$

where $h_l^{(+)}$ are spherical Haenkel functions, $\alpha_1^2 = \frac{2m}{\hbar^2}(\lambda + E)$ and $\beta_1^2 = \frac{2m}{\hbar^2}(\lambda - E)$. These solutions correspond to the bound quasiparticle spectrum.

For $E > -\lambda$ the spectrum is continuous and the solutions are:

$$u_{lj} = C[cos(\delta_{lj})j_l(\alpha_1 r) - sin(\delta_{lj})n_l(\alpha_1 r)] ,$$
$$v_{lj} = Dh_l^{(+)}(\beta_1 r) , \qquad (13)$$

where n_l are spherical Neumann functions and δ_{lj} is the phase shift corresponding to the angular momentum (lj).

The constants entering in the wave functions above are fixed by the continuity conditions of the solutions and of their derivatives at $r = a$ and the normalization. In what follows we discuss only the continuous solutions, i.e., for energies $E > -\lambda$.

Of particular interest are the values of energies for which the wave functions have maximum localizations inside the potential well. These are the regions close to quasiparticle resonances, which can be defined as complex outgoing solutions of HFB equations. In HFB approach one distinguishes two types of quasiparticle resonances. One type corresponds to the single-particle resonances of the mean field. For the pairing correlations an important role is played by those single-particle resonances which are close to the particle threshold and have relatively high angular momentum.

Another type of quasiparticle resonances corresponds to bound single-particle states. These resonant states, which appear due to the non-diagonal matrix elements of the pairing field, are specific of the HFB approach. The resonant states corresponding to deep hole states have small widths. The states with very small widths can be eventually treated as quasibound states, normalized to unity in the same volume as the bound quasiparticle states.

A special attention is paid usually to the continuum $s_{1/2}$ states [6, 7]. Apart from the deep hole $s_{1/2}$ states, which in HFB become narrow quasiparticle resonances with high quasiparticle energies, in drip line nuclei one may also find a loosely bound $s_{1/2}$ single-particle state. In continuum HFB approach this state appears often as a broad quasiparticle resonance (see example below), close to the continuum edge, and its role in pairing correlations cannot be distinguished from the rest of non-resonant $s_{1/2}$ continuum. This is different from a HF-BCS approach, where the contribution of the loosely bound single-particle $s_{1/2}$ state to pairing correlations is not mixed with the rest of background single-particle continuum.

The structure of the continuum discussed above is essentially the same for general, self-consistent HFB calculations. In this case the HFB equations for the continuum spectrum are integrated by starting far from the nucleus with the solution given by Equ. 13, which is propagated towards the matching point by the Numerov method. For each quasiparticle energy one calculates, by matching conditions, the phase shift. Then the energies (widths) of quasiparticle resonances are found from the energies where the derivative of the phase shift reaches its maximum (half of its maximum) value. This information is used afterwards to fix an appropriate energy grid (i.e., dense in the energy region of a resonance) for the calculation of the continuum contribution to the particle and pairing densities (see Equs. 9).

5.   HF-BCS Approximation

The HF-BCS approximation is obtained by neglecting in the HFB equations the non-diagonal matrix elements of the pairing field. This means that in the HF-BCS limit one neglects the pairing correlations induced by the pairs formed in states which are not time-reversed partners.

Particularly simple are the HF-BCS equations which include the effect of resonant continuum [8, 9]:

$$\Delta_i = \sum_j \langle i, \bar{i} | V | j, \bar{j} \rangle u_j v_j +$$

$$+ \sum_\nu \langle i, \bar{i} | V | \nu \epsilon_\nu, \overline{\nu \epsilon_\nu} \rangle \int_{I_\nu} g_\nu^c(\epsilon) u_\nu(\epsilon) v_\nu(\epsilon) d\epsilon ,$$

$$\Delta_\nu(\epsilon) = (g_\nu^c(\epsilon)/g_\nu(\epsilon)) (\sum_j \langle \nu \epsilon, \overline{\nu \epsilon} | V | j, \bar{j} \rangle u_j v_j$$

$$+ \sum_{\nu\prime} \langle \nu \epsilon_\nu, \overline{\nu \epsilon_\nu} | V | \nu\prime \epsilon_{\nu\prime}, \overline{\nu\prime \epsilon_{\nu\prime}} \rangle \int_{I_{\nu\prime}} g_{\nu\prime}^c(\epsilon\prime) u_{\nu\prime}(\epsilon\prime) v_{\nu\prime}(\epsilon\prime) d\epsilon\prime)$$

$$\equiv (g_\nu^c(\epsilon)/g_\nu(\epsilon)) \Delta_\nu . \tag{14}$$

where $g_\nu^c(\epsilon) = \frac{2j_\nu+1}{\pi} \frac{d\delta_\nu}{d\epsilon}$, $g_\nu(\epsilon)$ is the total level density and δ_ν is the phase shift of angular momentum $(l_\nu j_\nu)$. In these equations the interaction matrix elements are calculated with the scattering wave functions at resonance energies and normalized inside the volume where the pairing interaction is active. The particle number condition is:

$$N = \sum_i v_i^2 + \sum_\nu \int_{I_\nu} g_\nu^c(\epsilon) v_\nu^2(\epsilon) d\epsilon . \tag{15}$$

The energy factor $g_\nu^c(\epsilon)$ takes into account the variation of the localization of scattering states in the energy region of a resonance (i.e., the width effect) and goes to a delta function in the limit of a very narrow width. For more details one can see Ref. [9].

6.   Application: Continuum Coupling in Different Approximations

In order to illustrate the approximations discussed above we take as an example the nucleus ^{84}Ni, for which HFB calculations with discretized continuum can be found in the literature [4].

Table 1 Energies and widths of quasiparticle resonant states. In the two last columns the total occupancies for each (lj) channel are shown both in full and in box calculations.

j	l	E(MeV)	Γ(MeV)	Total occ. (full)	Total occ. (box)
1/2	0	1.276	0.742		
1/2	0	20.834	0.178		
1/2	0	43.392	10^{-6}		
				2.305	2.303
1/2	1	7.968	0.352		
1/2	1	33.451	0.105		
				1.994	1.995
3/2	1	9.720	0.60		
3/2	1	34.975	0.108		
				1.999	2.001
3/2	2	2.439	0.654		
3/2	2	22.026	0.069		
				1.168	1.171
5/2	2	1.843	0.072		
5/2	2	25.626	0.006		
				1.567	1.559
5/2	3	8.862	0.906		
				1.002	1.003
7/2	3	15.328	2.247		
				1.011	1.012
7/2	4	3.596	0.037		
				0.099	0.101
9/2	4	5.674	0.0007		
				0.960	0.959
11/2	5	5.381	0.050		
				0.069	0.07

In all calculations discussed below the HF mean field is calculated with the Skyrme interaction SIII. The pairing force is given by Equ. 6 and the parameters are the ones used in Ref. [4].

First, we present the results given by the continuum HFB equations solved with proper boundary conditions, as defined in the previous sections. These results, referred below as "full" continuum calculations, are compared to the HFB calculations performed with box boundary conditions.

In Tab. 1 are shown the quasiparticle resonant states, of hole and particle type. For each (lj) channel are shown also the total occupancy obtained in full and box calculations. One can notice the large widths of quasiparticle states corresponding to bound single-particle states close to the Fermi energy. The widths are obtained from the phaseshift behavior around $\pi/2$ except for the three states $s_{1/2}$, $p_{1/2}$ and $p_{3/2}$ close to the Fermi energy for which the widths are extracted from the occupancy profiles.

The results for the pairing field, pairing density and particle density are shown in Figs. 1–3. In the full and box HFB calculations we have used for each (lj) channel the same energy cut-off parameter, i.e. the absolute value of the depth of effective mean field potential corresponding to that (lj) [2]. From Fig. 2 one can see that the box calculations slightly overestimate the pairing density in the region of $5-6$ fm. This overestimation of pairing correlations is also seen by comparing the pairing energies (full: -18.757 MeV; box: -19.530 MeV) and total binding energies (full: -652.857 MeV; box: -653.315 MeV).

Next, we compare the results given by the HF-BCS approximation presented in the previous section, to the HFB calculations of Ref. [4]. In Ref. [4] the HFB equations are diagonalized in a basis formed by the single-particle states selected by a box of a finite radius. The single-particle continuum is cut at 5 MeV. This energy cutoff is different from the HFB box calculations presented above, where the energy cutoff is much higher. Up to 5 MeV, one finds three single-particle resonances, $d_{3/2}$, $g_{7/2}$ and $h_{11/2}$, which are treated in the HF-BCS approximation given by Equs. 14. The HFB and HF-BCS results for the pairing field are shown in Fig. 4. In the same figure is shown the HF-BCS result obtained by replacing each single-particle resonance by a discrete state, normalized in the same volume as the one used for the bound states. As seen from Fig. 4, this result is closer to HFB calculations. This similarity is due to the fact that in the HFB calculations each resonant state is actually represented by a unique state normalized in the box. In the HF-BCS calculations one notices that the pairing field is quite sensitive to the way the resonant states are treated. This is because in this case the widths of the resonant states affect rather strongly their occupation probabilities [9]. When the effect of the width of resonant states is included in HF-BCS equations the pairing correlations are decreasing, which produces a reduction of the total binding energy by about 400 keV. This is a trend similar with the one observed in continuum

HFB calculations presented above. Thus, one can conclude that, in order to take fully into account the effect of continuum upon pairing correlations, one needs to solve the HF-BCS and HFB equations with proper boundary conditions.

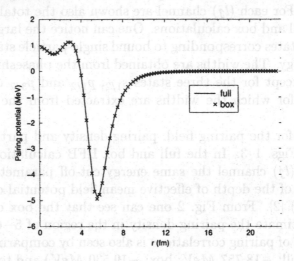

Figure 1   Pairing fields in full (solid line) and box (x) calculations.

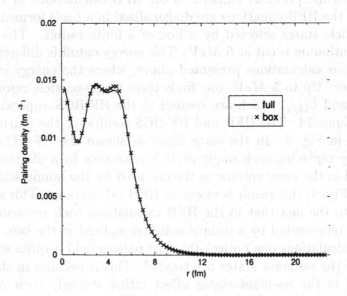

Figure 2   Pairing densities in full (solid line) and box (x) calculations.

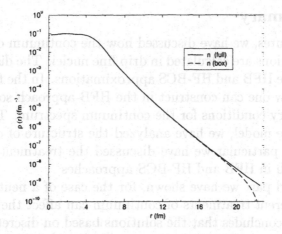

Figure 3 Particle densities in full (solid line) and box (dashed line) calculations.

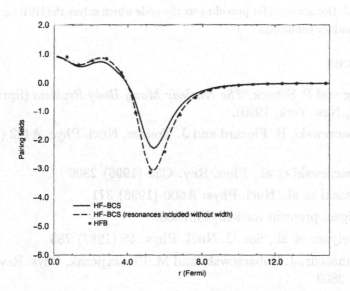

Figure 4 Pairing fields calculated in different approximations: HFB of Ref. [4] (stars), resonant HF-BCS with widths (solid) and without widths (dashed).

7.   Summary

In these lectures, we have discussed how the continuum coupling and pairing correlations are calculated in drip line nuclei. The discussion was restricted to the HFB and HF-BCS approximations. In the first part, we have shown how one can construct in the HFB approach solutions with proper boundary conditions for the continuum spectrum. Then, for the case of a simple model, we have analyzed the structure of quasiparticle continuum. In particular we have discussed the treatment of resonant continuum both in HFB and HF-BCS approaches.

In the second part we have shown, for the case of a neutron rich nucleus, how different treatments of continuum can affect the pairing correlations. One concludes that the solutions based on discretized continuum (box boundary conditions) can overestimate the pairing correlations when compared with proper continuum HFB or HF-BCS calculations. This may have consequences on the predictions of drip lines.

Acknowledgments

We thank J. Dobaczewski for providing us the code which solves the HFB equations with box boundary conditions.

References

[1] P. Ring and P. Schuck, *The Nuclear Many-Body Problem* (Springer-Verlag, New York, 1990).

[2] J. Dobaczewski, H. Flocard and J. Treiner, Nucl. Phys. A422 (1984) 103

[3] J. Dobaczewski et al., Phys. Rev. C53 (1996) 2809

[4] J. Terasaki et al., Nucl. Phys. A600 (1996) 371

[5] A. Bulgac, preprint nucl-th/9907088.

[6] S.T. Belyaev et al., Sov. J. Nucl. Phys. 45 (1987) 783

[7] K. Bennaceur, J. Dobaczewski and M. Ploszajjczak, Phys. Rev. C60 (1999) 2809

[8] N. Sandulescu, R.J. Liotta and R. Wyss, Phys. Lett. B394 (1997) 6

[9] N. Sandulescu, N. Van Giai and R.J. Liotta, Phys. Rev. C61 (2000) 061301(R)

The two nuclear superfluids in a rotating field: Competition between T = 0 and T = 1 pairing

Ramon Wyss

Royal Institute of Technology, KTH
KTH-Frescati, Frescativ. 24, 104 05 Stockholm, Sweden
wyss@msi.se

Keywords: Superfluidity, $N = Z$ nuclei, isoscalar and isovector pairing, rotational motion, Wigner energy, symmetry energy, delayed band crossing

Abstract In $N = Z$ nuclei one may encounter two fundamentally different pairing modes, both composed of neutron proton pairs, but of either isoscalar or isovector type. In this lecture, we discuss the role of isoscalar pairing with respect to the Wigner energy and we address the competition of isoscalar and isovector pairing in rotating nuclei. Both the additional binding in $N = Z$ nuclei as well as the recently observed shift of the crossing frequencies can find a microscopical explanation via the isoscalar, $T = 0$, pairing correlations.

1.   Introduction

An open issue in todays nuclear structure physics relates to the question of proton-neutron correlations in $N \approx Z$ nuclei. Although everybody agrees upon that the $T = 0$ neutron-proton interaction is of dominant importance, there is quite a controversy on the question whether these correlations may lead to a pair-condensate analogous to the well studied like particle pairing. Before one is addressing this question, one needs to define the frame in which to discuss this issue. The concept of collective motion, associated with an intrinsic frame, is inherent to the mean-field. In this model, it is a meaningful question to ask whether the nucleus is deformed or just soft with respect to changes in deformation. In a similar fashion, it is meaningful to ask whether there exist a static pairing gap. This question can be answered directly since it is related to the solution of the BCS gap-equation: When pairing correlations are

D. N. Poenaru et al. (eds.), Nuclei Far from Stability and Astrophysics, 139–149.
© 2001 *Kluwer Academic Publishers. Printed in the Netherlands.*

strong enough, there will be a non trivial solution, i.e. a deformed state in gauge-space.

In nuclei with same number of protons and neutrons, $N = Z$, the wave functions of neutrons and protons are essentially identical. One may therefore expect a superfluid state, that in addition to the well know Cooper pairs of like particles, i.e. neutron-neutron (nn) and proton-proton (pp), contains neutron-proton (np) pairs. Already iso-spin symmetry requires the presence of such a mixed pair-condensate. However, since neutron and protons are different fermions, they can occupy identical states. Therefore, there will be two fundamentally different kinds of np Cooper pairs, which form either symmetric ($T = 1$) or antisymmetric ($T = 0$) combinations in iso-space. The present lecture addresses the different role of these $T = 1$ and $T = 0$ pairs.

In order to deal with all different pairing-modes on the same footing, the BCS-theory requires to be extended. The first generalizations of the BCS-equations as to encompass $T = 1$ nn-, pp- and np-pairing were presented already in the mid-sixties [1–4]. A detailed review of the generalized BCS-theory, where both $T = 0$ and $T = 1$ are treated can be found in Goodmans work [5]. Most of these mean-field calculations that include neutron-proton pair correlations lead in general to the existence of a non-vanishing $n - p$ pair gap [5]. This implies that one expects indeed a neutron-proton superfluid to be present in $N = Z$ nuclei. However, it has been difficult to find clear signals associated with collective np-pairing. In this lecture I will discuss a few aspects of the observables, related to coherent proton-neutron pairing correlations. I will start with the discussion on the symmetry energy and then review aspects of high spin states in $N = Z$ nuclei.

2.   Symmetry Energy and Wigner Energy

In the study of masses it was noticed long ago, that nuclei along the $N = Z$ line are more bound than their even-even neighbors, see e.g. [6]. This additional binding energy has been phrased as the Wigner energy, in relation to the Wigner SU(4) model. In this model, the binding energy can be expressed analytically, corresponding to $a_a T(T + 4)$ [7]. Apparently, according to this formula, states with $T = 1$ are by a factor $5a_a$ less bound compared to the $T = 0$ states. However, analyzing masses within the single-j model yields instead a dependence proportional to $a_a T(T + 1)$ dependence [8]. Experimental values are in agreement with the single-j shell model, but not with Wigners SU(4) [9, 10]. A modern mass formula will indeed have to be based on a term proportional to $T(T + 1)$, in order to account for the binding energies. A specific property of these kind of mass formulae is that they are discontinuous along the $N = Z$ line, i.e. exhibit a clear cusp.

All these analysis are done within the shell model or shell model like Hamiltonian. In contrast, the mean-field model will give a symmetry energy that is proportional to $(N - Z)^2$ in accordance to the well known Bethe-Weizsäcker mass formula. The Fermi-gas model has this form of $N - Z$ dependence. Similarly, in the Woods-Saxon potential, the isovector potential is proportional to $V_o\kappa(N - Z)/A$. Again, this part of the potential is active for $(N - Z)$ nucleons, yielding an additional binding proportional to $(N - Z)^2/A$. Last not least, in the Hartree-Fock model, the energy scales with the square of the densities, $E \approx TrTr\rho v\rho$, where v correspond to the the two-body interaction, Tr stands for the trace, and ρ the density matrix. None of the mean-field models will exhibit such a cusp in the masses at the $N = Z$ line.

In this lecture, my starting point is the mean-field. This implies that quantities like the pairing gaps are well defined. It also means that the symmetry-energy in this model will be proportional to $(N - Z)^2$. Since the ground-state of all even-even nuclei has $T = T_z = 1/2(N - Z)$, the symmetry energy is proportional to T^2 in mean-field models.

What is now the effect when one allows both $T = 1$ and $T = 0$ pairing to be present in the mean-field? To address this question, we start from a single particle field, generated by a Woods-Saxon potential [11] and a two body residual pairing interaction of seniority type.

$$\hat{H}^\omega = \hat{h}_{WS} + G_{T=1}\hat{P}_1^\dagger\hat{P}_1 + G_{T=0}\hat{P}_0^\dagger\hat{P}_0 - \omega\hat{j}_x. \tag{1}$$

The cranking term $-\omega\hat{j}_x$ is used to generate states at given angular momentum I, and its role will be discussed later on. In order to avoid the spurious phase transition encountered in the BCS-approximation, we employ approximate particle number projection by means of the Lipkin-Nogami method. The model is discussed in detail in Ref. [12]. The pairing action is now determined by the two strength parameters $G_{T=0}$ and $G_{T=1}$. From shell model interactions one can infer that the strength of the $T = 0$ force is larger than the $T = 1$. We will start with investigating how the ground state energy is changing with the parameter x which determines the ratio of the $T = 0$ and $T = 1$ correlations, respectively, where $G_{T=0} = x \ G_{T=1}$.

Since the BCS-approach is strictly variational, the system will choose the pairing mode which yields the lowest energy. This means that when $x < 1$ one will encounter only $T = 1$ pairing and when $x > 1$, there will be $T = 0$-pairing only, see also the extensive discussion in Ref. [12]. Note that in the BCS-approximation, there is nothing preventing both pairing modes to be present simultaneously. It is entirely the effect of the seniority pairing (simple pair-counting) that causes this exclusion. More complicated forces can give a mixed $T = 0$ and $T = 1$ phase [13, 14]. If one projects on good particle number, or is doing an exact calculation, then again, the two phases of the pairing interaction, $T = 0$ and $T = 1$ will mix [12, 15].

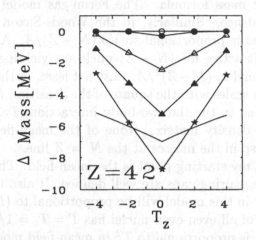

Figure 1 The additional binding energy, Δ, is depicted as a function of $N - Z = T_z$ (The standard definition is $T_z = 1/2(N - Z)$). The strength of the $T = 0$ pairing force $G_{T=0}$, ranges from $x = 1.0$ to $x = 1.4$ revealing the mass gain with increasing $T = 0$ pairing correlations.

Interestingly, the effect of the $T = 0$ pairing is to lower the ground-state energy. Since at the same time, $T = 0$ pairing is active only when $N \approx Z$, a microscopic explanation of the Wigner energy is provided: The mass-defect in $N = Z$ nuclei, the Wigner energy, finds a microscopic explanation via the $T = 0$ pairing energy, that provides the additional binding, not present in standard mean-field models. In other words, a generalized pairing theory, including both $T = 0$ and $T = 1$ cures the missing binding energy, related to the linear term $|T|$ in mass-formulae. The effect is nicely demonstrated in Fig. 1. We calculate here the additional binding energy, coming from the $T = 0$ pairing correlations for the case of $Z = 42$, i.e. Mo-isotopes. The strength of the $G_{T=0}$ pairing interaction is determined as above, from $G_{T=0} = x \, G_{T=1}$, where x ranges from 1.0 to 1.4. Of course, when $x = 1.0$, no energy is gained, since there is no additional binding from the $T = 0$ pairs. However, when $x = 1.1$ we already notice a slight increase in the pairing energy and therefore additional binding. With increasing x-value, the additional binding is increasing correspondingly. The solid curve, without symbols corresponds to the Wigner-energy from the mass-formula of Ref. [6]. An x-value between 1.30 and 1.40 would correspond to the experimental mass-defect. Note that it is the additional $T = 0$ pairing energy that accounts for the additional binding. As discussed more in detail in [12], the $T = 1$ np-pairing energy does not contribute at all to enhance pairing correlations.

The figure also reveals the pronounced drop of the $T = 0$ pairing with increasing $|T_z|(= T)$. Apparently, the $T = 0$ correlations are peaked at $N = Z$ and once $N - Z = 4$, little of the correlations is left. The reason for this effect is also simple to understand: When we add two more neutrons (or protons) to an $N = Z$ nucleus, this implies that the additionally occupied level cannot participate in the np-pairing, since the np-pairs cannot scatter into that level. The $N = Z + 2$ states thus corresponds to 2 qp-state with respect to the $T = 0$ pairing, and the $N = Z + 4$ to a 4 qp-state and so on. From normal pairing theory, we are well familiar with the drop in pairing correlations with increasing seniority, and this is just the analogue in iso-space, see also [12].

3. The Role of T = 0 Correlations at High Angular Momenta

Let us now turn to the issue of high angular momenta. States of high spin are treated by means of the cranking approximation, where the cranking frequency ω is adjusted as to fulfill the requirement of $< j_x > = sqrt(I(I + 1))$, see Equ. 1 and Ref. [12] The response of the $T = 1$ pairing to rotation has been studied since long and is well understood. Nucleons moving in time-reversed orbits feel the Coriolis force, that acts to align the angular momentum along the rotational axis. Hence, the rotational motion tends to counteract the pairing correlations. With increasing rotational frequency, the Coriolis force will finally win and strongly reduce $T = 1$ pairing. This is the analogue to the well-known Meissner-effect [16]. For the case of $T = 0$ pairing, the situation is different. $T = 0$ pairs can couple their intrinsic angular momenta parallel $\alpha\alpha$, as well as anti-parallel, $\alpha\tilde{\alpha}$. The $\alpha\tilde{\alpha}$, $T = 0$ pairing responds to the rotational motion in exactly the same manner as the $T = 1$ pairing. Indeed, for both cases, nucleons move in time reversed orbits and for the Coriolis force it does not matter whether these are proton pairs or a proton-neutron pair. On the other hand, the $\alpha\alpha$ pairs can align their angular momenta along the rotational axis. Such pairs will not be broken by the Coriolis force.

Interestingly, recent experiments in $N = Z$ nuclei reveal that the frequency at which the nucleons align their angular momenta, the crossing frequency ω_c, is shifted compared to isotopes in which $N = Z+2$ [17, 18].

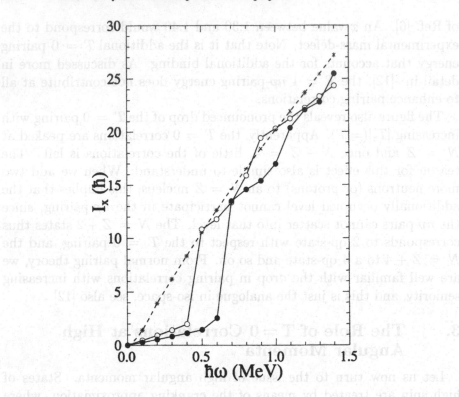

Figure 2 The angular momentum, I_x, as a function of the rotational frequency, $\hbar\omega$. Three different sets of calculations are depicted: $T = 1$ (o), $T = 0$, $\alpha\alpha$ pairing ($\star$) and $T = 0$, $\alpha\bar{\alpha}$ pairing ($\bullet$).

To clarify the role played by $T = 0$ and $T = 1$ pairing in $N = Z$ nuclei, we calculate the effect for a representative case like ^{72}Kr. The results are depicted in Fig. 2 where we show three different types of calculations. The standard calculations are done with $T = 1$ pairing only. At $\hbar\omega = 0.4\ MeV$, we see a sharp rise in the calculated angular momentum, I_x. This is the effect we mentioned above. The Coriolis force is strong enough to break a pair of nucleons, moving in time reversed orbits and to align their angular momenta. Since we have the same number of protons and neutrons, this occurs for both kind of nucleons, resulting in a total increase of the angular momenta by $\approx 10\ \hbar$. If the pairing field is dominated by $\alpha\alpha$ pairs, the picture is completely different. Such pairs can smoothly align their angular momenta with increasing frequency. The calculated spins, I_x, increase linearly with frequency and there is no sudden alignment, see Fig. 2. As discussed in Ref. [12], the moment of inertia in this case actually may *exceed* the rigid body moment of inertia, if one defines the rigid body moment as the one calculated without pairing correlations at all. There is thus a profound change with respect to the response to rotational motion in the presence of $T = 0$, $\alpha\alpha$ pairing.

Last not least, let us consider the case of $T = 0$, $\alpha\tilde{\alpha}$ pairing. The effect of this pairing mode is to shift the crossing frequency by $\Delta\hbar\omega \approx 0.2\ MeV$, see Fig. 2. Apparently, in the presence of a coherent $\alpha\tilde{\alpha}\ T = 0$ pairing field, the alignment is considerably delayed. This effect is entirely analogous to the mass defect of $N = Z$ nuclei. There the $T = 0$ pairing increases the mass-defect since the pairing correlations are stronger than in non $N = Z$ nuclei. For the case of angular momentum alignment, the increase of pairing correlations requires a larger Coriolis force, and hence the pair-breaking mechanism is shifted to higher frequencies. The same effect is also obtained in single-j shell calculations, where one can solve the Hamiltonian exactly, see [15].

The question still remains how to determine the balance of $\alpha\tilde{\alpha}$- and $\alpha\alpha$-pairing. We know that the total strength of the $T = 0$ pairing needs to be larger than the $T = 1$ in order to account for the Wigner energy. Hence, one can determine the strength parameter x from the mass-difference along the $N = Z$ line, analogous to the determination of the $T = 1$ pairing strength via odd-even mass differences [12]. In the presence of time-reversal symmetry, there is no difference between $\alpha\tilde{\alpha}$- and $\alpha\alpha$-pairing. But as seen clearly in Fig. 2, in the presence of a rotating field, time-reversal symmetry is broken and the two $T = 0$ modes respond entirely different.

The effect of the $\alpha\alpha$-pairing on rotational motion has been discussed for the case of ^{48}Cr. Similar to the situation in ^{72}Kr shown in Fig. 2, there is no sudden alignment but only a smooth increase in angular momentum, I. On the other hand, from experiment we know that there is a backbend [19]. This means that the force balance is in favor for the $\alpha\tilde{\alpha}$-pairing. For the case of ^{72}Kr, as we discussed above, using a strength of the $x = 1.3$, corresponding to the Wigner energy mass defect, will result in a shift of the crossing frequency of the amount shown in Fig. 2.

One can understand that $\alpha\tilde{\alpha}$-pairing is stronger than $\alpha\alpha$. In the language of the shell-model, the $\alpha\alpha$-pairing corresponds to the coupling of two particles with given j to a total angular momentum of $J = 2j$, whereas the $\alpha\tilde{\alpha}$-pairing corresponds to the coupling to $J = 1$, i.e. low J. In the $N = 3$ shell, above ^{56}Ni, we deal with states of many different j-values, like $p_{1/2}$, $p_{3/2}$, $f_{5/2}$ and $g_{9/2}$. Whereas all these states can couple to angular momentum $J = 1$, the coupling to $J = 2j$ is totally fragmented. Therefore, one indeed expects a stronger $\alpha\tilde{\alpha}$-pairing field [15].

The dominance of the $\alpha\tilde{\alpha}$-pairing at low angular momentum, does not mean that the $\alpha\alpha$-pairing is unimportant. Recent work showed that this pairing mode may become important at high angular momenta, since there most of the $\alpha\tilde{\alpha}$-pairing, be it $T = 1$ or $T = 0$ is quenched [12, 13]. In other words, if pairing will be important at high angular momenta, it will be the $T = 0$ $\alpha\alpha$-pairing.

What are the expected signatures of this pairing mode? In a Skyrme-HFB study of bandtermination in ^{48}Cr [13], it was shown that the non-collective yrast levels were strongly affected by the presence of $\alpha\alpha$-pairing. This pairing mode mixes in aligned high-spin components in the wave-function, resulting in enhanced quadrupole collectivity. States above band-termination are affected by $T = 0$ pairing. A possible signature may therefore be to measure the spectrum above and the decay into the terminating state. However, this may not be an easy task. But also the collective rotational states can be modified in the presence of $T = 0$ pairing. The $\alpha\alpha$-pairing allows particles to be scattered into higher lying orbits. Since with increasing frequency, orbits with high-j are approaching the Fermi-surface, they will be partially occupied, resulting in an increase of the moments of inertia.

The situation is nicely demonstrated for the case of ^{88}Ru, see Fig. 3. In this nucleus there is a low lying, energetically very favored SD structure, with deformation $\beta_2 \approx 0.6$. This SD structure crosses the normal deformed states already at spins $I \approx 30$. The calculations with no pairing show a smooth decrease in the moment of inertia, due to the limited angular momentum available in finite nuclei. Once particles align they cannot continue to contribute to the angular momentum, anymore. In the presence of $T = 1$ pairing, the alignment of expecially the $h_{11/2}$ particles is delayed (with respect to unpaired calculations) and maximizes at $\hbar\omega \approx 0.6\ MeV$, see Fig. 3. After that alignment, the moment of inertia drops quite strongly and approaches that of the calculations without pairing. This is expected, since the $T = 1$ pairing is almost quenched. In contrast, the calculations with $T = 0$ $\alpha\alpha$-pairing show a steep increase at $\hbar\omega = 0.3\ MeV$. This is the frequency, where the balance of $T = 1$ and $T = 0$ pairing is in favor of the latter and we encounter a phase transition from $T = 1$ to $T = 0$. The frequency where the phase transition occurs depends sensitively on the strength of the $T = 0$ pairing. For this calculations, we choose a strength of $x = 1.1$. Note, however, that with this undercritical strength, undercritical in the sense that at $\hbar\omega = 0.0\ MeV$ only $T = 1$ pairing is present, the moments of inertia at high angular momenta is not decreasing in the manner like for $T = 1$ or unpaired calculations. Indeed, the moment of inertia in the presence of $T = 0$ makes the nucleus more rigid! Once the strength

x becomes smaller than 0.9, the effect disappears. For more details, see also the discussion in [12]. Whether one will observe this kind of 'super'-rigid rotations remains to be seen and will establish the strength of $\alpha\alpha$-pairing.

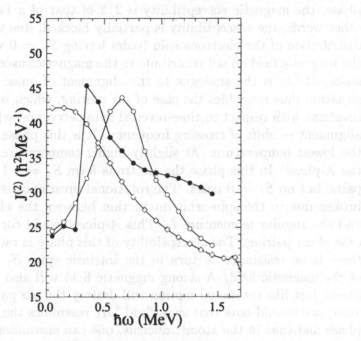

Figure 3 The kinematical, J^2, moments of inertia for the SD-band in ^{88}Ru. Calculations with no pairing ($\diamond$) are compared to calculations with standard $T = 1$ ($\circ$) and the $T = 0$, $\alpha\alpha$ ($\bullet$) pairing. The sharp rise at $\hbar\omega = 0.3$ MeV is due to the pairing phase transition from $T = 1$ to $T = 0$.

There is an interesting analogy to superfluid ^{3}He. The superfluid phases of ^3He are nicely discussed in e.g. Ref. [20]. It was realized early on in ^3He, that the paired electrons carry orbital angular moment $L = 1\hbar$ with respect to the nucleus at the center, in addition to the internal spin of the electrons, which is $S = 1\hbar$. One is thus dealing with the case of p-wave pairing, which is analoguous to the situation in $N = Z$ nuclei where the nucleons are in an $S = 1$ state for the case of $T = 0$ pairing. In ^3He one encounters two different phases, A and B, which occur for different temperatures (and pressures). The effect of a magnetic field on superfluid ^3He is exactly the same as the one of a rotational field on an atomic nucleus in a superfluid phase. Time-reversal symmetry is broken in the intrinsic system and the electrons (nucleons) try to orient their angular momenta along the direction of the magnetic (rotating) field.

It turns out that the spin properties of the two phases in ^3He remind of the two phases of $T = 0$ pairing: In phase B, the so called Balian-Wertheimer phase, all electron spins are isotropically oriented (We have equal probability of finding electrons with $S_z = +1$, 0 and -1). In this phase, the magnetic susceptibility is 2/3 of that of a Fermi-liquid. In other words, the susceptibility is partially blocked, due to the isotropic distribution of the electrons spin (pairs having $S_z = 0$ with respect to the magnetic field do not contribute to the magnetic susceptibility). The susceptibility is the analogue to the alignment of quasi-particles. The situation thus resembles the case of $\alpha\tilde{\alpha}$ pairing, which is also isotropic, invariant with respect to time-reversal symmetry, and which resists the alignment — shift of crossing frequency. It is this phase that occurs at the lowest temperature. At slightly higher temperature, we encounter the A-phase. In this phase the electrons form $S_z = -1$ and $S_z = +1$ pairs, but no $S_z = 0$ pairs. The rotational invariance is spontaneously broken due to the spin-orbit interaction between the electron spins S and the angular momentum L. This A-phase would correspond to the case of $\alpha\alpha$ pairing. The susceptibility of this phase is rather high, since there is no resistance to turn in the intrinsic spins S_z along the axis of the magnetic field. A strong magnetic field will also destroy the B-phase, just like rotational motion will destroy the $\alpha\alpha$ pairing. Last not least, one should note that superfluid ^4He resembles the $T = 1$ pairing phase and that in the atomic nucleus, one can encounter a *single* wave function, in which all phases, like A and B of ^3He together with the ^4He phase, are mixed [14].

4.   Conclusions

The $N = Z$ nuclei form an unique laboratory to explore different competing excitations. In a single wave function, both $T = 1$ and $T = 0$ correlations may coexist. The $T = 0$ correlations contribute to the binding energy but only in close vicinity of the $N = Z$ line. It accounts in a microscopic fashion for the Wigner energy. Two different pairing modes may be present for $T = 0$ pairing. The $\alpha\tilde{\alpha}$ mode, which has similar properties like the standard isovector pairing. It can be viewed as being responsible for the shift of the crossing frequency. In contrast the $\alpha\alpha$ pairing mode results in enhanced collectivity, where the moment of inertia may exceed the case of an unpaired nucleus. From simple considerations one can expect the $\alpha\tilde{\alpha}$ mode being stronger but the exact balance between the two requires more experimental data on high spin states in $N = Z$ nuclei. Nice analogies with superfluid ^3He can be established. Important experiments to directly measure the collectivity of $T = 0$ correlations would imply pair transfer. This will require radioactive ion-beams, which will become available soon. The atomic nucleus exhibits the unique feature among fermionic systems, where different pairing phases can compete in a single system. Rotations are an important probe to establish the balance of the $T = 0$ and $T = 1$ correlations.

Acknowledgments

This work is done in collaboration with W. Satuła from Warsaw University and is supported by the Swedish Natural Science Research Council (NFR) and the Göran Gustafsson Foundation. R.W. thank the (Department of Energy's) Institute of Nuclear Theory at the University of Washington for its hospitality and the Department of Energy for partial support during the completion of this work.

References

[1] A. Goswami, Nucl. Phys. 60 (1964) 228

[2] A. Goswami and L.S. Kisslinger, Phys. Rev. 140 (1965) B26

[3] P. Camiz, A. Covellho and M. Jean, Nuovo Cimento 36 (1965) 663

[4] P. Camiz, A. Covellho and M. Jean, Nuovo Cimento 42 (1966) 199

[5] A.L. Goodman, Adv. Nucl. Phys. 11 (1979) 263

[6] W.D. Myers and W. Swiatecki, Nucl. Phys. 81 (1966) 1

[7] E.P. Wigner, Phys. Rev. 51 (1937) 106

[8] I. Talmi, Rev. Mod. Phys. 34 (1962) 704

[9] J. Jänecke, Nucl. Phys. 73 (1965) 97

[10] N. Zeldes, in Handbook of Nuclear Properties, Ed. by D. Poenaru and W. Greiner, Clarendon Press, Oxford, 1996, p. 13

[11] S. Ćwiok, J. Dudek, W. Nazarewicz, J. Skalski and T. Werner, Comp. Phys. Comm. 46 (1987) 379

[12] W. Satuła and R. Wyss, A676 (2000) 120

[13] J. Terasaki, R. Wyss, P.-H. Heenen, Phys. Lett. B437 (1998) 1

[14] A.L. Goodman, Phys. Rev. C60 (1999) 014311

[15] J. Sheikh and R. Wyss, Phys. Rev. 62 (2000) 051302(R)

[16] W. Meissner and R. Ochsenfeld, Naturwiss. 21 (1933) 787

[17] G. de Angelis, et. al., Phys. Lett B415 (1997) 217

[18] K. Lister, Proceedings of the Int. Workshop on $N = Z$ Nuclei, Pingst 2000, Lund, to be published

[19] S. M. Lenzi et al. Z. Phys. A354 (1996) 117

[20] D. M. Lee, Rev. Mod. Phys. 69 (1997) 645

Ternary and Multicluster Cold Fission

Dorin N. Poenaru

Horia Hulubei National Institute of Physics and Nuclear Engineering
P.O. Box MG-6, RO-76900 Bucharest, Romania
poenaru@ifin.nipne.ro

Keywords: ternary fission, cold fission, multicluster fission

Abstract   Recently obtained experimental results on particle accompanied cold fission of ^{252}Cf suggest the existence of a short-lived quasi-molecular state. Within our three-center phenomenological model, described in this lecture, we found a possible explanation based on a new minimum in the deformation energy at a separation distance very close to the touching point. Half-lives of some quasimolecular states which could be formed in ^{10}Be and ^{12}C accompanied fission of ^{252}Cf are roughly estimated to be the order of 1 ns, and 1 ms, respectively. By extending our unified approach of cold fission, cluster radioactivity and α-decay to multifragment fission we obtain informations on the most probable mechanism of this complex process. Besides fission into two or three fragments, a heavy nucleus spontaneously break into four, five or six nuclei. Examples are presented for cold fission of ^{252}Cf, accompanied by following emitted clusters: 2α, $\alpha +^{10}Be$, 3α, $\alpha + ^{6}He + ^{10}Be$, and 4α.

1.   Introduction

Various fission processes [1, 2] are very complex phenomena dominated by dynamics of a large collective motion. Despite a longstanding effort of research and development, they continue to surprise us. Let us recollect some historical mile-stones. In 1929 the fine structure in α decay, in 1939 and 1940 the induced and spontaneous fission, in 1946 the α accompanied (ternary) fission, in 1980 prediction of cluster radioactivities which have been experimentally confirmed since 1984, in 1981 the cold fission, in 1989 the fine structure in ^{14}C decay, in 1998 the α- and ^{10}Be accompanied cold fission as well as the double and triple fine structure in a binary and ternary fission, respectively.

D. N. Poenaru et al. (eds.), Nuclei Far from Stability and Astrophysics, 151–162.
© *2001 Kluwer Academic Publishers. Printed in the Netherlands.*

We contributed to the development of *cluster radioactivity (CR)* by publishing our predictions in a paper [3] which is now the most cited one in this field, and by our theoretical investigations (see the books [2, 4] and references therein) leading to estimated half-lives which have been experimentally confirmed since 1984. In Physics and Astronomy Classification Scheme (PACS) one has now a new field — *23.70.+j Heavy-particle decay*. Following conclusions have been drawn. All nuclides with $Z > 40$ are metastable with respect to CR, but the main region of emitters is above $Z = 86$ (trans-francium nuclei), with daughters around ^{208}Pb. Usually α-decay is the main competitor. Half-lives short enough ($T < 10^{30}$ s) and relatively large branching ratios ($b = T_\alpha/T > 10^{-17}$) can be measured. Both fission-like and α-like models can explain CR. Our fission models allow a unified description of α-decay, CR, and cold fission. The measurable quantity (decay constant) $\lambda = \nu S P_s$ is a product of three model-dependent expressions. Preformation, S, can be calculated quasiclassically as an internal barrier penetrability in a fission model. Universal curve $\log T = f(\log P_s)$ of even-even nuclei, is a single straight line for a given decay mode. "Geiger-Nuttal" plot $\log T = f(1/\sqrt{Q})$ gives much scattered points (one line for each Z number of the parent). From about 150 decay modes predicted within our analytical superasymmetric fission (ASAF) model, the following have been experimentally confirmed: ^{14}C, ^{20}O, ^{23}F, $^{24-26}Ne$, $^{28,30}Mg$, and $^{32,34}Si$. Cold-fission ($TKE \simeq Q$) can be interpreted as CR. In a region of U, Pu, Cm, Cf it is a rare process. Bimodal fission of Fm, Md, No is comparable with the usual (hot) fission mechanism. The best cold fissioning nucleus, with a CF stronger than α-decay, should be ^{264}Fm. For cold fission processes the doubly magic ^{132}Sn plays the role of the ^{209}Pb in CR.

Many properties of binary fission process have been explained [5] within the liquid drop model (LDM); others like the asymmetric mass distribution of fragments and the ground state deformations of many nuclei, could be understood only after adding the contribution of shell effects [6, 7] which proved also to be of vital importance for CR.

The total kinetic energy (TKE) of fragments, in the most frequently detected binary or ternary fission mechanism, is smaller than the released energy (Q) by about $25-35$ MeV, which is used to produce deformed and excited fragments. These then emit neutrons (each with a binding energy of about 6 MeV) and γ-rays. There is also a rare "cold" fission mechanism, in which the TKE exhausts the Q-value, hence no neutrons are emitted, and the fragments are produced in or near their ground-state. The first experimental evidence for cold binary fission in which its TKE exhaust Q was reported [8] in 1981. Larger yields were measured [9] in trans-Fm ($Z \geq 100$) isotopes, where the phenomenon was called bimodal fission.

From time to time (at best once per about 1000 fission events) a spontaneous (or induced) fission of a nucleus leads to three fragments, usually one light particle (which is more frequently ^4He or some Be, C, or O isotope) and two large fragments. Since 1946, when "long-range" α particles accompanying fission were observed for the first time [10], progress made in the field of ternary fission has been reviewed many times (for example in Refs. [11–13]). Attempts to study various aspects of this phenomenon were made both by using the liquid drop model (LDM) [14] and the three-center shell model [15] Many other theoretical approaches have also been developed [16–21].

Our knowledges about fission phenomena were substantially enriched during the last years, as a consequence of experiments [22] performed by using triple γ coincidences in a modern large array of γ-ray detectors (GAMMASPHERE). A particularly interesting feature, observed [23, 24] in ^{10}Be accompanied cold fission of ^{252}Cf is related to the width of light particle γ-ray spectrum. The 3.368 MeV γ line of ^{10}Be, with a lifetime of 125 fs is not Doppler-broadened, as it should be if it would be emitted when ^{10}Be is in flight. The absence of Doppler broadening is related to a trapping of ^{10}Be in a potential well of nuclear molecular character.

We would like to extend to cold ternary fission the unified approach of cold binary fission, cluster radioactivity, and α-decay. A three-center phenomenological model, able to explain qualitatively the recently obtained experimental results will be presented below. A new minimum of a short-lived molecular state appears in the deformation energy at a separation distance very close to the touching point. This minimum allows the existence of a short-lived quasi-molecular state, decaying into three final fragments. Half-lives of some quasimolecular states which could be formed in ^{10}Be and ^{12}C accompanied fission of ^{252}Cf are roughly estimated to be the order of 1 ns, and 1 ms, respectively.

Long ago it was shown [25], on the basis of the liquid drop model, that for increasingly heavier nuclei fission into three, then four, and even five fragments becomes energetically more favorable than binary fission. Swiatecki took into consideration only equally sized fragments and ignored any shell effect. The "true" (in which the fragments are equally sized) ternary or quaternary spontaneous cold fission has not been experimentally discovered until now. On the other side the multi-fragmentation process taking place at high excitation energies, well over the potential barrier, is intensively studied. From the analysis of different configurations of fragments we conclude that the most favorable mechanism in such a decay mode should be the cluster emission from an elongated neck formed between the two heavy fragments.

2.   Shape Parametrization

The shape parametrization with one deformation parameter as follows has been suggested from the analysis [26] of different aligned and compact configurations of fragments in touch. A lower potential barrier for the aligned cylindrically-symmetric shapes with the light particle between the two heavy fragments, is a clear indication that during the deformation from an initial parent nucleus to three final nuclei, one should arrive at such a scission point. We increase continuously the separation distance, R, between heavy fragments, while the radii of heavy fragment and of light particle are kept constant, R_1 constant, $R_3 = $ constant. We adopt the following convention: $A_1 \geq A_2 \geq A_3$. The hadron numbers are conserved: $A_1 + A_2 + A_3 = A$.

At the beginning (the neck radius $\rho_{neck} \geq R_3$) one has a two-center evolution until the neck between the fragments becomes equal to the radius of the emitted particle, $\rho_{neck} = \rho(z_{s1}) |_{R=R_{ov3}} = R_3$. This Equ. defines R_{ov3} as the separation distance at which the neck radius is equal to R_3. By placing the origin in the center of the large sphere, the surface equation in cylindrical coordinates is given by:

$$\rho_s^2 = \begin{cases} R_1^2 - z^2 & , \quad -R_1 \leq z \leq z_{s1} \\ R_2^2 - (z - R)^2 & , \quad z_{s1} \leq z \leq R + R_2 \end{cases} \tag{1}$$

Then for $R > R_{ov3}$ the three center starts developing by decreasing progressively with the same amount the two tip distances $h_1 + h_{31} = h_{32} + h_2$. Besides this constraint, one has as in the binary stage, volume conservation and matching conditions. The R_2 and the other geometrical quantities are determined by solving numerically the corresponding system of algebraic equations. By assuming spherical nuclei, the radii are given by $R_j = 1.2249 A_j^{1/3}$ fm $(j = 0, 1, 3)$, $R_{2f} = 1.2249 A_2^{1/3}$ with a radius constant $r_0 = 1.2249 \ fm$, from Myers-Swiatecki's variant of LDM. Now the surface equation can be written as

$$\rho_s^2 = \begin{cases} R_1^2 - z^2 & , \quad -R_1 \leq z \leq z_{s1} \\ R_3^2 - (z - z_3)^2 & , \quad z_{s1} \leq z \leq z_{s2} \\ R_2^2 - (z - R)^2 & , \quad z_{s2} \leq z \leq R + R_2 \end{cases} \tag{2}$$

and the corresponding shape has two necks and two separating planes. Some of the important values of the deformation parameter R are the initial distance $R_i = R_0 - R_1$, and the touching-point one, $R_t = R_1 + 2R_3 + R_{2f}$. There is also R_{ov3}, defined above, which allows one to distinguish between the binary and ternary stage.

3.   Deformation Energy

According to the LDM, by requesting zero energy for a spherical shape, the deformation energy, $E^u(R)\, E^0$, is expressed as a sum of the surface and Coulomb terms

$$E^u_{def}(R) = E^0_s[B_s(R) - 1] + E^0_C[B_C(R) - 1] \tag{3}$$

where the exponent ^{u} stands for uniform (fragments with the same charge density as the parent nucleus), and ^{0} refers to the initial spherical parent. In order to simplify the calculations, we initially assume the same charge density $\rho_{1e} = \rho_{2e} = \rho_{3e} = \rho_{0e}$, and at the end we add the corresponding corrections. In this way we perform one numerical quadrature instead of six. For a spherical shape $E^0_s = a_s(1 - \kappa I^2)A^{2/3}$; $I = (N - Z)/A$; $E^0_C = a_c Z^2 A^{-1/3}$, where the numerical constants of the LDM are: $a_s = 17.9439\ MeV$, $\kappa = 1.7826$, $a_c = 3e^2/(5r_0)$, $e^2 = 1.44\ MeV \cdot fm$.

The shape-dependent, dimensionless surface term is proportional to the surface area:

$$B_s = \frac{E_s}{E^0_s} = \frac{d^2}{2} \int_{-1}^{+1} \left[y^2 + \frac{1}{4}\left(\frac{dy^2}{dx}\right)^2 \right]^{1/2} dx \tag{4}$$

where $y = y(x)$ is the surface equation in cylindrical coordinates with -1, +1 intercepts on the symmetry axis, and $d = (z'' - z')/2R_0$ is the seminuclear length in units of R_0. Similarly, for the Coulomb energy [27] one has

$$B_c = \frac{5d^5}{8\pi} \int_{-1}^{+1} dx \int_{-1}^{+1} dx' F(x, x') \tag{5}$$

$$
\begin{aligned}
F(x, x') = {} & a_\rho^{-1}\{yy_1[(K - 2D)/3] \cdot \\
& \left[2(y^2 + y_1^2) - (x - x')^2 + \frac{3}{2}(x - x')\left(\frac{dy_1^2}{dx'} - \frac{dy^2}{dx}\right) \right] + K \cdot \\
& \left\{ y^2 y_1^2/3 + \left[y^2 - \frac{x - x'}{2}\frac{dy^2}{dx} \right]\left[y_1^2 - \frac{x - x'}{2}\frac{dy_1^2}{dx'} \right] \right\} \}
\end{aligned}
\tag{6}
$$

K, K' are the complete elliptic integrals of the 1st and 2nd kind

$$K(k) = \int_0^{\pi/2} (1 - k^2 \sin^2 t)^{-1/2} dt; \quad K'(k) = \int_0^{\pi/2} (1 - k^2 \sin^2 t)^{1/2} dt \tag{7}$$

and $a_\rho^2 = (y + y_1)^2 + (x - x')^2$, $k^2 = 4yy_1 / a_\rho^2$, $D = (K - K')/k^2$.

The new minimum, at a separation distance $R = R_{min-t} > R_{ov3}$, is the result of a competition between the Coulomb and surface energies. At the beginning ($R < R_{min-t}$) the Coulomb term is stronger, leading to a decrease in energy, but later on ($R > R_{min-t}$) the light particle formed in the neck posses a surface area increasing rapidly, so there is also an increase in energy up to $R = R_t$.

After performing numerically the integrations, we add the following corrections: for the difference in charge densities reproducing the touching point values; for experimental masses reproducing the Q_{exp}-value at $R = R_i$, when the origin of energy corresponds to infinite separation distances between fragments, and the phenomenological shell corrections δE

$$E_{LD}(R) = E_{def}^u(R) + (Q_{th} - Q_{exp})f_c(R) \qquad (8)$$

where $f_c(R) = (R - R_i)/(R_t - R_i)$, and

$$Q_{th} = E_s^0 + E_C^0 - \sum_1^3 (E_{si}^0 + E_{Ci}^0) + \delta E^0 - \sum_1^3 \delta E^i \qquad (9)$$

The correction increases gradually (see Fig. 1) with R up to R_t and then remains constant for $R > R_t$. The barrier height increases if $Q_{exp} < Q_{th}$ and decreases if $Q_{exp} > Q_{th}$. In this way, when one, two, or all final nuclei have magic numbers of nucleons, Q_{exp} is large and the fission barrier has a lower height, leading to an increased yield. In a binary decay mode like cluster radioactivity and cold fission, this condition is fulfilled when the daughter nucleus is ^{208}Pb and ^{132}Sn, respectively.

4.   Shell Corrections and Half-Lives

Finally we also add the shell terms

$$E(R) = E_{LD}(R) + \delta E(R) - \delta E^0 \qquad (10)$$

Presently there is not available any microscopic three-center shell model reliably working for a long range of mass asymmetries, hence we use a phenomenological model, instead of the Strutinsky's method, to calculate the shell corrections. The model is adapted after Myers and Swiatecki [7]. At a given R, we calculate the volumes of fragments and the corresponding numbers of nucleons $Z_i(R)$, $N_i(R)$ ($i = 1, 2, 3$), proportional the volume of each fragment. We add for each fragment the contribution of protons and neutrons

$$\delta E(R) = \sum_i \delta E_i(R) = \sum_i [\delta E_{pi}(R) + \delta E_{ni}(R)] \qquad (11)$$

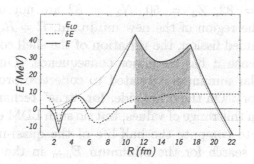

Figure 1 The liquid drop model, E_{LD}, the shell correction, δE, and the total deformation energies, E, for the ^{10}Be accompanied cold fission of ^{252}Cf with ^{146}Ba and ^{96}Sr heavy fragments. The new minimum appears in the shaded area from R_{ov3} to R_t.

which are given by

$$\delta E_{pi} = Cs(Z_i); \quad \delta E_{ni} = Cs(N_i) \tag{12}$$

where

$$s(Z) = F(Z)/[(Z)^{-2/3}] - cZ^{1/3} \tag{13}$$

$$F(n) = \frac{3}{5}\left[\frac{N_i^{5/3} - N_{i-1}^{5/3}}{N_i - N_{i-1}}(n - N_{i-1}) - n^{5/3} + N_{i-1}^{5/3}\right] \tag{14}$$

in which $n \in (N_{i-1}, N_i)$ is either a current Z or N number and N_{i-1}, N_i are the closest magic numbers. The constants $c = 0.2$, $C = 6.2\ MeV$ were determined by fit to the experimental masses and deformations. The variation with R is calculated [28] as

$$\delta E(R) = \frac{C}{2}\left\{\sum_i [s(N_i) + s(Z_i)]\frac{L_i(R)}{R_i}\right\} \tag{15}$$

where $L_i(R)$ are the lengths of the fragments along the axis of symmetry, at a given separation distance R. During the deformation, the variation of separation distance between centers, R, induces the variation of the geometrical quantities and of the corresponding nucleon numbers. Each time a proton or neutron number reaches a magic value, the correction energy passes through a minimum, and it has a maximum at midshell (see Fig. 1). The first narrow minimum appearing in the shell correction energy δE in Fig. 1, at $R = R_{min1b} \simeq 2.6\ fm$, is the result of almost simultaneously reaching the magic numbers $Z_1 = 20$, $N_1 = 28$, and $Z_2 = 82$, $N_2 = 126$. The second, more shallower one around $R =$

$R_{min2b} \simeq 7.2 \ fm$ corresponds to a larger range of R-values for which $Z_1 = 50$, $N_1 = 82$, $Z_2 = 50$, $N_2 = 82$ are not obtained in the same time. In the region of the new minimum, $R = R_{min-t}$, for light-particle accompanied fission, the variation of the shell correction energy is very small, hence it has no major consequence. One can say that the quasimolecular minimum is related to collective properties (liquid-drop like behavior). On the other side, for "true" ternary process both minima appear in this range of values, but no such LDM effect was found there. In order to compute the half-life of the quasi-molecular state, we have first to search for the minimum E_{min} in the quasimolecular well, from R_{ov3} to R_t, and then to add a zero point vibration energy: $E_{qs} = E_{min} + E_v$.

The half-life, T, is expressed in terms of the barrier penetrability, P, which is calculated from an action integral, K, given by the quasi-classical WKB approximation

$$T = \frac{h \ln 2}{2E_v P}; \quad P = exp(-K) \tag{16}$$

where h is the Planck constant, and

$$K = \frac{2}{\hbar} \int_{R_a}^{R_b} \sqrt{2\mu[E(R) - E_{qs}]} \, dR \tag{17}$$

in which R_a, R_b are the turning points, defined by $E(R_a) = E(R_b) = E_{qs}$ and the nuclear inertia is roughly approximated by the reduced mass $\mu = m[(A_1 A_2 + A_3 A)/(A_1 + A_2)]$, where m is the nucleon mass, $\log[(h \ln 2)/2] = -20.8436$, $\log e = 0.43429$.

The results of our estimations for the half-lives of some quasimolecular states formed in ^{10}Be and ^{12}C accompanied fission of ^{252}Cf are of the order of 1 ns and 1 ms, respectively. For example $\log T(s) = -8.89$ for ^{10}Be accompanied fission with $^{138}Te + ^{104}Mo$ fragments and -3.25 for $^{12}C + ^{140}Te + ^{100}Zr$. Consequently the new minimum we found can qualitatively explain the quasimolecular nature of the narrow line of the ^{10}Be γ-rays.

5.   Multi-Cluster Fission

Basic condition to be fulfilled in the natural process, $^A Z \rightarrow \sum_1^n {}^{A_i} Z_i$, concerns the energy release

$$Q = M - \sum_1^n m_i \tag{18}$$

which should be positive and high enough. We took the masses (in units of energy), entering in the above equation, from the updated compilation of measurements [30]. When $n = 2$ one has a binary process, $n = 3$ means a ternary fragmentation, $n = 4$ — a quaternary fission, $n = 5$ — a division into five fragments, $n = 6$ — a division into six fragments, etc.

In a first approximation, one can obtain an order of magnitude of the potential barrier height by assuming spherical shapes of all the participant nuclei. This assumption is realistic if the fragments are magic nuclei. For deformed fragments it leads to an overestimation of the barrier. By taking into account the prolate deformations, one can get smaller potential barrier height, hence better condition for multicluster emission. We use the Yukawa-plus-exponential (Y+E) double folded model [31] extended [32] for different charge densities. In the decay process from one parent to several fragments, the nucleus deforms, reaches the touching configuration, and finally the fragments became completely separated. The interaction energy of the fragments at the touching point configuration is given by:

$$E_t = \sum_i^n \sum_{j>i}^n E_{ij} \; ; \; E_{ij} = E_{Cij} + E_{Yij} \tag{19}$$

where $E_{ij}(i, j = 1, 2, ...n)$ is the sum of Coulomb and Y+E potential energies, given below for spherical shapes. The origin of potential energy is taken at an infinite separation distance of fragments. In such a way one has initially a potential energy equal to the Q-value. By assuming a uniform distribution of the electric charge, the electrostatic interaction energy of nonoverlapping spherical nuclei $A_i Z_i$ and $A_j Z_j$ is the same as if all charges were concentrated at the sphere centers. Consequently one has

$$E_{Cij} = e^2 Z_i Z_j / R_{ij} \tag{20}$$

where R_{ij} is the distance between centers, which becomes $R_{tij} = r_0(A_i^{1/3} + A_j^{1/3})$ at the touching point. Within Myers-Swiatecki's liquid drop model there is no contribution of surface energy to the interaction of separated fragments; deformation energy has a maximum at the touching point configuration. The proximity forces acting at small separation distances (within the range of strong interactions) give rise in the Y+EM to a term expressed as follows

$$E_{Yij} = -4 \left(\frac{a}{r_0}\right)^2 \sqrt{a_{2i} a_{2j}} \left[g_i g_j \left(4 + \frac{R_{ij}}{a}\right) - g_j f_i - g_i f_j\right] \frac{\exp(-R_{ij}/a)}{R_{ij}/a} \tag{21}$$

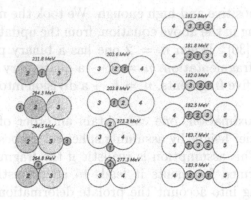

Figure 2 Aligned and compact configurations for α accompanied cold fission (left hand side) and for $\alpha + {}^{10}Be$ accompanied cold fission (middle). On the right hand side there are aligned configurations with three clusters between light and heavy fragment for $\alpha + {}^{6}He + {}^{10}Be$ accompanied cold fission. The parent is ${}^{252}Cf$, and the heavy fragment is a doubly magic ${}^{132}Sn$. The corresponding touching point energies are shown.

where

$$g_k = \frac{R_k}{a} \cosh\left(\frac{R_k}{a}\right) - \sinh\left(\frac{R_k}{a}\right) \; ; \; f_k = \left(\frac{R_k}{a}\right)^2 \sinh\left(\frac{R_k}{a}\right) \qquad (22)$$

in which R_k is the radius of nucleus $A_k Z_k$, $a = 0.68$ is diffusivity parameter, and a_{2i}, a_{2j} are expressed in terms of model constants a_s, κ and nuclear composition parameters I_i and I_j, $a_2 = a_s(1 - \kappa I^2)$, $a_s = 21.13 \; MeV$, $\kappa = 2.3$, $I = (N - Z)/A$, $R_0 = r_0 A^{1/3}$, $r_0 = 1.16 \; fm$ is the radius constant, and e is the electron charge.

Different kinds of aligned and compact configurations of fragments in touch are shown in Fig. 2. On the left hand side there are three aligned fragments on the same axis, in the following order of three partners: 213, 123, and 132 (or 231) and one compact configuration (in which every partner is in touch with all others). Touching point energy for "polar emission" (123 or 132) is much higher than that of an emission from the neck (213), which explains an experimentally determined low yield of polar emission compared to the "equatorial" one. As it should be, compact configuration posses maximum total interaction energy, hence it has a lowest chance to be observed. For examples given in Fig. 2 the energy of compact configuration compared to that of aligned configuration is higher by 33 MeV for ternary fission, 73 MeV for quaternary fission, and 86 MeV for fission into five fragments.

An important conclusion can be drawn, by generalizing this result, namely: *multiple clusters $1, 2, 3, \ldots$ should be formed in a configuration of nuclear system in which there is a relatively long neck between light* *(n − 1)* and heavy *(n)* fragment. Such shapes with long necks in fission have been considered [29] as early as 1958. For "true" ternary fission, in two ^{84}As plus ^{84}Ge, $E_t - Q = 98$ MeV! Despite the larger Q-value (266 MeV), a very large touching point energy explains why this split has a low chance to be observed.

On the right-hand side of Fig. 2, we ignore aligned configurations in which heavy fragments are not lying at the two ends of a chain. By arranging in six different manners the α-particle, ^6He, and ^{10}Be clusters between two heavy fragments from a cold fission of ^{252}Cf, the difference in energy is relatively small. Nevertheless, 43125 configuration gives a lowest touching point energy. Touching point energy of a compact configuration, which is not shown in Fig. 2, is higher by 86 MeV.

When the parent nucleus is heavier ($^{252,254}Es$, $^{255,256}Fm$, $^{258,260}Md$, $^{254,256}No$, ^{262}Lr, $^{261,262}Rf$, etc), multi cluster emission could be stronger. While minimum energy of the most favorable aligned configuration of fragments in touch, when at least one cluster is not an alpha particle, becomes higher and higher with increasing complexity of partners, the same quantity for multi-alphas remains favorable, hence we expect that multi-alpha emission could be more easily detected.

One can estimate, very tentatively, an order of magnitude of emission probability for 2α accompanied fission of trans-uranium nuclei, by comparing results for α and Be accompanied fission. Both Q-values and mass numbers are very similar, hence we expect to have a similar probability — four to five orders of magnitude lower than that of a single α accompanied fission, which is measurable. In conclusion, we suggest experimental searches for multi-alpha accompanied fission, starting with 2α.

References

[1] D.C. Hoffman, T.M. Hamilton and M.R. Lane, in *Nuclear Decay Modes*, Institute of Physics Publishing, Bristol (1996) ch. 10, p. 393

[2] D.N. Poenaru and W. Greiner, in *Nuclear Decay Modes*, Institute of Physics Publishing, Bristol (1996) ch. 6, p. 275

[3] A. Săndulescu, D.N. Poenaru and W. Greiner, Sov. J. Part. Nucl. (1980) 528

[4] D.N. Poenaru and W. Greiner, in *Handbook of Nuclear Properties*, Oxford University Press (1996), ch. 5, p. 131

[5] N. Bohr and J. Wheeler, Phys. Rev. 55 (1939) 426

[6] V.M. Strutinsky, Nucl. Phys. A95 (1966) 420

[7] W.D. Myers and W.J. Swiatecki, Nucl. Phys. A81 (1966) 1

[8] C. Signarbieux et al., J. Phys. Lett. (Paris), 42 (1981) L437

[9] E.K. Hulet et al., Phys. Rev. Lett. 56 (1986) 313

[10] L. W. Alvarez, as reported by G. Farwell, E. Segrè, and C. Wiegand, Phys. Rev. 71 (1947) 327. T. San-Tsiang, R. Chastel, H. Zah-Way and L. Vigneron, C. R. Acad. Sci. Paris 223 (1946) 986

[11] I. Halpern, Annu. Rev. Nucl. Sci. 21 (1971) 245

[12] C. Wagemans, in *Particle Emission from Nuclei, Vol. III* CRC Press, Boca Raton, Florida (1989), ch. 3, p. 63

[13] M. Mutterer and J.P. Theobald, in *Nuclear Decay Modes*, Institute of Physics Publishing, Bristol (1996), ch. 12, p. 487

[14] H. Diehl and W. Greiner, Nucl. Phys. A229 (1974) 29

[15] J. Hahn, H.J. Lustig and W. Greiner, Z. Nat. 32a (1977) 215

[16] Y. Boneh, Z. Fraenkel and I. Nebenzahl, Phys. Rev. 156 (1967) 1305

[17] G.M. Raisbeck and T.D. Thomas, Phys. Rev. 172 (1968) 1272

[18] N. Cârjan and B. Leroux, Phys. Rev. C22 (1980) 2008

[19] O. Tanimura and T. Fliessbach, Z. Phys. A328 (1987) 475

[20] V.A. Rubchenya and S.G. Yavshits, Z. Phys. A329 (1988) 217

[21] R. Schäfer and T. Fliessbach, J. Phys. G 21 (1995) 861

[22] J.H. Hamilton et al., Prog. Part. Nucl. Phys. 35 (1995) 635

[23] A. V. Ramayya et al., Phys. Rev. Lett. 81 (1998) 947

[24] P. Singer et al., in *Dynamical Aspects of Nuclear Fission (Proc. Conf., Častá-Papiernička, Slovakia)*, JINR Dubna (1996) p. 262

[25] W.J. Swiatecki, in *Second U.N. Int. Conf. on the peaceful uses of atomic energy* Geneva (1958) p. 651

[26] Poenaru, D.N. et al. Phys. Rev. C59 (1999) 3457

[27] D.N. Poenaru and M. Ivaşcu, Comp. Phys. Comm. 16 (1978) 85

[28] H. Schultheis and R. Schultheis, Phys. Lett. B37 (1971) 467

[29] D.L. Hill in *Proc. of the 2nd U N Int. Conf. on the Peaceful Uses of Atomic Energy*, United Nations, Geneva (1958), p. 244

[30] G. Audi and A.H. Wapstra, Nucl. Phys. A595 (1995) 409

[31] H.J. Krappe, J.R. Nix and A.J. Sierk, Phys. Rev. C20 (1979) 992

[32] D.N. Poenaru, M. Ivaşcu and D. Mazilu, Comput. Phys. Commun. 19 (1980) 205

[33] D.N. Poenaru, W. Greiner and R.A. Gherghescu, Atomic Data Nucl. Data Tables, 68 (1998) 91

Fusion and Quasifission of Heavy Nuclei

G.G. Adamian[a], N.V. Antonenko[b], A. Diaz Torres[a], S.P. Ivanova[b],
W. Scheid[a] and V.V. Volkov[b]

[a] Institut für Theoretische Physik, Justus-Liebig-Universität, Giessen, Germany
[b] Joint Institute for Nuclear Research, Dubna, Russia
Werner.Scheid@theo.physik.uni-giessen.de

Keywords: Fusion, quasifission, superheavy nuclei, evaporation residue cross sections, dinuclear system concept

Abstract   Evaporation residue cross sections for superheavy nuclei and quasifission distributions in heavy ion collisions are described within the dinuclear system concept. This concept assumes that the compound nucleus is formed by a transfer of nucleons between two touching nuclei. The calculated cross sections and quasifission distributions agree well with experimental data with exception of that for $Z = 118$.

1.   Introduction

The experimental synthesis of superheavy elements and heavy nuclei far from the line of stability stimulates the study of the mechanism of fusion in heavy ion collisions at low energies. Models for calculation of fusion cross sections can be discriminated by the assumptions on the dynamics of the fusion process. The important collective degrees of freedom are the relative internuclear distance R between the nuclear centers and the mass asymmetry coordinate η defined as $\eta = (A_1 - A_2) / (A_1 + A_2)$ where A_1 and A_2 are the mass numbers of the clusters. Here, $\eta = 0$ means a symmetric system and $\eta = \pm 1$ the fused system.

Depending on the main degree of freedom used for the description of fusion, two different types of models for fusion can be distinguished. The first type of models assumes a melting of nuclei along the internuclear distance [1]. These models describe the fusion of not too heavy nuclei. The second type is based on the dinuclear system concept [2, 3]. We use the dinuclear system (DNS) model for calculating production cross sections of superheavy nuclei. Here, the approaching nuclei get captured

D. N. Poenaru et al. (eds.), Nuclei Far from Stability and Astrophysics, 163–172.
© 2001 *Kluwer Academic Publishers. Printed in the Netherlands.*

into an excited dinuclear configuration. Then the dinuclear system develops in the mass asymmetry coordinate by diffusion in time and can simultaneously decay by quasifission. When the system crosses the inner fusion barrier in the mass asymmetry coordinate, a compound nucleus is formed. This system can approach the ground state by neutron evaporation in competition with fission, which is expressed by the survival probability.

In this paper we show by comparison with experimental data that the calculated evaporation residue cross sections for the production of actinides and superheavy nuclei support the basic assumption of the dinuclear system concept that the nuclei do not directly melt together, but form the compound nucleus by transferring nucleons in a dinuclear configuration of touching nuclei. Further we discuss distributions of the fragment masses and charges in quasifission which give information on the dynamics and time-development of the dinuclear system.

2.   Evaporation Residue Cross Section for Fusion

The evaporation residue cross section can be written as a sum over the contributions of partial waves

$$\sigma_{ER}(E_{c.m.}) = \sum_J \sigma_c(E_{c.m.}, J) P_{CN}(E_{c.m.}, J) W_{sur}(E_{c.m.}, J). \tag{1}$$

The first factor is the partial capture cross section for the transition of colliding nuclei over the entrance (Coulomb) barrier with a probability $T(E_{c.m.}, J)$. The second factor in Equ. 1 is the probability of complete fusion that the system approaches an excited compound nucleus, starting from the touching configuration. This probability includes the competition with quasifission after capture. The last factor in Equ. 1 is the surviving probability of the fused system and regards the fission and neutron evaporation of the excited compound nucleus. The contributing angular momenta in σ_{ER} are limited by W_{sur} with $J_{max} \approx 10 - 20$ in the case of highly fissionable superheavy nuclei. For small angular momenta we can approximate this cross section as follows

$$\sigma_{ER}(E_{c.m.}) = \sigma_c(E_{c.m.}) P_{CN}(E_{c.m.}, J = 0) W_{sur}(E_{c.m.}, J = 0) \tag{2}$$

with $\sigma_c(E_{c.m.}) = \pi \lambda^2 (2J_{max} + 1) T(E_{c.m.}, J = 0)$. For reactions leading to superheavy nuclei, values of $J_{max} = 10$ and $T(E_{c.m.}, J = 0) = 0.5$ are chosen for bombarding energies $E_{c.m.}$ near the Coulomb barrier.

3.   Adiabatic Treatment of Internuclear Motion

Models describing the fusion process as an internuclear melting of nuclei often use adiabatic potential energy surfaces (PES). These potentials are calculated with the macroscopic-microscopic method of the Strutinsky formalism. Here we apply the microscopic two-center shell model with the following coordinates: elongation $\lambda = \ell/(2R_0)$, where ℓ is the length of the system and $2R_0$ the diameter of the compound system, neck parameter ε with $\varepsilon = 0$ showing no neck and $\varepsilon \approx 1$ showing necked-in shapes, mass asymmetry η and deformation parameters $\beta_i = a_i/b_i$ which are ratios of semiaxes of the clusters $i = 1$ and 2. The adiabatic potential is obtained as

$$U(\lambda, \varepsilon, \eta, \eta_z, \beta_i) = U_{LDM} + \delta U_{shell} + \delta U_{pairing}, \qquad (3)$$

where U_{LDM} is the liquid drop potential, δU_{shell} the shell correction part originating from the TCSM and $\delta U_{pairing}$ the pairing energy part. In Fig. 1 we show such an adiabatic potential energy surface for the system $^{110}Pd + {}^{110}Pd$ [4]. Calculations of fusion probabilities [4] were carried out in adiabatic PES for symmetric and near symmetric heavy ion systems, e.g. for the $^{110}Pd + {}^{110}Pd$ - system. We found that the fusion probabilities obtained with this adiabatic model are much larger than the values derived from experimental data. We conclude that an adiabatic treatment of the potential energy surface is not adequate for the description of fusion of heavy ions. This statement depends also on the choice of the mass parameters which can be calculated with an extended temperature-dependent cranking formula [5].

4.   Dynamical Diabatic Description

Because the adiabatic PES can not be applied for the description of fusion, we investigated the dynamics of the transition between an initially diabatic (sudden) interaction potential and an adiabatic potential during the fusion process. The main question is the survival of the dinuclear configuration evolving in the mass asymmetry.

In the calculations we used the maximum symmetry method for the Hamiltonian of the diabatic TCSM [6]. Diabatic TCSM levels do not show avoided level crossings. Avoided level crossings can be removed from the adiabatic TCSM by eliminating the symmetry-violating parts from the adiabatic Hamiltonian of the TCSM. The diabatic potential can be written

$$U_{diab}(\lambda, \varepsilon) = U_{adiab}(\lambda, \varepsilon) + \delta U_{diab}(\lambda, \varepsilon), \qquad (4)$$

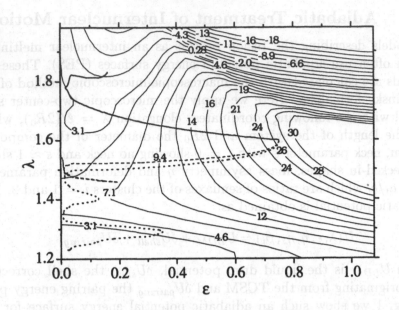

Figure 1 Adiabatic potential energy contours drawn in the (λ, ε)-plane for the system $^{110}Pd + {}^{110}Pd$ with shell corrections and $\beta_i = 1$. The dynamical trajectories starting from the touching configuration with initial kinetic energies of 0, 40 and 60 MeV are shown by solid, dashed and dotted curves, respectively.

where

$$\delta U_{diab} = \sum_\alpha \epsilon_\alpha^{diab} n_\alpha^{diab} - \sum_\alpha \epsilon_\alpha^{adiab} n_\alpha^{adiab}. \tag{5}$$

Here, ϵ_α^{diab}, ϵ_α^{adiab} and n_α^{diab}, n_α^{adiab} are the single particle energies and occupation numbers of the diabatic and adiabatic levels, respectively. The initial occupation probabilities n_α^{diab} are determined by the config-uration of the separated nuclei. They depend on time since the excited diabatic levels get deexcited:

$$\frac{dn_\alpha^{diab}(\lambda, \varepsilon, t)}{dt} = -\frac{1}{\tau(\lambda, \varepsilon, t)} (n_\alpha^{diab}(\lambda, \varepsilon, t) - n_\alpha^{adiab}(\lambda, \varepsilon)), \tag{6}$$

where τ is a relaxation time determined by a mean single particle width depending on the diabatic occupation numbers

$$(\tau(\lambda, \varepsilon, t))^{-1} = \sum_\alpha \bar{n}_\alpha^{diab} \Gamma_\alpha / (2\hbar \sum_\alpha \bar{n}_\alpha^{diab}). \tag{7}$$

Diabatic potentials show a strong increase with decreasing elongation λ. Their repulsive character screens smaller values of the elongation and hinders the DNS to melt into the compound nucleus. Fig. 2 shows the diabatic potential for the system $^{110}Pd + {}^{110}Pd$ as a function of the elongation λ for the initial time and the lifetime t_0 of the DNS. The nuclei are considered as spherical with a neck parameter $\varepsilon = 0.75$. Probabilities for fusion are calculated with Kramers expressions in the coordinates λ and η. These expressions result from quasistationary solutions of the Fokker-Planck equation and are written as integrals over the fusion rates.

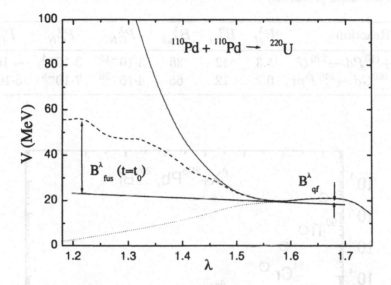

Figure 2 The diabatic time-dependent potentials (solid and dashed curves) and the adiabatic potential (dotted curve) for $^{110}Pd + {}^{110}Pd$ as a function of λ.

$$P_{CN}^{\lambda \ or \ \eta} = \int_0^{t_0} \Lambda_{fus}^{\lambda \ or \ \eta} dt \qquad (8)$$

The fusion rates $\Lambda_{fus}^{\lambda \ or \ \eta}$ are proportional to $\exp(-B_{fus}^{\lambda \ or \ \eta}/T)$. Here, B_{fus}^{λ} and B_{fus}^{η} are the corresponding barriers for fusion and $T = \sqrt{E^*/a}$ the local thermodynamic temperature with the excitation energy E^* of the system. In Tab. 1 we list examples for barriers and probabilities for forming compound nuclei [7]. Our analysis with the diabatic dynamics yields the result that the fusion of heavy nuclei along the internuclear distance in the coordinates R or λ is very unprobable, especially in the case of fusion of equal nuclei. These facts, demonstrated with the examples of Tab. 1, strongly support our standpoint that the correct model for fusion of heavy nuclei is the dinuclear system model where fusion is described by the transfer of nucleons in the touching configuration.

The results are also supported by microscopical investigations [8], based on the SU(3) - symmetry with harmonic oscillator wavefunctions for the nucleons, which showed a structural forbiddenness for the motion to smaller relative distances.

Table 1 Quasifission barrier (in MeV), inner barriers (in MeV) in η and λ calculated with the dynamical diabatic potential, fusion probabilities in λ and η, and experimental fusion probability.

Reaction	B_{qf}^{λ}	B_{fus}^{η}	B_{fus}^{λ}	P_{CN}^{λ}	P_{CN}^{η}	P_{fus}^{exp}
$^{110}Pd + ^{110}Pd \rightarrow ^{220}U$	1.3	12	36	$3 \cdot 10^{-15}$	$3 \cdot 10^{-4}$	$\sim 10^{-4}$
$^{86}Kr + ^{160}Gd \rightarrow ^{246}Fm$	0.2	12	65	$4 \cdot 10^{-26}$	$7 \cdot 10^{-5}$	$5 \cdot 10^{-5}$

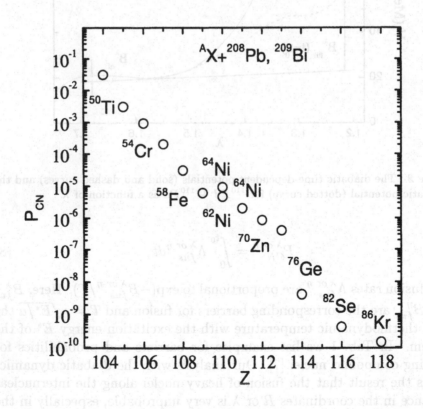

Figure 3 Calculated fusion probabilities P_{CN} for cold fusion in (HI,1n) lead-based reactions for the projectiles indicated.

5.   Cross Sections for Superheavy Nuclei

The evaporation residue cross sections, calculated according to Equ. 1, depend on the surviving probabilities. W_{sur} is essentially given by the ratio between the width for neutron decay to the width for fission of the excited compound nucleus which is the main decay process. The probability is obtained [9, 10] with a statistical model and varies moderately with the shell structure of the compound nucleus between 10^{-4} and 10^{-2} for lead-based fusion with the evaporation of one neutron.

The available experimental data can be well reproduced with the dinuclear system concept with exception of the reaction $^{86}Kr + ^{208}Pb \rightarrow ^{293}118 + 1n$. We calculated a list of evaporation residue cross sections for lead- and actinide-based reactions [9, 11]. The strong decrease of the

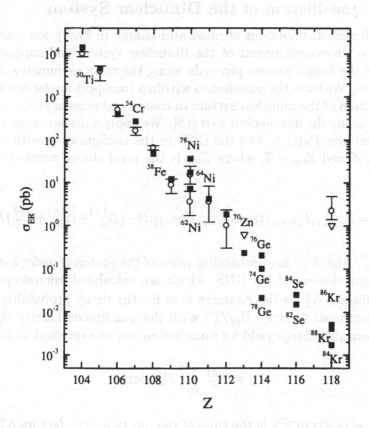

Figure 4 Evaporation residue cross sections for ^{208}Pb- and ^{209}Bi-based fusion. The full squares and the open circles are calculated and experimental data, respectively. The open triangles give experimental upper limits (GSI).

fusion cross sections with the charge number of the fused system is caused by the decrease of the probability P_{CN} for forming the compound nucleus (see Figs. 3 and 4). For example, for the reaction $^{70}Zn + ^{208}Pb \rightarrow {}^{277}112 + 1n$ we found $P_{CN}=1 \cdot 10^{-6}$, $\sigma_{ER}=1.8$ pb (exp. 1.0 pb) and for the reaction $^{86}Kr + ^{208}Pb \rightarrow {}^{293}118 + 1n$ we obtained $P_{CN} = 1.5 \cdot 10^{-10}$, $\sigma_{ER} = 5.1$ fb (exp. 2.2 pb and < 1.0 pb). From the extrapolation of the experimental results for the elements 102 – 112 one concludes for the reaction $^{86}Kr + ^{208}Pb \rightarrow {}^{293}118 + 1n$ that the value σ_{ER} is of the order of 0.01 pb in close agreement to our result.

The elements 114 and 116 were synthesized in the actinide based hot fusion reactions $^{48}Ca + ^{244}Pu$ and $^{48}Ca + ^{248}Cm$ at JINR in Dubna. The earlier calculated values of σ_{ER} [11, 12] for these reactions are in agreement with the experimental data of about 1 pb.

6.   Quasifission of the Dinuclear System

Quasifission distributions of mass and charge in heavy ion reactions reveal the time-development of the dinuclear system and support the idea that the fusion process proceeds along the mass asymmetry degree of freedom. We treat the quasifission within a transport model describing the evolution of the dinuclear system in charge and mass asymmetry and its decay along the internuclear axis [13]. We apply a master-equation for the probability $P_Z(t)$ to find the DNS in the configuration with charge numbers Z and $Z_{tot} - Z$, where Z_{tot} is the total charge number of the system:

$$\frac{\partial P_Z(t)}{\partial t} = \Delta^{(-)}_{(Z+1)}P_{(Z+1)}(t)+\Delta^{(+)}_{(Z-1)}P_{(Z-1)}(t)-(\Delta^{(+)}_Z+\Delta^{(-)}_Z+\Lambda^{qf}_Z)P_Z(t).$$
(9)

Here, $\Delta^{(+)}_Z$ and $\Delta^{(-)}_Z$ are probability rates of the proton transfer between the nuclear clusters of the DNS, which are calculated microscopically. The coefficient Λ^{qf}_Z is the Kramers rate for the decay probability in R and proportional to $\exp(-B_{qf}/T)$ with the quasifission barrier $B_{qf}(\eta)$. The measurable charge yield for quasifission can be expressed as follows:

$$Y_Z = \Lambda^{qf}_Z \int_0^{t_0} P_Z(t)dt,$$
(10)

where $t_0 = (3\text{-}4)\text{x}10^{-20}$s is the time of the reaction. The factors Λ^{qf}_Z and $P_Z(t)$ in Equ. 10 are considered separately because the characteristic time for nucleon transitions between the nuclei is much shorter than the decay time of the DNS. Similar formulas are true for P_A and Y_A as functions of the mass number.

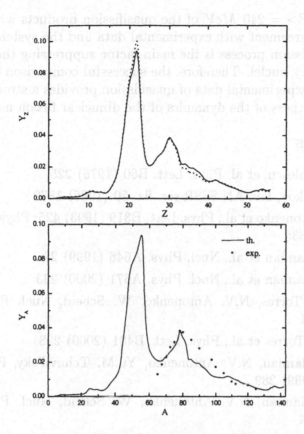

Figure 5 Charge (upper part) and mass (lower part) yields of the quasifission products as a function of Z and A of the light fragment, respectively, for the hot fusion reaction $^{48}Ca + {}^{238}U \rightarrow {}^{286}112$ and an excitation energy of the initial DNS of $E^* = 5$ (dotted curve) and 15 (solid curve) MeV. The experimental data [14] are shown by solid points for $E^* = 10$ MeV corresponding to an excitation energy of the compound nucleus of 33 MeV.

Fig. 5 shows the calculated and experimental charge and mass yields Y_Z and Y_A as functions of the charge and mass of the light fragment, respectively, for the hot fusion reaction $^{48}Ca + {}^{238}U \rightarrow {}^{286}112$. The calculated mass distribution is in good agreement with the measured one [14]. It should be noted that, near the initial combination, quasifission events overlapping with the products of deep inelastic collisions were not taken into account in the experimental data because of the difficulty to discriminate these data. Regarding the deformation of the nuclei in the $^{48}Ca + {}^{238}U$ reaction, we find that the calculated average total kinetic energy $<TKE> = 240$ MeV of the quasifission products with $70 < A < 120$ is in agreement with experimental data and the systematics.

The quasifission process is the main factor suppressing the complete fusion of heavy nuclei. Therefore, the successful comparison of the theoretical and experimental data of quasifission provides a strong support for the correctness of the dynamics of the dinuclear fusion model.

References

[1] A. Sandulescu, et al. Phys. Lett. B60 (1976) 225

[2] V.V. Volkov, Izv. AN SSSR ser. fiz. 50 (1986) 1879

[3] N.V. Antonenko et al., Phys. Lett. B319 (1993) 425; Phys. Rev. C51 (1995) 2635

[4] G.G. Adamian et al., Nucl. Phys. A646 (1999) 29

[5] G.G. Adamian et al., Nucl. Phys. A671 (2000) 233

[6] A. Diaz-Torres, N.V. Antonenko, W. Scheid, Nucl. Phys. A652 (1999) 61

[7] A. Diaz-Torres et al., Phys. Lett. B481 (2000) 228

[8] G.G. Adamian, N.V. Antonenko, Yu.M. Tchuvil'sky, Phys. Lett. B451 (1999) 289

[9] G.G. Adamian, N.V. Antonenko, W. Scheid, Nucl. Phys. A678 (2000) 24

[10] G.G. Adamian et al., Phys. Rev. C (2000) in print

[11] G.G.Adamian et al., Nucl. Phys. A633 (1998) 409; Nuovo Cimento A110 (1997) 1143

[12] E.A. Cherepanov, preprint JINR, E7-99-27(1999)

[13] A. Diaz-Torres et al., *Quasifission process in the dinuclear system*, preprint, Giessen (2000)

[14] M.G. Itkis et al., *Proc. 7ᵗʰ Int. Conf. on Clustering Aspects of Nuclear Structure and Dynamics* (Rab, Croatia, 1999), ed. M. Korolija, Z. Basrak, R. Caplar (World Scientific, Singapore, 2000) p. 386; preprint JINR E15-99-248(1999)

Binary and Ternary Fission Yields of ^{252}Cf

J.K. Hwang[a], C.J. Beyer[a], A.V. Ramayya[a], J.H. Hamilton[a],
G.M. Ter Akopian[b,c], A.V. Daniel[b,c], J.O. Rasmussen[d], S.-C. Wu[d],
R. Donangelo[d,e], J. Kormicki[a], X.Q. Zhang[a], A. Rodin[b], A. Formichev[b],
J. Kliman[b], L. Krupa[b], Yu. Ts. Oganessian[b], G. Chubaryan[f],
D. Seweryniak[g], R.V.F. Janssens[g], W.C. Ma[h], R.B. Piercey[h],
and J.D. Cole[i]

[a] Department of Physics, Vanderbilt University, Nashville, TN37235, USA

[b] Flerov Laboratory for Nuclear Reactions, Joint Institute for Nuclear Research, Dubna, Russia

[c] Joint Institute for Heavy Ion Research, Oak Ridge, TN37831, USA

[d] Lawrence Berkeley National Laboratory, Berkeley, CA 94720, USA

[e] Instituto de Física, Universidade Federal do Rio de Janeiro, C.P. 68528, 21945-970 Rio de Janeiro, Brazil

[f] Cyclotron Institute, Texas A and M University, Texas 77843-3366, USA

[g] Argonne National Laboratory, Argonne, IL 60439, USA

[h] Department of Physics, Mississippi State University, MS39762, USA

[i] Idaho National Engineering and Environmental Laboratory, Idaho Falls, ID83415, USA

ramayya1@ctrvax.vanderbilt.edu

Keywords: Binary and ternary fission, ^{252}Cf, cube data, Gammasphere

Abstract The spontaneous fission of ^{252}Cf has been studied via $\gamma - \gamma - \gamma$ coincidence and $\gamma - \gamma$-light charged particle coincidence with Gammasphere. The yields of correlated $Mo - Ba$ pairs in binary fission with $0 - 10$ neutron emission have been remeasured with an uncompressed cube. The previous hot fission mode with $8 - 10$ neutron emission seen in the $Mo - Ba$ split is found to be smaller than earlier results but still present. New $0n$ binary SF yields are reported. By gating on the light charged particles detected in $\Delta E - E$ detectors and $\gamma - \gamma$ coincidences with Gammasphere, the relative yields of correlated pairs in alpha ternary fission with zero to 6n emission are observed for the first time. The peak occurs around the α $2n$ channel. A number of correlated pairs are identified in ternary fission with ^{10}Be as the LCP. We observed only cold, $0n$ ^{10}Be and little, if any, hot, xn ^{10}Be channel.

D. N. Poenaru et al. (eds.), Nuclei Far from Stability and Astrophysics, 173–184.

1.   Introduction

Studies of prompt gamma rays emitted in spontaneous fission with large detector arrays have given new insights into the fission process [1–5]. From $\gamma - \gamma - \gamma$ coincidence studies of the prompt γ-ray emitted in the spontaneous fission of ^{252}Cf with Gammasphere, yields of individual correlated pairs in binary and ternary fission were determined for the first time for 0 to 10 neutron emission. [2–4]. Earlier we reported an ultra hot fission mode in the $Mo - Ba$ split in the ^{252}Cf SF [2]. By using an uncompressed $\gamma - \gamma - \gamma$ cube, problems in the fission data analysis from complexity in the spectra in the $8-10$ neutron emission yields were overcome. The new $Mo - Ba$ yields show a reduced yield for the ultra hot mode. These data also allowed the extraction of more accurate zero neutron emission yields in cold binary fission. A new experiment was carried out in which the light charged particles (LCP) involved in ternary fission were detected in a LCP-$\gamma - \gamma$ coincidence mode. From gating on the alpha particles and a gamma ray, the relative 0 to $6n$ yields associated with α ternary fission were extracted. Gamma rays associated with new correlated pairs in coincidence with high energy ^{10}Be particles also were identified. These data give new insights into fission. In contrast to α ternary SF, only the $0n$ mode is observed for ^{10}Be.

2.   A new Determination of the Ba-Mo Yield Matrix for ^{252}Cf

We carried out pioneering work on the quantitative determination of yield matrices, using $\gamma-\gamma$ and $\gamma-\gamma-\gamma$ coincidence data to extract yields of particular fragment pairs in the SF of ^{252}Cf [2,3,4]. One interesting finding was that $\approx 14\%$ of the ^{252}Cf barium-molybdenum split goes via a "hot fission" mode, where, as many as 10 neutrons are emitted [2,3]. This latter feature stimulated some theoretical speculations and also some skepticism, since the hot fission mode (called Mode 2) has been reported only in the $Ba-Mo$ pairs in ^{252}Cf and not in ^{248}Cm spontaneous fission [6]. There have been some theoretical efforts to understand how this hot fission could arise [4,7,8]. In the present work we used our 1995 Gammasphere data, taken by the GANDS95 collaboration [2]. The analysis was carried out with uncompressed triple coincidence spectra. This differs from the previous analyses where either uncompressed double coincidence spectra or compressed triple coincidence spectra were used. In both of these methods one faces problems, because of the vast number of γ-rays in the spectrum and particulary because of the degeneracy of several γ-rays in the $8-10$ neutron emission yields for $Mo - Ba$. In

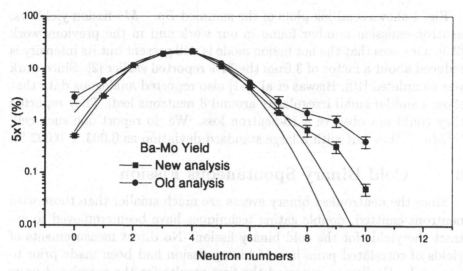

Figure 1   Mo − Ba yields vs. neutron numbers.

this new analysis, using the uncompressed 3D data, we remeasured the pair yields of barium ($Z = 56$) with molybdenum ($Z = 42$) partners. Because ^{104}Mo and ^{108}Mo have $2^+ \to 0^+$ transitions that are too close in energy to resolve and their $4^+ \to 2^+$ transitions are barely resolvable with peak-fitting routines, we have generally chosen to double-gate on the Ba fragments and measure the $2^+ \to 0^+$ intensities in the Mo partners (and $4^+ \to 2^+$ where the $2^+ \to 0^+$ are unresolvable.) The barium double gates are on the $4 \to 2 \to 0$ cascade and the $3 \to 2 \to 0$ cascade, the latter being significant in the heavier bariums where octupole deformation is reported [9]. The odd-A nuclei are special cases discussed in a separate publication. Their yields in our triple-coincidence analysis fall rather smoothly into the yield patterns of their even-even neighbors. In the yield calculations we have taken into account that Compton suppression is not complete and that, also, Compton scattering on the walls of the chamber and into a detector occur and that true continuum gammas are simultaneously present. Rather than using one of the existing gamma efficiency curves for Gammasphere, as determined off-line with radioactive standards in singles mode, we checked the efficiency curves with rotational cascades in the actual experiment, double-gating on two transitions high in the rotational band and measuring the intensities of the lower transitions in the band. Thus, these efficiency measurements involved coincidence efficiencies and take into account Compton suppression, "time-walk," and other factors at the high count rates of the actual experiment.

Fig. 1 shows semi-log plots of the summed $Ba - Mo$ fission yields vs. neutron-emission number found in our work and in the previous work [2,3]. One sees that the hot fission mode is still present but its intensity is reduced about a factor of 3 from the 14% reported earlier [2]. Since work was completed [10], Biswas et al. [11] also reported analogous data that show a similar small irregularity around 8 neutrons lost. They reported they could not observe a 10-neutron loss. We do report one such cell, ^{104}Mo- ^{138}Ba, but with a large standard deviation as 0.003 ± 0.002 %.

3.   Cold Binary Spontaneous Fission

Since the neutronless binary events are much smaller than those with neutrons emitted, double gating techniques have been employed to extract the yields for the cold binary fission. No direct measurements of yields of correlated pairs in cold binary fission had been made prior to our work. Earlier we reported the first results for the correlated pairs in cold binary fission in ^{252}Cf [1,3] and ^{242}Pu [12]. Subsequently we extracted additional and more accurate yields of cold binary fission [4] from Gammasphere data with 72 detectors.

Table 1 Average cold binary fission yields from gates on two light fragment and two heavy fragment transitions

	A_L / A_H	Y_{exp}	Y_{the}	$Y_{the}^{(ren)}$
Zr/Ce	100/152	0.010(2)	0.38	0.004
	102/150	0.020(4)	2.82	0.033
	103/149	0.030(6)	4.21	0.049
	104/148	0.010(2)	1.03	0.012
Mo/Ba	104/148	0.010(2)	0.47	0.005
	106/146	0.040(8)	0.61	0.007
	107/145	0.070(14)	3.07	0.036
	108/144	0.030(6)	7.45	0.087
Tc/Cs	109/143	0.090(18)	11.03	0.128
Ru/Xe	110/142	0.060(12)	3.78	0.044
	111/141	0.10(2)	7.12	0.083
	112/140	0.020(4)	0.59	0.007
	114/138	0.020(4)	1.17	0.014
Pd/Te	116/136	0.050(20)	2.35	0.027

The cold binary fission yields are shown in Tab. 1 along with the theoretical values predicted by Sandulescu et al. [13]. In Tab. 1, the first reports of the cold binary fission of an odd-Z - odd-Z fragmentation is shown for the Tc and Cs pair and of an even Z- odd A pair. The over all agreement with theory is generally good, including the predicted enhancement of the odd Z and odd A cases.

4.   Light Charged Particle Ternary Fission

More recently, another experiment was performed incorporating charged particle detectors to detect ternary particles in coincidence with γ rays in Gammasphere. The energy spectrum of charged particles emitted in the spontaneous fission of ^{252}Cf was measured by using two $\Delta E - E$ Si detector telescopes installed at the center of the Gammasphere array at Argonne National Laboratory.

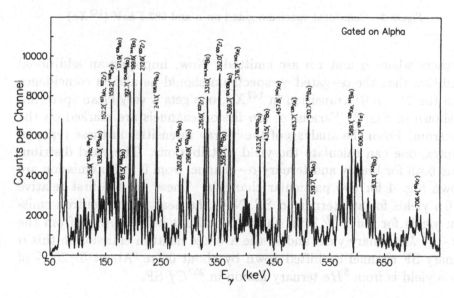

Figure 2 α gated spectrum

With the position resolution of the strip detector (4 mm wide strips and 1 mm resolution along each strip), the $\Delta E - E$ telescopes provided unambiguous Z and A identification for all the light charged particles of interest. The energy calibration of the telescopes was performed with ^{224}Ra and ^{228}Th radioactive sources. The γ-ray spectrum in coincidence with ternary α-particles is shown in Fig. 2. In this spectrum, one can easily see the γ-transitions for various partner nuclei where a ternary α-particle is emitted. For example, Xe and Mo isotopes are

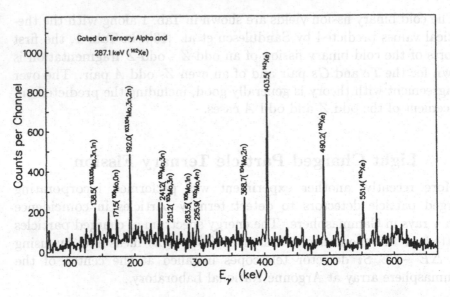

Figure 3   Coincident spectrum gated on α and 287.1 keV (^{142}Xe)

partners where α and xn are emitted. Now, imposing an additional condition that the α−gated γ−spectrum should be also in coincidence with the $2^+ \to 0^+$ transition in ^{142}Xe, one gets a very clean spectrum as shown in Fig. 3. Various α, xn fission channels are marked on the spectrum. From the analysis of the γ−ray intensities in these types of spectra, one can calculate the yield distributions. The yield distributions both for binary and ternary α−channel from 0 to 6n emission are shown Fig. 4 for two particular channels. These are the first relative $0-6n$ yields for any ternary α SF. Note the peak of the neutron emission yields for Ba-α-^{102}Zr is shifted up about half an AMU from the $Ba - ^{106}Mo$ binary yield and so the average neutron emission in this α ternary SF channel is shifted down by about 0.4n. About 5−20 % of the α yield is from ^5He ternary fission in ^{252}Cf SF.

5.   Identification of the Cold ^{10}Be Ternary SF Pairs of ^{252}Cf

Ternary fission is very rare process that occurs roughly only once in every 500 spontaneous fissions (SF) dominated by α ternary fission. Roughly, the ^{10}Be particles are emitted once per 10^5 spontaneous fissions. The maximum yield in the binary spontaneous fission is located around 3 to 4 neutrons. We now find that the α ternary fission is, mostly, accompanied by $\approx$ 2 to 3 neutrons.

In neutronless ternary spontaneous fission (SF), the two larger frag-

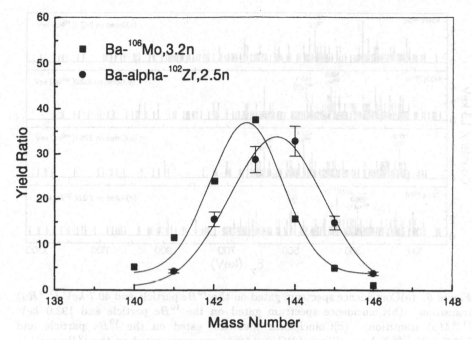

Figure 4 Yield spectrum for α

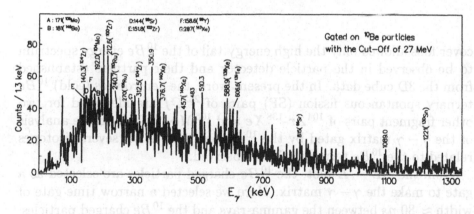

Figure 5 Gamma spectrum gated on ^{10}Be particles. Several peaks are marked with several isotopes related to the present work. But, it is hard to assign the right isotopes to each peak because of the expected γ-ray multiplicity in this spectrum.

ments have very low excitation energy and high kinetic energies. Experimentally it is not easy to identify the γ transitions of the cold or hot ^{10}Be ternary SF pair because it is very rare process. The first case of neutronless ^{10}Be ternary spontaneous fission (SF) in ^{252}Cf was reported from $\gamma - \gamma - \gamma$ coincidence spectrum where the pairs are ^{96}Sr and ^{146}Ba without neutrons emitted [5]. In our LCP-$\gamma - \gamma$ data, the

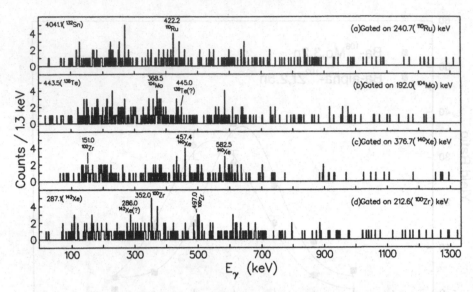

Figure 6   (a)Coincidence spectrum gated on the ^{10}Be particle and 40.7 *keV* (^{110}Ru) transition.   (b)Coincidence spectrum gated on the ^{10}Be particle and 192.0 *keV* (^{104}Mo) transition.   (c)Coincidence spectrum gated on the ^{10}Be particle and 376.7 *keV* (^{140}Xe) transition.   (d)Coincidence spectrum gated on the ^{10}Be particle and 212.6 *keV* (^{100}Zr) transition.

cover foils allowed only the high energy tail of the ^{10}Be energy spectrum to be observed in the particle detector and their partners established from the 3D cube data. In the present work, the neutronless (cold) ^{10}Be ternary spontaneous fission (SF) pairs of ^{252}Cf are identified for two other fragment pairs of ^{104}Zr-^{138}Xe and ^{106}Mo-^{136}Te from the analysis of the $\gamma - \gamma$ matrix gated by the ^{10}Be particles. Also, several isotopes related to the ^{10}Be ternary SF are observed.

From the $\Delta E - E$ plot, the ^{10}Be charged particles are selected as a gate to make the $\gamma - \gamma$ matrix. Here we selected a narrow time gate of width $\approx$ 80 *ns* between the gamma-rays and the ^{10}Be charged particles. Also, we did not subtract the background spectrum from the full projection of the $\gamma - \gamma$ matrix because of poor statistics. The high efficiency of Gammasphere enables coincidence relationships to be established even with the low statistics data associated with a small ^{10}Be ternary SF yield. The gamma spectrum gated on the ^{10}Be charged particles is shown in Fig. 5. Several peaks are marked with isotopes related to the present work. It is sometimes hard to assign the right isotopes from the energies of the peaks alone because the same energy transition may be present in one or more isotopes. For example, the strong 212 *keV* peak in the spectrum can come from several sources such as ^{100}Zr, $^{111,113}Rh$,

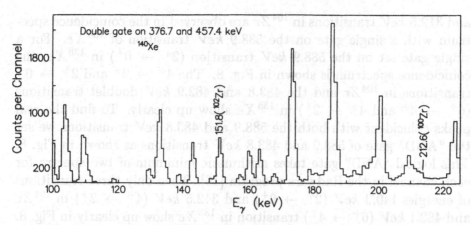

Figure 7 Coincidence spectrum double-gated on 376.7 and 457.4 keV transitions in ^{140}Xe. See 151.0 (^{102}Zr) and 212.0 (^{100}Zr).

and ^{147}La. So next, in the ^{10}Be gated $\gamma - \gamma$ matrix we set a gate on the 212.6 keV energy in ^{100}Zr. There we can see the 352.0 ($4^+ \rightarrow 2^+$) and 497.0 ($4^+ \rightarrow 2^+$) keV transitions in ^{100}Zr. Four examples are shown in Fig. 6 to identify ^{110}Ru (or ^{108}Ru), ^{104}Mo (or ^{108}Mo), ^{140}Xe and ^{100}Zr, respectively. But the identification of the gamma-transitions belonging to these partner fragments is not clear in those spectra. From the $\gamma - \gamma - \gamma$ cube we could clearly establish coincidence for ^{100}Zr-^{142}Xe and ^{102}Zr-^{140}Xe. Also, by double gating on the 376.7 and 457.3 γ-rays in ^{140}Xe (Fig. 7), we can see clearly the zero neutron channel ^{102}Zr and probably the ^{100}Zr $2n$ channel which is weaker by a factor of $5-10$ if present. The identification of several isotopes related with the ^{10}Be emission is made by the observation of two or three transitions in coincidence belonging to each isotope and from the $\gamma - \gamma - \gamma$ cube. All isotopes and the related γ transitions identified in the present work are tabulated in Tab. 2. In Tab. 2, partner fragments pertaining to the cold (neutronless) channel are shown some of which are confirmed as noted. Quadrupole deformations for each isotope are taken from Refs. [14,15]. From these examples, we can see that the statistics of the coincident spectrum with a single gate on the lowest gamma transition does not depend on the statistics of the gated peak shown in Fig. 1 because of complexity of the gamma-ray multiplicity and the enhanced population of the lowlying levels in the ^{10}Be SF.

Two fragment pairs, ^{138}Xe-^{104}Zr and ^{136}Te-^{106}Mo with no neutrons emitted show γ rays produced from both of pair fragments in the ^{10}Be gated coincidence spectrum with a single γ gate. In other words, the 171.6 keV transition of ^{106}Mo is observed in the coincidence spectrum with a single gate on the 606.6 keV transition of ^{136}Te. Also, the 140.3

and 312.5 keV transitions in ^{104}Zr are observed in the coincidence spectrum with a single gate on the 588.9 keV transition of ^{138}Xe. For a single gate set on the 588.9 keV transition ($2^+ \to 0^+$) in ^{138}Xe, the coincidence spectrum is shown in Fig. 8. The $4^+ \to 2^+$ and $2^+ \to 0^+$ transitions in ^{104}Zr and the 483.8 and 482.9 keV doublet transitions ($6^+ \to 4^+$ and $4^+ \to 2^+$) in ^{138}Xe show up clearly. To find the real peaks coincident with both the 588.9 and 483.8 keV transitions we set the "AND" gate of 588.9 and 483.8 keV transitions as shown in Fig. 8. This logical "AND" gate takes arithmetic minimum of two spectra for each channel in the Radware program [16]. Then only three transitions of energies 140.3 keV ($2^+ \to 0^+$) and 312.5 keV ($4^+ \to 2^+$) in ^{104}Zr and 482.1 keV ($6^+ \to 4^+$) transition in ^{138}Xe show up clearly in Fig. 8. Although the three peaks in Fig. 4c contain only two counts, the background is less than 0.01/channel in Fig. 8. The 140.3 and 312.5 keV

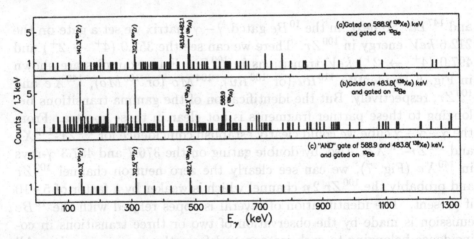

Figure 8 (a)Coincidence spectrum gated on the ^{10}Be particle and 588.9 keV (^{138}Xe) transition. (b)Coincidence spectrum gated on the ^{10}Be particle and 483.8 keV (^{138}Xe) transition. (c)Coincidence spectrum with "AND" gate of 588.9 and 483.8 keV (^{138}Xe) transitions, and gated on ^{10}Be. This logical "AND" gate takes arithmetic minimum of two spectra for each channel in the Radware program [16].

transitions do not exist in the level scheme of ^{138}Xe. Since the $\gamma - \gamma$ matrix is gated by ^{10}Be particles, 140.3 and 312.5 keV transitions belong to the partner nucleus ^{104}Zr. However, 109.0 and 146.8 keV transitions in ^{103}Zr ($^{10}Be + 1n$ channel) and the 151.8 and 326.2 keV transitions in ^{102}Zr ($^{10}Be + 2n$ channel) do not show up clearly. In another case of ^{136}Te-^{106}Mo, also, the $^{10}Be+1n$ and $^{10}Be+2n$ channels are not observed. This could be caused by the larger feeding to ground state but more likely by small yields in the $^{10}Be + n$ and $^{10}Be + 2n$ channels. The hot fission mode can excite the fragments up to higher level energies than the cold fission. Therefore, SF yields of the $^{10}Be + n$ and $^{10}Be + 2n$ channels have to be smaller than the neutronless (cold) ^{10}Be SF yield. The present results indicate that the cold (neutronless) process is dominant in the ternary SF accompanying a heavy third particle such as ^{10}Be with high kinetic energy.

In our work, we are gating only on the high kinetic energy part of the ^{10}Be particles. The ^{104}Zr isotope is highly deformed with a β_2 value of around 0.4 [15,16] and the ^{138}Xe nucleus is very spherical. Therefore,

Table 2 Fragments identified from the coincidence relationship between γ-rays and ^{10}Be ternary particle. * : identified in $\gamma - \gamma - \gamma$ data and ** : LCP-$\gamma - \gamma$ data.

Identified Isotopes (β_2 [14,15])	Observed γ rays (keV)	Partner isotopes (β_2 [14,15])
$^{100}_{40}Zr$ (0.321)	212.6, 352.0, 497.0	$^{142}_{54}Xe$ (0.145])*
$^{102}_{40}Zr$ (0.421)	151.8, 326.2	$^{140}_{54}Xe$ (0.1136)*
$^{104}_{40}Zr$ (0.381)	140.3, 312.5	$^{138}_{54}Xe$ (0.0309)**
$^{104}_{42}Mo$ (0.325)	192.0, 368.5	$^{138}_{52}Te$ (0.000)
(or $^{108}_{42}Mo$)		(or $^{134}_{52}Te$)
$^{106}_{42}Mo$ (0.353)	171.6 with 606.6 (^{136}Te)	$^{136}_{52}Tc$ (0.000)**
$^{110}_{44}Ru$ (0.303)	240.7, 422.2	$^{132}_{50}Sn$ (0.000)
(or $^{108}_{44}Ru$)		(or $^{134}_{50}Sn$)
$^{136}_{52}Te$ (0.000)	606.6, 424.0	$^{106}_{42}Mo$ (0.353)**
$^{138}_{54}Xe$ (0.0309)	588.9, 483.8, 482.1	$^{104}_{40}Zr$ (0.381)**
$^{140}_{54}Xe$ (0.1136)	376.7, 457.4, 582.5	$^{102}_{40}Zr$ (0.421)*

the ^{10}Be particle seems to be emitted from the breaking of $^{148}Ce = {}^{138}Xe + {}^{10}Be$ at scission which would enhance the ^{10}Be kinetic energy. Increased deformation at the scission point increases excitation energy for the third ternary particle and two heavy fragments. Therefore the possibility of observing the excited levels in both the fragments increases when both of them are deformed at scission point such as ^{104}Zr (deformed)-^{148}Ce ($^{138}Xe + {}^{10}Be$) (deformed). Actually, the neutronless binary fission yield for ^{148}Ce-^{104}Zr pair is as high as 0.05(3) per 100 SF of ^{252}Cf [17]. These cases are very similar to the one we reported earlier for the pair ^{96}Sr (spherical shape) and ^{146}Ba (deformed shape) [5].

In the α ternary fission we see the cold, zero, neutron fission but $2n$ and $3n$ channels are much stronger. However, for the cold ^{10}Be ternary SF pairs identified from the $\gamma - \gamma$ matrix gated on ^{10}Be charged particles and the 3D data, we find the zero neutron channel clearly much stronger than $1n$ and $2n$. This is a very unique discovery in the study of the cold (zero neutron) fission processes.

Acknowledgments

Research at Vanderbilt University and Mississippi State University is supported in part by the U.S. Department of Energy under Grants No. DE-FG05-88ER40407 and DE-FG05-95ER40939. Work at Idaho National Engineering and Environmental Laboratory is supported by the U.S. Department of Energy under Contract No. DE-AC07- 76ID01570. Work at Argonne National Laboratory is supported by the Department of Energy under contract W-31-109-ENG-38.

References

[1] J.H. Hamilton et al., J. Phys. G20 (1994) L85

[2] G.M. Ter-Akopian et al., Phys. Rev. Lett. 77 (1996) 32

[3] G.M. Ter-Akopian et al., Phys. Rev. C55 (1997) 1146

[4] A.V. Ramayya et al., *Heavy elements and related new phenomena, Volume I* World Scientific (1999) 477

[5] A.V. Ramayya et al., Phys. Rev. Lett. 81 (1998) 947

[6] N. Schultz, private communication

[7] Yu.U. Pyatkov et al., Nucl. Phys. A624 (1997) 140

[8] R. Donangelo et al., Int. J. Mod. Phys. E7 (1998) 669

[9] J.H. Hamilton et al., Proc. Nuclear Structure 98, Gatlinburg (1999) p. 473

[10] S.-C. Wu et al., Phys. Rev. C62 (2000) 041601

[11] D.C. Biswas et al., Eur. Phys. J. A7 (2000) 189

[12] Y.X. Dardenne et al., Phys. Rev. C54 (1996) 206

[13] A. Sandulescu et al., Int. J. Modern Phys. E7 (1998) 625

[14] S. Raman et al., Atomic Data and Nucl. Data Tables 36 (1987) 1

[15] P. Moller et al., Atomic Data and Nucl. Data Tables 59 (1995) 185

[16] D.C. Radford, Nucl. Instr. Meth. Phys. Res. A361 (1995) 297

[17] J.H. Hamilton et al., Prog. in Part. and Nucl. Phys. 35 (1995) 635

Recent Experiments on Particle-Accompanied Fission

Manfred Mutterer

Institut für Kernphysik, Technische Universität Darmstadt
Schlossgartenstrasse 9, D-64289 Darmstadt, Germany
mutterer@hrz1.hrz.tu-darmstadt.de

Keywords: Nuclear fission, particle-accompanied fission

Abstract   Particle-accompanied fission, known also as ternary fission, is a process where, close to the instant of scission, a light charged particle is
ejected from the space between the emerging fission fragments. Since
only a few fission events in a thousand are such ternary fragmentations, precise ternary fission experiments are generally hard to perform.
The present report is primarily concerned with a recent experimental
study on the ternary fission of $^{252}Cf(sf)$ which was carried out with the
Darmstadt-Heidelberg 4π $NaI(Tl)$ Crystal Ball spectrometer as highly
efficient γ-ray and neutron detector which was additionally equipped
with detectors for fission fragments and ternary particles. This advanced
multiparameter study was able to cover besides the most abundant α-
particle-accompanied fission process also much rarer fission modes with
emission of various ternary particles, from tritons to carbon nuclei. The
present report will be complimented with the results from other recent
experiments, and by an outlook to future work.

1.   Introduction and Fundamentals

From classical hydrodynamics it is known that when a cylindrical
laminar flow of liquid decays by a Rayleigh instability [1] into a chain
of individual drops usually small droplets appear between them. Similarly, when chopping-up a soap bubble tiny bubbles remain between
bigger spherical ones. In nuclear physics, an analogous phenomenon
of a particle-associated fragmentation of an actinide nucleus was discovered in 1946 after exposing uranium-loaded photographic plates to
a neutron flux [2]. The present article deals with this so-called light-
charged-particle (LCP) accompanied fission process, also being known
as ternary fission (TF).

D. N. Poenaru et al. (eds.), Nuclei Far from Stability and Astrophysics, 185–196.
© *2001 Kluwer Academic Publishers. Printed in the Netherlands.*

For fission processes at low excitation, as near-barrier fission induced by thermal neutrons and sub-barrier spontaneous fission of actinides, only a few events in a thousand are ternary fragmentations. Among them, the probability of a splitting of the heavy nucleus into three fragments of nearly equal masses (so-called "symmetric tripartition" or "true ternary" fission) is extremely rare, with measured upper limits in the order of 10^{-9} compared to binary fission [3, 4]. All ternary fragmentations investigated so far can be attributed to particle-accompanied fission processes leading to a pair of fission fragments with mass distributions closely resembling binary fission ones, but which are accompanied by emission of a third light nucleus. These ternary fissions occur with a comparably moderate yield of $3-6 \times 10^{-3}$, in about 90 % of cases with the emission of α-particles.

The main emission direction of ternary α-particles approximately orthogonal to the direction of motion of the fragments, and the high mean ternary α-particle energy of about 16 MeV, were soon qualitatively understood by imaging the formation of the α-particle from the neck joining the fragments at the instant of scission where the common Coulomb field of the fragments has strong focusing strength. Thus, many ternary fission studies have aimed at getting experimental access to intermediate stages of the fission process, esp. for probing fragment scission-point configurations and prescission dynamics. This goal has not been fully met to date, though considerable progress has been achieved in recent years by multiparameter studies on the three-body kinematics with the detection system DIOGENES [5], by thorough measurements of LCP probabilities and energy spectra with LOHENGRIN (e.g., [6, 7]), and by high-resolution triple γ-ray correlation experiments with GAMMA-SPHERE (e.g., [8]).

The present report is primary concerned with a recent experiment which has aimed at a more detailed study on the kinematics in LCP-accompanied spontaneous ^{252}Cf fission, including registration of prompt fission γ-rays and neutrons. References to other closely related work will be given, but no attempt is made at giving a comprehensive survey. Review papers on TF were published most recently by Wagemans (1991) [9], and Mutterer and Theobald (1996) [10].

2.   Experiment

In the experiment a combination of high-efficiency angle-sensitive detectors is applied for determining not only the kinematical parameters of the reaction, i.e. kinetic energies and mutual emission angles of the ternary particle and the correlated pair of fission fragments, but also the excitation state of the products by registering simultaneously neutrons and γ-rays from their disintegration. The measurement was performed at the MPI-K, Heidelberg, using the Darmstadt-Heidelberg Crystal Ball (CB) spectrometer [11, 12]. The CB is a homogeneous 4π detection system consisting of a dense spherical package of 162 large (20 cm long) $NaI(Tl)$ crystals of high detection efficiency ($\geq$ 90 % for γ- rays and $\simeq$ 60 % for fission neutrons) [13]. Neutrons registered in NaI mainly via the $(n, n')\gamma$ reaction were separated from the prompt γ-rays by time-of-flight. The ^{252}Cf sample (4×10^3 $fissions/s$) and the detector system "CODIS" for fragments and particles [14] were mounted at the center of the CB. A Frisch-gridded 4π twin ionization chamber (IC) measures the energies of both fission fragments and their emission angles. The LCPs were measured by a ring of 12 $\Delta E - E$ telescopes made from ΔE ICs and silicon PIN diodes, which surrounds one half of the fragment IC. The solid angle Ω_{LCP} of the telescopes was $\pi/4$. Angular resolutions were $\leq$ 5° fwhm for fission fragments and light particles.

The set of measured parameters has allowed to determine, for each fission event, the following quantities and their mutual correlations: fragment masses and kinetic energies; multiplicity and angular distribution of fission neutrons; multiplicity, energy and angular distributions of fission γ-rays; energy, nuclear charge (mass), and emission angle of the LCP from ternary fission. A total number of 1.2×10^6 ternary fission events accompanied by α-particles were registered, simultaneously 8×10^4 events with ^3H, 2.5×10^4 events with ^6He, 2.5×10^3 events with the emission of Li and Be each, and 70 (130) events with B (C) nuclei. At the same time, 7×10^7 binary fission events were recorded, by reducing the data acquisition rate by a factor of 1/64.

3.   Results and Discussion

3.1.   Fragment Excitation Energy and LCP Emission Probability

From the kinematical data obtained with CODIS we could deduce, for the first time, fission-fragment energy and mass distributions, and the fragment total excitation energies TXE, for the rare ternary fission modes with the emission of LCPs heavier than He isotopes. It turns out that LCP emission proceeds in expense of a considerable amount of fragment excitation energy (35 MeV, on average, for binary ^{252}Cf fission), with the required energy increasing with LCP mass and energy. As an example, the average TXE decreases from 27 MeV to 15 MeV when instead of an α-particle a ternary C-isotope is emitted. In this sense

TF with emission of heavier LCPs features a rather cold large-scale re-arrangement of nuclear matter.

The relation between the probability for emission of different LCPs and the corresponding fragment TXE losses is depicted in Fig. 1. There is an obvious discontinuity between the data for ternary α-particles and tritons which was already noted by Halpern [15] (so-called "H-isotopes anomaly"). The present results constitute a hint for the occurrence of preformed α-clusters in the fissioning nucleus, which could take up energy from fission dynamics in the saddle-to-scission stage, as is postulated by the cluster emission model of α-accompanied fission [16]. The significance of α-cluster preformation in α-TF was recently discussed also by Serot and Wagemans [17]. For emitting the majority of other LCPs the main clue might rather be the dynamics of neck rupture, what is the physics picture behind the early empirical model of Halpern [15]. Similarly, Pik-Pichak [18] has described ternary particle emission near scission by adiabatic first-order perturbation theory.

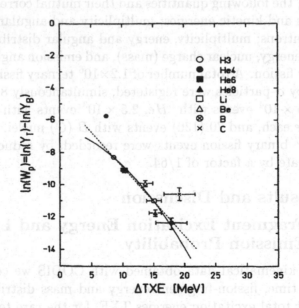

Figure 1 Emission probability (logarithmic scale) of light charged particles, as a function of the difference between mean total excitation energies TXE in binary and ternary fission. Average TXE in binary fission is 35 MeV. The dotted line is a linear fit through the data points without ^4He and C. The solid line is a quadratic fit through the same data points.

3.2. Neutron Multiplicity and Angular Distribution

For most LCPs the results on the excitation energy TXE deduced from the kinematics are found to be in close correspondence with the mean number of neutrons $\bar{\nu}$ registered by the CB. However, slightly enhanced neutron yields emerge in the cases of the $^{4,6,8}He$ isotopes, and of Li [10]. The explanation was found in the formation of short-lived neutron-unstable "primary" ternary particles which disintegrate by neutron emission, either from the ground state or an excited state, into the registered "secondary" ones.

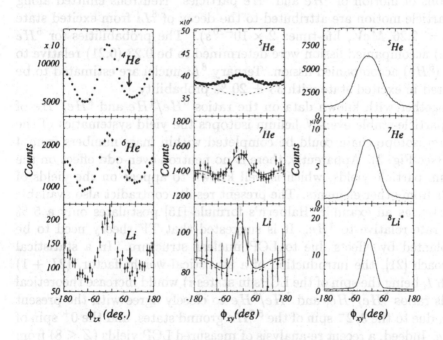

Figure 2 Different steps in the analysis of the neutron angular correlations in ^5He, ^7He and Li accompanied fission: (left-hand side) Projections w_{zx} of measured neutron intensity on the zx plane in the coordinate system xyz. Arrows indicate mean LCP emission angles. (center) respective w_{xy} distributions perpendicular to the fission axis, with the background from prompt fission neutrons (dashed line) included. (right-hand side): calculated angular distributions w_{xy}, without (dotted lines) and with account of experimental resolution (full lines)

For the relatively abundant ternary ^4H, ^6He and Li particles, the good counting statistics achieved has allowed to evaluate distinct angular distributions of the emitted neutrons [19]. Fig. 2 displays neutron emission probabilities as a function of the azimuthal angles ϕ in a coordinate system xyz, with the polar axis in the direction of motion of the light fission fragment, and zero ϕ-angle in the direction of the LCP. There are enhanced neutron yields along the direction of these particles ($\phi = 0°$), the angular width of these contributions getting narrower in the cases of ^6He and Li. The result for ^4He qualitatively agrees with former work [20] in which yields and energy spectra of neutrons at angles of about $0°$ and $180°$ with respect to the α-particles were measured. The comparison with trajectory calculations, shown in the right-hand patterns of Fig. 2, unambiguously identifies ternary emission of the short-lived ^5He and ^7He nuclei (life-times: $1 \times 10^{-21}s$, and $4 \times 10^{-21}s$, respectively) by the angular distribution of their decay neutrons centered at the directions of motion of ^4He and ^6He particles. Neutrons emitted along Li-particle motion are attributed to the decay of ^8Li from excited state ($E^* = 2.26\ MeV$, life-time: $2 \times 10^{-20}s$). The probabilities for ^5He (^7He) accompanied fission were determined to be 0.23 (0.21) relative to ^4He (^6He) accompanied fission. Ternary ^8Li nuclei are estimated to be emitted in excited state with $33 \pm 20\ \%$ probability.

Together with known data on the ratios $^6He/^4He$ and $^8He/^4He$ of the particle-stable even-N helium isotopes the yield systematics of the helium isotope chain could be completed within mass numbers from 4 to 8 (see Fig. 3). Apparently, there is no neutron even-odd effect on the helium particle yields, which is well known to appear on the yields of LCPs from other elements. The present results contradict also available theories; as an example, Halpern's formula [15] postulates only a 5 % ^5He rate relative to ^4He. It is suggested that TF theory need to be supplanted by effects due to LCP nuclear structure. In a statistical approach [21], the introduction of a statistical-weight factor, $(2I_i + 1)$ (with I_i being the spin of the LCPs in states i) would increase theoretical yields ratios $^5He/^4He$, and $^7He/^6He$ to closely agree with the present data, due to the $3/2^-$ spin of the $^{5,7}He$ ground states, and the 0^+ spin of $^{4,6}He$. Indeed, a recent re-analysis of measured LCP yields ($Z \leq 8$) from thermal-neutron induced fission of ^{235}U, with inclusion of LCP spins and the population of LCP excited states (shown at the right-hand side of Fig. 3), has confirmed this assumption [19].

In the meanwhile other decay channel from particle-unstable LCPs have been identified, such as the ^8Be decay into two α-particles, and the decay of 4.63 MeV excited ^7Li into an α and a triton [22]. These sequential decays mimic a real quaternary fission process, with two LCPs being coincidently emitted. The study of quaternary fission requires highly intense fission sources, what has been perceived in a recent TF

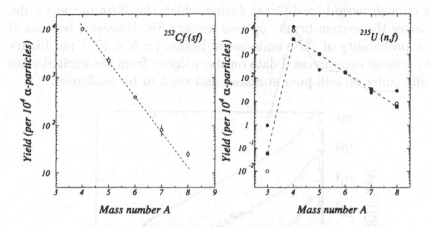

Figure 3 (left side): Yields of helium isotopes ($4 \leq A \leq 8$) in ternary fission of ^{252}Cf. The ^6He and ^8He yields are from literature (dashed line, to guide the eye). (right side): Theoretical analysis of helium isotope yields ($3 \leq A \leq 8$) in ternary fission of $^{235}U(n_{th}, f)$, with (squares) and without (points) taking LCP excited states and spins into account. Experimental values (open symbols) are normalized to 10^4 α-particles.

study on triple correlations in $^{233,235}U$ fission, induced by an intense polarized cold neutron beam at the high-flux reactor of ILL, Grenoble [23].

3.3. Gamma-Ray Emission from Ternary Particles

From the γ-ray pulse-height distributions, resulting from summing up the spectra from all crystals of the CB, we have got evidence for the formation of Be nuclei in an excited level which decays by γ-ray emission. Fig. 4 shows the γ-ray spectrum for Be-accompanied fission, in comparison with the spectra for binary, α- and Li-accompanied fission, respectively. In the Be-TF γ-ray spectrum the 3.368 MeV γ-emission from the 0.18 ps disintegration of the first excited level in ^{10}Be is clearly identified. Amazingly, an important fraction of the ^{10}Be radiation appears as a non-Doppler broadened γ-line in the γ-ray spectrum summed over the full solid angle, indicating γ-decay from a source at rest. A narrow line was also registered in a recent high-resolution triple γ-ray correlation experiment with GAMMASPHERE [8]. The absence of Doppler broadening observed in both experiments constitutes a hint for a possible occurrence of a specific long-lived triple nuclear configuration at scission, i.e. a molecular-type of nuclear structure with a ternary ^{10}Be particle held in the potential well between the two fission fragments for a long enough period ($\simeq 10^{-12}s$) during which the ^{10}Be nucleus γ decays before the system breaks up into ternary SF. However, because of the low probability of ^{10}Be emission in fission ($\simeq 6 \times 10^{-5}$ per binary fission) present experimental data on the γ-decay from the excited state naturally suffer on still poor statistics and need to be confirmed.

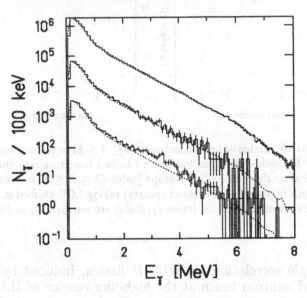

Figure 4 Gamma-ray spectra for α-particle, *Li*- and *Be*- accompanied fission (top to bottom). The spectrum for *Li* is multiplied by a factor of 20. Dashed lines represent the shape of the spectrum for binary fission. Error bars are statistical errors.

3.4.   Gamma-Ray Anisotropy

The anisotropy of the γ-ray emission with respect to the fission axis yields information about fragment spin orientation. Previous work found the well known anisotropy in binary fission of ^{252}Cf [24] to be either removed [25] or characteristically changed [26] when a ternary α-particle accompanies fission. Different possibilities for a correlation of the ternary particle emission process with collective vibrational modes of the fissioning nucleus were proposed for explaining the experimental findings.

Due to the high detection efficiency and granularity of the CB, we could measure with good accuracy the full angular distribution of the γ-rays with respect to the fission axis, both for the binary and the LCP-accompanied fission modes. For γ-energy bins from 0.1 to 1.5 *MeV*, the results on binary and α-accompanied fission [27, 28] are depicted in Fig. 5, along with theoretical curves. No significant difference between binary and ternary fission in the angular distribution patterns is ob-

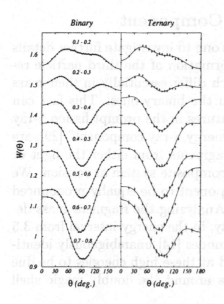
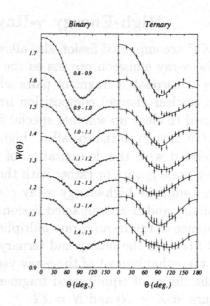

Figure 5 Angular distributions $W(\theta)$ of γ-rays with respect to the light fission fragment motion, in the laboratory system. Gamma-ray energy intervals are given in MeV. The scale is correct for the lowest plots; others are shifted consecutively by 0.1 units. Solid lines are fitted curves, calculated with assuming the γ multipolarity to be ≤ 2.

served. Our binary fission value for the average anisotropy ratio of both fragments in their respective rest frames [27], $A = W(0°) / W(90°) - 1$, is 0.0872 ($\pm$ 0.0002 statistical error), in fair agreement with Ref. [24], while the present result of $A = 0.0903$ ($\pm$ 0.0019) for α-particle accompanied fission is in obvious contradiction to published data, A = 0.015(22) [26] and A = 0.03(2) [25].

The present observation means that the emission of ternary α-particles, although it seems to take away some amount of fragment spin, does not influence or even destroy the spin alignment. The analysis of the γ-ray anisotropy for other LCPs (tritons, ^6He, and Li and Be nuclei), which was possible for the first time with the present data, confirms our observation made for the α-particles. We thus conclude that the difference in γ-ray anisotropy between the ternary and binary fission modes is very small. We have also analyzed the projection of the γ-ray angular distribution onto the plane perpendicular to the (light) fragment direction, as a function of the azimuthal angle ϕ. The γ-ray emission was found to be isotropic, within 1 % of error, with respect to ϕ, indicating that there is no correlation of the fragment spin orientation with the emission direction of ternary particles.

3.5.   High-Energy γ-Ray Component

LCP accompanied fission also allows one to investigate further details of the γ-ray emission process as the formation of the third particle results in correlated fragment pairs which differ essentially in their mass (and nuclear charge) composition from the binary ones. This fact can be used to clear up whether special features in the prompt fission γ-ray spectrum, such as the so-called "high-energy γ-ray component" [29], are correlated with the de-excitation of fragments from either the light or heavy mass group, or rather with the composite system at scission. We have found the high-energy γ-ray component to be equally pronounced in binary and α-accompanied fission. Analyzing the fragment mass dependence of the mean γ-ray multiplicity, in the energy interval from 3.5 to 8 MeV, in the binary and ternary modes [14] unambiguously identified the enhancement of the γ-ray yield at these high energies to be due to the decay of equilibrated fragments around the doubly magic shell closure at $Z = 50$ and $N = 82$.

4.   Summary and Conclusions

The use of highly efficient angle-sensitive detectors, namely the CODIS detection system for fission fragments and LCPs, and the Crystal Ball for fission γ-rays and neutrons, has provided a considerably deeper insight into the rare process of particle-accompanied fission, discovering also previously unexploited features. For the first time it could be determined to what extent the fragment TXE is reduced when LCPs with nuclear charges up to carbon are emitted. This information is important for a correct modeling of ternary scission configurations.

Formation of particle-unstable species, such as ^5He and ^7He, was found to be a well pronounced source of surplus neutrons. These neutrons contribute, albeit little, to the ^{252}Cf neutron spectrum used as a reference standard in nuclear measurements. The dominant part of surplus neutrons comes from ^5He, with about one neutron in every 1500 binary fission events.

The production of LCPs in rather highly excited states decaying by neutron emission (in the case of lithium) or by γ-ray emission (in the case of beryllium), is a new finding opening up new possibilities for studying experimentally the energetics of the fissioning system at scission. Future experimental studies on the population of excited states in ternary particles seem to be very promising since, compared to the high densities of levels in the primary fission fragments, level densities in the light ternary particles are low and the level structures are well known.

The measured anisotropy of γ-ray emission for various TF modes shows, contrary to previous experimental studies, that the emission of ternary particles at scission does not influence the alignment of fragment spins. The processes of ternary particle emission and fragment spin formation proceed close to scission but, as it seems, independently from each other. Evidently, angular momentum generation in fission fragments is still a challenge to theory (see, e.g. [30]), and even more sophisticated experimental studies might be needed for gaining further insight into the intimate details of the nuclear dynamics at scission.

As regards the formation of long-lived complex nuclear molecules in particular ternary scission-point configurations which the γ-decay data in ^{10}Be accompanied ^{252}Cf(sf) seem to suggest it is important to say that the observed neutron decay from excited ternary $^8Li^*$ nuclei gives no hint for a comparably large hindrance of LCP emission. The measured neutron angular distribution tells that the break-up of $^8Li^*$, being faster than the γ-decay of $^{10}Be^*$ by seven orders of magnitude, occurs after LCP separation. If the hypothesis of nuclear molecule formation in Be-accompanied fission holds, this phenomenon seems to depend on LCP structure, and/or presumably occurs more likely with LCPs of nuclear charges $Z > 3$, and masses $A > 8$. The very low yields of these processes make further experiments with spontaneous fission sources rather complicated. Neutron induced reactions at high neutron fluxes would be an alternative.

Acknowledgments

This work was partly supported by the BMBF, Bonn (06DA461, 06DA913 and 06TU669). Fruitful discussions with F. Gönnenwein, D. Schwalm and Yu.N. Kopatch are kindly acknowledged.

References

[1] J.M. Rayleigh, Proc. London Math. 10 (1878) 4

[2] T. San-Tsiang et al., C. R. Acad. Sc. 223 (1946) 986

[3] R. Stoenner, and M. Hillman, Phys. Rev. 71 (1966) 716

[4] P. Schall et al., Phys. Lett. B191 (1987) 339

[5] P. Heeg, PhD Thesis, TH Darmstadt, 1990; P. Heeg et al., Proc. Conf. 50 Years with Nuclear Fission, Gaithersburg (1989), Vol I, p. 299.

[6] M. Hesse et al., Proc. Int. Conf. DANF96, Častá Papiernička (1996), p. 238.

[7] F. Gönnenwein et al., Proc. Seminar on Fission Pont d'Oye IV, Habay-la-Neuve (1999) p. 59

[8] A.V. Ramayya et al., Phys. Rev. Lett. 81 (1998) 947; J.H. Hamilton et al., Acta Physica Slovaca 49 (1999) 31

[9] C. Wagemans, in *The Nuclear Fission Process*, CRC Press, Boca Radon (1991), p. 580.

[10] M. Mutterer and J.P. Theobald, in *Nuclear Decay Modes*, Institute of Physics Publ., Bristol (1996) p. 487

[11] M. Mutterer et al., Proc. Int. Conf. DANF96, Častá Papiernička (1996) p. 250.

[12] P. Singer et al., Proc. Int. Conf. DANF96, Častá Papiernička (1996) p. 262.

[13] V. Metag et al, Lecture Notes in Physics 178 (1983) 163

[14] P. Singer et al., Z. Phys. 359 (1997) 41

[15] I. Halpern, Ann. Rev. Nucl. Sci. 21 (1971) 245

[16] N. Cârjan, J. Physique 37 (1976) 1279; S. Oberstedt and N. Cârjan, Z. Physik 344 (1992) 59

[17] O. Serot and C. Wagemans, Proc. Seminar on Fission Pont d'Oye IV, Habay-la-Neuve (1999) p. 45

[18] A. Pik-Pichak, Phys. Atom. Nuclei 57 (1994) 906

[19] M. Mutterer et al., Proc. Conf. Fission and Properties of Neutron-Rich Nuclei, St. Andrews (1999) p. 316

[20] E. Cheifetz et al. Phys. Rev. Lett. 29 (1972) 805

[21] G. Valskii, Sov. J. Nucl. Phys. 24 (1976) 140

[22] P. Jesinger et al., Proc. Sem. ISINN-8, Dubna (2000), to be publ.

[23] P. Jesinger et al., Proc. Seminar on Fission Pont d'Oye IV, Habay-la-Neuve (1999) p. 111; also Nucl. Inst. Meth. A 440 (2000) 618

[24] K. Skarsvåg, Phys. Rev. C22 (1980) 638

[25] O.I. Ivanov, Sov. J. Nucl. Phys. 15 (1972) 620

[26] W. Pilz and W. Neubert, Z. Phys. A338 (1991) 75.

[27] Yu.N. Kopach et al., Phys. Rev. Lett. 82 (1999) 303

[28] M. Mutterer et al., Proc. Seminar on Fission Pont d'Oye IV, Habay-la-Neuve (1999) p. 95

[29] A. Hotzel et al., Z. Phys. A356 (1996) 299

[30] I.N. Mikhailov and P. Quentin, Proc. Conf. Fission and Properties of Neutron-Rich Nuclei, St. Andrews (1999) p. 384

Fusion Valleys for Synthesis of Superheavy Nuclei

Radu A. Gherghescu^a and W. Greiner^b

^{a}*Horia Hulubei National Institute of Physics and Nuclear Engineering*
P.O. Box MG-6, RO-76900 Bucharest, Romania
radu@th.physik.uni-frankfurt.de

^{b}*Institut für Theoretische Physik, Robert Mayer Str. 8-10*
Frankfurt am Main, Germany
greiner@th.physik.uni-frankfurt.de

Keywords: Superheavy nuclei, fusion, fission

Abstract   Nuclear deformation energy of the superheavy nucleus $^{304}120$ is studied by using the macroscopic-microscopic method. Deformation coordinate space provide asymmetric ellipsoidal shapes for the two target-projectile nuclei. The Yukawa-plus-exponential model gives the macroscopic energy. Shell effects are obtained by Strutinsky method. The main ingredient in the method is the use of a deformed asymmetric two-center shell model which provides the interacting levels within the overlapping configuration of the two fusioning partners. A complete set of spin-orbit strength parameters has been obtained leading to magic numbers $Z = 120$ for protons and $N = 184$ for neutrons. As a consequence, two fusion valleys are visible on the potential energy surface, shaped around double magicity of ^{132}Sn and ^{208}Pb.

1.   Introduction

At present time there is a boost in interest for superheavy nuclei, motivated by the successful synthesis of the heaviest elements. According to different theoretical approaches, the next spherical magic numbers could be $Z = 114$, 120 or 126. The superheavy nucleus with $Z = 120$ is the next to be synthesized. The proton number 120 is obtained recently as magic in a relativistic mean-field calculation [1].

There is always the crucial practical problem of how to produce superheavy elements. Since the ratio of neutrons to protons increases with the

D. N. Poenaru et al. (eds.), Nuclei Far from Stability and Astrophysics, 197–207.
© 2001 *Kluwer Academic Publishers. Printed in the Netherlands.*

mass number A, one cannot easily combine nuclei from the known region, because the compound system would be severely neutron-deficient.

This work focus on a suitable theoretical model to provide the projectile-target in favorable combinations, and to allow the transition from two different systems to one. Possible fission channels are also investigated for the compound system. Consequently, the steps which have to be followed within this approach are:

• the choice of a deformation space capable to describe the transition from two deformed or spherical independent nuclei to a final superheavy one. The calculations will focus on $^{304}120$ as a possible double-magic nucleus.

• calculation of the potential deformation energy within the chosen space of deformation. This step will be accomplished using the macroscopic-microscopic method. Minimization of the total energy with respect to all coordinates will provide a first glance on the process, namely a shape of the static barriers of fission and fusion.

• the dynamic study of the process involves calculation of the mass inertia tensor which couples different deformation coordinates. Then minimization of the action integral will lead to the fusion and fission paths and the dynamic barriers.

2.   Surface Parametrization

A first set of deformation parameters which is considered is exemplified in Fig. 1. Calculations are restricted to axially symmetric shapes. Here we have a typical fission like shape where the fragments are spherical. R_1 and R_2 are their radii, R is the distance between centers. The two spheres are smoothly joined by a neck obtained by rolling a third sphere of radius R_3 around the symmetry axis. All these three parameters can be varied independently. One can achieve for example fusion like shapes by considering zero neck radius, R_3 up to very elongated shapes (large R_3) until the system exits the barrier. Certain situations when one or both fragments are deformed are worth to be investigated. Such an approach yield to cold valleys due to deformed shell structure, even though these valleys are less pronounced than for spherical fragments. Therefore the two ratios of the possible asymmetric ellipsoidal target-projectile nuclei are also considered in the space of deformation, providing overlapping fusion like shapes as in Fig. 2.

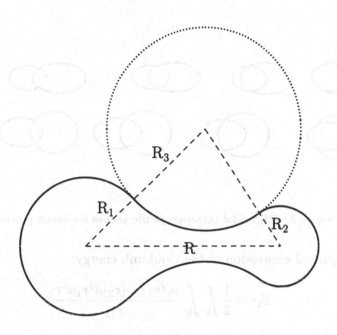

Figure 1 Shape parametrization for two spherical fragments smoothly joined by a neck region. R_1, R_2 are the radii of the two fragments, whereas R_3 is the neck radius.

3.   Deformation Potential Energy

The macroscopic-microscopic method has been used to calculate the total energy of deformation, which reads:

$$E_{total}(R) = E_{LDM} + \delta E_{shell} \tag{1}$$

The macroscopic part, E_{LDM} contains the two only shape dependent components: the Coulomb interaction energy and the nuclear energy, calculated within the Yukawa-plus-exponential model.

The most important, event though not the largest, part of the energy of a superheavy nucleus is the shell correction energy δE_{shell}. This last one is computed by Strutinsky method. The new feature in this approach is the use of a deformed asymmetric two-center shell model, which provides the input energy levels for the two interacting potential schemes.

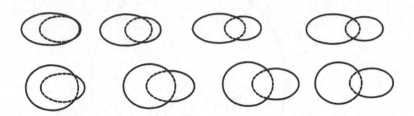

Figure 2 Ellipsoidal target-projectile shapes for fusion processes

The general expression of the Coulomb energy:

$$E_C = \frac{1}{2} \int_V \int_V \frac{\rho_e(\mathbf{r}_1)\rho_e(\mathbf{r}_2) d^3r_1 d^3r_2}{r_{12}} \tag{2}$$

becomes for two overlapping nuclei:

$$E_C = \frac{\rho_{1e}^2}{2} \int_{V_1} d^3r_1 \int_{V_1} \frac{d^3r_2}{r_{12}} + \rho_{1e}\rho_{2e} \int_{V_1} d^3r_1 \int_{V_2} \frac{d^3r_2}{r_{12}} + \tag{3}$$
$$\frac{\rho_{2e}^2}{2} \int_{V_2} d^3r_1 \int_{V_2} \frac{d^3r_2}{r_{12}}$$

where $r_{12} = |\mathbf{r}_1 - \mathbf{r}_2|$. The first and last term represent the self-energies of the two fragments, and the intermediate one — their interaction energy.

The same is true for the nuclear energy:

$$E_Y = -\frac{a_2}{8\pi^2 r_0^2 a^4} \int_V \int_V \left(\frac{r_{12}}{a} - 2\right) \frac{\exp(-r_{12}/a)}{r_{12}/a} d^3r_1 d^3r_2 \tag{4}$$

where we get three terms, in a way similar with the Coulomb energy

$$E_Y = -\sum_{i=1}^{2} \frac{a_{2i}}{8\pi^2 r_0^2 a^4} \int_{V_i} d^3r_1 \int_{V_i} \left(\frac{r_{12}}{a} - 2\right) \frac{\exp(-r_{12}/a)}{r_{12}/a} d^3r_2 -$$
$$\frac{2\sqrt{a_{21}a_{22}}}{8\pi^2 r_0^2 a^4} \int_{V_1} d^3r_1 \int_{V_2} \left(\frac{r_{12}}{a} - 2\right) \frac{\exp(-r_{12}/a)}{r_{12}/a} d^3r_2 \tag{5}$$

The *a* parameter accounts for the finite range of nuclear forces.

4. The Deformed Asymmetric Two-Center Shell Model

A specialized microscopic model has to be developed that allows nuclear shapes to be adapted to the fusion process, and this is the deformed asymmetric two-center shell model. A two-center shell model, pioneered for the first time by the Frankfurt school [2], is essential for the description of fusion. It shows that the shell structure of the two participating target-projectile nuclei is visible far beyond the barrier into the fusioning nucleus.

The Hamiltonian is expressed as a sum of the following terms

$$H = H_{osc} + V_{ls} + V_{l^2} \tag{6}$$

where in cylindrical coordinates H_{osc}:

$$H_{osc} = -\frac{\hbar^2}{2m_0}\left[\frac{\partial^2}{\partial\rho^2} + \frac{1}{\rho}\frac{\partial}{\partial\rho} + \frac{1}{\rho^2}\frac{\partial^2}{\partial\phi^2} + \frac{\partial^2}{\partial z^2}\right] + V(\rho, z) \tag{7}$$

is the oscillator term. The model is based on a double-center deformed asymmetric oscillator potential:

$$V(\rho, z) = \frac{m_0}{2}\begin{cases} \omega_{\rho 1}^2\rho^2 + \omega_{z1}^2(z + z_1)^2 &, \quad z < z_0 \\ \omega_{\rho 2}^2\rho^2 + \omega_{z2}^2(z - z_2)^2 &, \quad z \geq z_0 \end{cases} \tag{8}$$

where z_1 and z_2 are the centers of the target and projectile nuclei and z_0 is the separation plane. Different frequencies are connected to the geometry of the interacting ellipsoids:

$$\omega_{zi} = \left(\frac{b_i}{a_i}\right)^{2/3}\omega_0 \; ; \; \omega_{\rho i} = \left(\frac{a_i}{b_i}\right)^{1/3}\omega_0 \tag{9}$$

where ω_0 is the frequency of a spherical nucleus having the same volume as the ellipsoidal one.

The oscillator Hamiltonian is not separable due to the inequality $\omega_{\rho 1} \neq \omega_{\rho 2}$. We choose a particular transition system with $\omega_{\rho 1} = \omega_{\rho 2} = \omega_{z1} = \omega_1$ and $\omega_{z2} = \omega_2$. This is the configuration of a sphere intersected with a vertical spheroid. Then we obtain three second order differential equations [3] and the solutions are:

$$\Phi_m(\phi) = \frac{1}{2\pi}\exp(im\phi) \tag{10}$$

$$R_{n_\rho m}(\rho) = \left[\frac{2\Gamma(n_\rho + 1)\alpha_\rho^2}{\Gamma(n_\rho + |m| + 1)}\right]^{1/2}\exp\left(-\frac{\alpha_\rho^2\rho^2}{2}\right)(\alpha_\rho\rho)^{|m|}L_{n_\rho}^{|m|}(\alpha_\rho^2\rho^2) \tag{11}$$

$$
Z_{\nu_i}(z) = \begin{cases} C_{\nu_1} \exp\left[-\frac{\alpha_{z1}^2}{2}(z+z_1)^2\right] \mathcal{H}_{\nu_1}[-\alpha_{z1}(z+z_1)] & , \quad z < z_0 \\[3mm] C_{\nu_2} \exp\left[-\frac{\alpha_{z2}^2}{2}(z-z_2)^2\right] \mathcal{H}_{\nu_2}[-\alpha_{z2}(z-z_2)] & , \quad z > z_0 \end{cases}
$$

$$(12)$$

where $L_{n_\rho}^{|m|}(x)$ are the Laguerre polynomials, $\mathcal{H}_\nu(x)$ the Hermite functions. The normalization constants C_{ν_1} and C_{ν_2} and the z-quantum numbers ν_1 and ν_2 are obtained from normalization conditions and continuity of the wave functions and their derivatives.

At this point we have a useful basis to help us calculate the energy levels for the real deformed two-center system.

The next step will be to obtain the single-particle levels of the two overlapping deformed oscillators. Thus one has to diagonalize the differences obtained by subtracting the already diagonalized potential from the real one, namely:

$$
\Delta V(\rho) = \frac{m_0}{2} \begin{cases} (\omega_{\rho 1}^2 - \omega_1^2)\rho^2 & , \quad z < z_0 \\[3mm] (\omega_{\rho 2}^2 - \omega_1^2)\rho^2 & , \quad z \geq z_0 \end{cases}
$$

$$(13)$$

and:

$$
\Delta V(z) = \frac{m_0}{2} \begin{cases} (\omega_{z1}^2 - \omega_1^2)(z+z_1)^2 & , \quad z < z_0 \\[3mm] (\omega_{z2}^2 - \omega_2^2)(z-z_2)^2 & , \quad z \geq z_0 \end{cases}
$$

$$(14)$$

5.   Spin-Orbit and l^2 Interaction

One knows that the spin-orbit and l^2 interaction terms generate the necessary level splitting to obtain the correct schemes of the individual fragments. The difference from one fragment to another even for overlapping nuclei comes from the strength parameters κ and μ of the interaction; these quantities take values according to different mass regions. Since we have two asymmetric partners in the usual spin-orbit and l^2 interaction potentials:

$$V_{ls} = -2\kappa\hbar\omega \mathbf{ls} \tag{15}$$

$$V_{l^2} = -\kappa\mu\hbar\omega \mathbf{l}^2 \tag{16}$$

κ and μ are no more constants but z-dependent:

$$\kappa\hbar\omega(z) = \kappa_1\hbar\omega_1 + (\kappa_2\hbar\omega 2 - \kappa_1\hbar\omega_1)\Theta(z-z_0) \tag{17}$$

$$\kappa\mu\hbar\omega(z) = \mu_1\kappa_1\hbar\omega_1 + (\mu_2\kappa_2\hbar\omega 2 - \mu_1\kappa_1\hbar\omega_1)\Theta(z-z_0) \tag{18}$$

Hence, to assure the hermiticity of the operators, one uses the anticommutators [4]:

$$V_{ls} = -\left\{ \kappa\hbar\omega(z), \frac{\Delta V_{osc} \times \mathbf{p}}{m_0\omega_0^2}\mathbf{s} \right\} \tag{19}$$

$$V_{l^2} = -\left\{ \kappa\mu\hbar\omega(z), \left(\frac{\Delta V_{osc} \times \mathbf{p}}{m_0\omega_0^2}\right)^2 \right\} \tag{20}$$

Then in the usual form of the spin-orbit and $\mathbf{l}^2$ operators:

$$\mathbf{ls} = \frac{1}{2}(\mathbf{l}^+\mathbf{s}^- + \mathbf{l}^-\mathbf{s}^+) + \mathbf{l}_z\mathbf{s}_z \tag{21}$$

$$\mathbf{l}^2 = \frac{1}{2}(\mathbf{l}^+\mathbf{l}^- + \mathbf{l}^-\mathbf{l}^+) + \mathbf{l}_z^2 \tag{22}$$

the creation and annihilation operators become:

$$\mathbf{l}^+ = -e^{i\phi}\left[m_0\omega_{\rho 1}^2\rho\frac{\partial}{\partial z} - m_0\omega_{z1}^2(z+z_1)\frac{\partial}{\partial\rho} - m_0\omega_{z1}^2 i\frac{z+z_1}{\rho}\frac{\partial}{\partial\phi} \right] \tag{23}$$

$$\mathbf{l}^- = e^{-i\phi}\left[m_0\omega_{\rho 1}^2\rho\frac{\partial}{\partial z} - m_0\omega_{z1}^2(z+z_1)\frac{\partial}{\partial\rho} + m_0\omega_{z1}^2 i\frac{z+z_1}{\rho}\frac{\partial}{\partial\phi} \right] \tag{24}$$

for the z_1 centered partner; for the z_2-centered one, we replace $z + z_1$ with $z - z_2$, $m_0\omega_{\rho_1}^2$ with $m_0\omega_{\rho_2}^2$ and $m_0\omega_{z1}^2$ with $m_0\omega_{z2}^2$.

Finally one diagonalizes the total operator:

$$\Delta V(\rho, z, \phi) = \Delta V(z) + \Delta V(\rho) + V_{ls} + V_{l^2} \tag{25}$$

The result is the level scheme of two overlapping deformed asymmetric target-projectile nuclei, from the separation starting point to the synthesized superheavy nucleus.

6.   Potential Energy Surfaces

Shell effects are vital for the survival of superheavy elements. In Fig. 3 the dotted line is the macroscopic energy as a function of elongation R for two different target-projectile pairs $^{208}Pb + {}^{96}Sr$ (left hand side) and $^{132}Sn + {}^{172}Yb$. As one can see, there is no macroscopic barrier. But if one adds the shell corrections (dashed line) a minimum appears and a barrier is created (solid line) which generates a ground state and allows the superheavy system to survive.

In such a way, the potential energy surfaces (PES) as they are calculated in the framework of the two-center shell model (Fig. 4) exhibit pronounced valleys, such that these valleys provide promising doorways to the fusion of superheavy nuclei for certain projectile-target combinations. As can be seen in Fig. 4, which was drawn as PES of $^{304}120$,

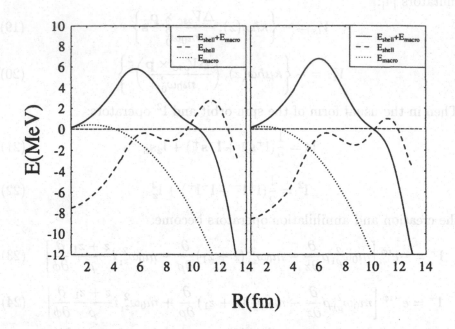

Figure 3 Survival of $^{304}120$ superheavy element as a consequence of the shell effects. Dotted line is the macroscopic energy, dashed line is the shell correction and the solid line is the total deformation energy.

with mass asymmetry η and elongation R as variables, two valleys are noticeable: one which is shaped around the double magicity of ^{208}Pb — the asymmetry valley, providing cluster decay type of disintegration [5], and a second one, carved by the double magicity of ^{132}Sn and its neighbors — the quasisymmetric valley.

If projectile and target approach each other trough those "cold" fusion valleys, they get only minimally excited and the barrier which has to be overcome (fusion barrier) is lowest as compared to neighboring projectile-target combinations.

The next step is to browse these potential valleys and study pairs of fusion partners with the lowest total deformation energy values. Minimization of energy within the deformation space will provide us the static barrier of the fusion (zero neck radius) or fission process. For a dynamical study, the mass inertia tensor is calculated with the Werner-Wheeler method. Then the action integral is minimized within the whole deformation space. As a result, one obtains the trajectories a certain pair will follow in order to synthesize the superheavy element.

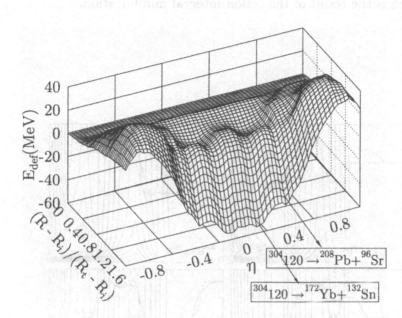

Figure 4 Potential energy surface as a function of mass asymmetry (η) and distance between centers R for $^{304}120$. The two valleys are visible around double magicity of ^{132}Sn and ^{208}Pb.

In Fig. 5 static (dashed lines) and dynamic paths (solid lines) are drawn on contour maps having the neck parameter R_3 and the elongation R as variables. Left hand side set contains pairs belonging to the quasisymmetric valleys, whereas right hand side set displays target-projectile within the asymmetric valley.

A prominent feature, noticeable especially for quasisymmetric pairs, is the abrupt decrease of the neck towards the almost separated fusion like shapes configuration. One can maybe conclude that even in the fusion process, pairs of target-projectile start overlapping with zero neck (intersected ellipsoids) but later on they develop a smooth necking region in between, which helps them follow the optimal dynamic path for synthesis of the superheavy element.

The difference between the static and the dynamic paths comes from different deformation coordinate couplings — via the mass tensor components. Each time a new degree of freedom is added in the deformation space, the corresponding component of the mass tensor drives the system on a dynamic trajectory which accounts for the new couplings with the new deformation component. The consequence of this fact is the difference between the static paths, which represent the result of energy minimization along all the deformation parameters and the dynamic path, which is the result of the action integral minimization.

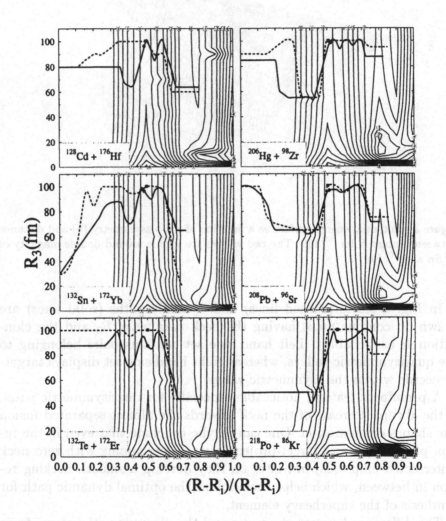

Figure 5 Contour maps with static barriers (dashed curves) and dynamic paths (solid curves) for tin valley pairs (left hand side plots) and lead valley pairs as a function of reduced elongation and neck radius R_3.

7. Conclusions

The message which this work intends to transmit, is that a deformed asymmetric two-center shell model is a major tool to approach fusion and fission phenomena.

On the potential energy surfaces, two valleys are shaped by two double-magic partners, ^{132}Sn and ^{208}Pb and their neighbors. These valleys are deep enough to settle the mass asymmetry parameter.

Besides the known used reactions having ^{208}Pb as one of the partners, these calculations propose, on the basis of the action integral minimization, hence on penetrability values comparison, quasisymmetric pairs for fusion, like:

$$^{128}Cd + {}^{176}Hf \rightarrow {}^{304}120$$
$$^{134}Te + {}^{170}Er \rightarrow {}^{304}120$$
$$^{132}Sn + {}^{172}Yb \rightarrow {}^{304}120$$

to synthesize the next superheavy element $^{304}120$.

References

[1] K. Rutz et al., Phys. Rev. C56 (1997) 238

[2] W. Greiner, J. Maruhn, 1996, *Nuclear Models*, Berlin: Springer Verlag, p275

[3] E. Badralexe, M. Rizea, A. Sandulescu, Rev. Roum. Phys. 19 (1974) 63

[4] J. A. Maruhn, W. Greiner, Z. Phys. 251 (1972) 431

[5] D. N. Poenaru, W. Greiner, 1996, *Nuclear Decay Modes*, Bristol-IOP, p. 275

Theoretical Description of
Ternary Cold Fission

A. Săndulescu[a], Ş. Mişicu[b], F. Carstoiu[a] and W. Greiner[b]

[a] National Institute of Physics and Nuclear Engineering, P.O. Box MG-6,
76900 Bucharest-Magurele, Romania

[b] Institut fűr Theoretische Physik der J. W. Goethe Universität,
D-60054, Frankfurt am Main, Germany
misicu@th.physik.uni-frankfurt.de

Keywords: Cold fission, cluster radioactivity, giant molecules

Abstract   The ternary cold fission of ^{252}Cf is described as the decay of a quasi-
bound giant trinuclear molecule, the fragments being in their ground
states with multipolar deformations. Topics like calculation of the half-
lives of the alpha-like trinuclear molecules, the collective modes for the
^{10}Be-like trinuclear molecules and the shift of the first 2^+ state of ^{10}Be
in the quasi-molecular configuration are addressed.

1.   Introduction

The ternary cold fission is a rare phenomenon consisting in the disintegration of a massive nucleus, such as ^{252}Cf, in two heavy nuclei and
a light particle, with a very small dissipation of energy on degrees of
freedom, other than the translational motion of the three nuclei. About
90 % of the light particles emitted in the ternary spontaneous fission are
α particles [1]. The rest is accounted by protons, neutrons, $^{3,5}He$, ^{10}Be
and a few others light particles.

Before scission takes place, and after preformation from the mother
nucleus was accomplished, there is a transient stage when the clusters
are in close vicinity. A very recent experiment on ^{10}Be-accompanied
cold fission of ^{252}Cf is supporting the idea that the time spent by the
three clusters in this transient stage is larger than 10^{-13} s which brings
into discussions the concept of a long living giant molecule. The present
work is dedicated to the theoretical treatment of this topic.

D. N. Poenaru et al. (eds.), Nuclei Far from Stability and Astrophysics, 209–220.

2.   Half-Lives of Trinuclear Molecules

A coplanar three cluster model consisting of two deformed fragments and a spherical α particle is considered. Three main coordinates were retained for the description of the dynamics during the rearrangement and penetration through the barrier processes: the coordinates (x_α, y_α) of the c.m. of the α particle, and a collective coordinate R describing the separation distance between the heavy fragments along the fission axis. We assume that the fission axis is conserved during the penetration process. No preformation factors were included in our description. The three body potential was computed with the help of a double folding potential generated by the M3Y-NN effective interaction [1]. We would like to stress that the interaction barriers are calculated quite accurately on the absolute scale and that the touching configuration for the heavy fragments is situated outside the barrier. Due to the lack of any explicit density dependence in the M3Y effective interaction, which leads to unphysically deep potentials for large density overlaps in the following we introduce a compression term in order to correct the short range dependence of the calculated interaction potentials which reads:

$$V(\boldsymbol{R}) = \int dr_1 dr_2 \rho_1(\boldsymbol{r}_1)\rho_2(\boldsymbol{r}_2)v(|\boldsymbol{s}|) + V_0 \int dr_1 dr_2 \tilde{\rho}_1(\boldsymbol{r}_1)\tilde{\rho}_2(\boldsymbol{r}_2)\delta(\boldsymbol{s}).$$

(1)

The terms entering in the above formula were discussed extensively in [2]. Our strategy is the following. We assume first that the movement of the three particles is so slow that the system adjusts adiabatically to stay in a configuration corresponding to the minimum in the three body potential. We distinguish three regions: the minimum where the system oscillates corresponding to a cold rearrangement process, the classically forbidden region (the barrier) and the asymptotic region (beyond the outer turning point) where the system decays into three fragments. In the minimum region we solved the classical equations of motion thus obtaining the oscillation time T_{osc}, the corresponding α particle trajectory and the inner turning point. The oscillation time is of the order of 0.6×10^{-21} s. and the barrier assault frequency of the order $\nu \sim 1.7 \times 10^{21} s^{-1}$. The zero point energy is estimated from the uncertainty principle giving $E_0 \simeq 0.9 \ MeV$ for highly deformed fragments and slightly larger for spherical splittings ($\sim 1.2 \ MeV$). In the classically forbidden region we solve the semiclassical trajectory equations for the relevant degrees of freedom [2]

$$x_i'' = \frac{M}{2(Q-V)}\left(\frac{x_i'}{m_1}\frac{\partial V}{\partial x_1} - \frac{1}{m_i}\frac{\partial V}{\partial x_i}\right),$$

(2)

with $M = m_1 + \sum m_j(x'_j)^2$ the effective mass (or the effective inertia) and $x' = \frac{dx}{dx_1}$. m_i are the masses for the corresponding degrees of
freedom, Q is the reaction energy. Mixed derivatives do not appear in
the effective inertia since in our case the mass tensor is diagonal. x_1 is
one of the relevant variables (supposed to vary monotonically) chosen to
parametrize the trajectory. In practice we found that the variable R is
the best candidate. The coupled differential equations (Equ. 2) describe
the motion of the system in the barrier region and satisfy the minimum
action principle. It can be shown that the set of equations (Equ. 2) is
fully equivalent with the classical equations of motion in imaginary time
$\tau = it$. The reduced action is given by

$$S_0 = \int \sqrt{2(V-Q)\left(m_1 + \sum_i m_i \left(\frac{dx_i}{dx_1}\right)^2\right)} \, dx_1, \qquad (3)$$

and the penetrability factor is simply $P = e^{-\frac{2S_0}{\hbar}}$. A close examination
of Equ. 2 shows that the semiclassical approximation is not valid on the
surface $Q = V$. In order to have a solution we must require that the
factor in parenthesis in Equ. 2 vanishes on that surface. This gives us
the so called *transversality conditions* or the *steepest descent equations*:

$$\frac{m_i x'_i}{\frac{\partial V}{\partial x_i}} = \frac{m_1}{\frac{\partial V}{\partial x_1}}, \qquad (4)$$

which are solved near the turning points.

Having solved the dynamical equations of motion, using the reaction
energies (Q_3) calculated from the experimental masses or in few cases
taken from the theoretical predictions of Möller and Nix [3] we evaluate
the penetrabilities as described above. The relative production yields
are evaluated from the simple formula

$$Y_L(A_L, Z_L) = \frac{P_L(A_L, Z_L)}{\sum_{L'} P_{L'}(A_{L'}, Z_{L'})} \qquad (5)$$

Evidently we assume that the preformation factors are the same for all
cold splittings and consequently in the expression (7) this factor can-
cels. The calculated yields (normalized to the experimental ones [4])
are displayed in Fig. 1. Full details of the calculation are given in [2].
Another remark concerns the energy and the emission angle of the α
particle. The energy is larger for large asymmetry mass splittings and
the emission angle is smaller. The average values, taking the calculated
dynamical yields as weighting factors are: $< E_{kL} > = 111.0 \pm 1.3 \, MeV$,
$< E_{kH} > = 80.6 \pm 4.5 \, MeV$, $< E_{k\alpha} > = 19.3 \pm 0.5 \, MeV$ and

$< \theta_\alpha > = -82.5 \pm 1.9°$. The quoted variances are calculated from the widths distributions given in Ref.[2]. We shall show below that these variances become much larger if quantum fluctuations are taken into account. According to Heeg [5], the mean experimental angle, measured with respect to the fission axis oriented towards the light fragment (opposite to our convention, see below) is 83°. The adiabatic scenario gives ternary yields only for splittings with one of the fragment spherical. The corresponding yields are evidently dominated by Q-value effects. Only high Q value channels contribute to the total yield. In the dynamical calculation there is a delicate interplay between deformation, Q-value and penetration path effects. Note that the theoretical yields are calculated strictly at zero excitation energy and we neglected the level density close to the ground state of the fragments and as a result the odd-even effects, noticeable in the experiment around the masses $A_L = 101,103,107$ are not well reproduced.

The lifetime for the decaying molecule is evaluated using the relation $T_{1/2} = \ln 2/\nu P$, where ν is the barrier assault frequency and P the penetrability.

Next we consider fluctuations in the entrance point since the calculated action along the penetration path should be minimized not only with respect to the path, but also with respect to the initial conditions. If we neglect the α particle coordinates, then the lowest quasi-molecular state of the system in the potential minimum is of the form $\Psi \sim \exp(-\frac{1}{2}\alpha(R - R_0)^2)$. From the estimated energy for the lowest state we have $E_0 = \frac{\hbar^2 \alpha}{2m}$ and the estimated value for the distribution width is $\frac{1}{2\sqrt{\alpha}} = 0.7 \ fm$. Therefore we started a Monte Carlo evaluation of the trajectories using as inner turning point $R_i = R_{t1} + \sigma \times gauss(seed)$ where R_{t1} is the turning point given by the minimization procedure, $gauss(seed)$ is a random Gaussian generator with zero mean value and variance 1. σ is the desired variance of the generated distribution, taken here of the order of 0.1 fm. For all generated R_i values we search for the solutions of the equation for energy conservation $V(R_i, x_\alpha(R_i), y_\alpha(R_i)) = Q_3$ at the entrance point. In practice, we relaxed this condition, searching for solutions $(x_{\alpha i}, y_{\alpha i})$ in a band $Q_3 \pm 50 \ keV$. No initial velocity distributions were considered in this simulation, since in our particular system of coordinates it is difficult to conserve the total linear momentum. In this way a number of some 8000 initial positions were generated. The calculation was completed by solving the dynamical equations in the barrier up to the outer turning point. From then on the system disintegrates. The final energy and angular distribution of the fragments were obtained by solving the classical equation

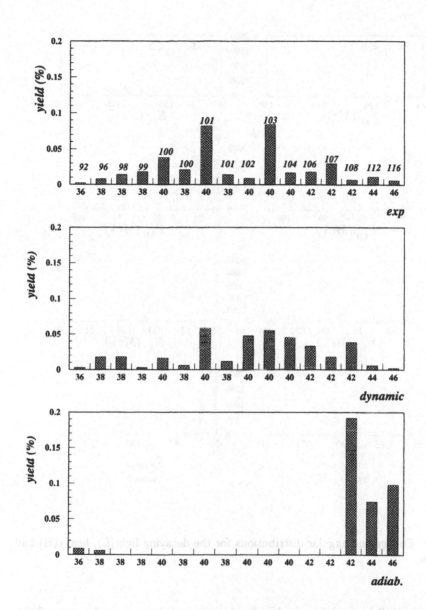

Figure 1 Experimental (upper panel) and calculated yields in the the dynamic model (middle panel) and adiabatic model(bottom). The fragmentation channel is indicated by the charge number (on the x axis) and by the mass number (on top of the stacks) of the light fragment.

of motion for a three body system. Since the exit point locates at rather large distances ($R_f \sim 18\ fm$, $y_{\alpha f} \sim 8\ fm$), the nuclear forces are com-

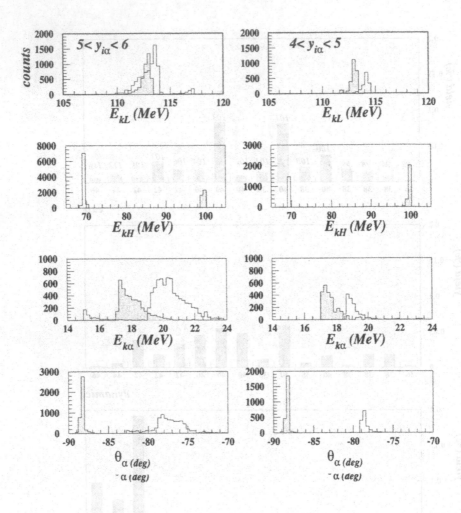

Figure 2 Energy and angular distributions for the decaying light(L), heavy(H) and α particle.

pletely negligible, and we used only the monopole-monopole part of the Coulomb interaction. We have shown [6] that quadrupole-monopole and other higher multipolarity terms have little influence on the final distributions. The results of the simulations are displayed in Fig. 2 for the splitting $^{252}Cf \rightarrow \alpha + {}^{92}Kr + {}^{156}Nd$. We have also considered the more symmetric splitting $^{252}Cf \rightarrow \alpha + {}^{116}Pd + {}^{132}Sn$ shown by shadowed histograms in Fig. 2. In the left panels we have selected all events in the band $5 \leq y_{\alpha i} \leq 6 \ fm$ and $4 \leq y_{\alpha i} \leq 5 \ fm$ in the right panels. Much lower initial values $y_{\alpha i}$ are corresponding to an α particle in overlap with the fragments and lead to trajectories traversing one of the heavy fragments and with penetrabilities of the order 10^{-100} or smaller and they are disregarded. The higher the initial position of the α particle, the wider the distribution in the final $y_{\alpha f}$ coordinate. Also the energy distribution of the light fragment (E_{kL}) is wider. The mean kinetic energy for the α particle is 21(18) MeV and the mean emission angle is $-78°$ ($-88°$) for large (small) mass asymmetries. The angle is negative since we have chosen a system of coordinates with the light fragment placed on the left side, and the α particle is deviated strongly by the heavy fragment, so the emission angle is always in the second quadrant ($\theta_\alpha \geq \pi/2$). For initial positions in the lower band (right panels in Fig. 2) all distributions are much narrower, showing a strong focusing effect. The final kinetic energy $E_{k\alpha}$ lowers to 19(17.5) MeV and the emission angle is $-79°$ ($-88°$). In the upper position band the penetrabilities are small ($\leq 10^{-22}$) and the lifetime is of the order of tens of seconds. In the lower band the penetrabilities increase (10^{-21}) and the lifetime varies between 1 and 10 seconds. As we must minimize the reduced action, one should choose the trajectory with maximum of penetrability. Therefore our model predicts lifetimes for the trinuclear molecule of the order of 1 second. Of course, all distributions will be much larger if averaged over all fragmentation channels. Note the strong dependence of all distributions on the mass asymmetry of the heavy fragments (and implicitly on Q value and deformation)(Fig. 2). This is especially evident for exit configuration ($R_f, y_{\alpha f}$) and for the kinematic variables of the fragments ($E_{k\alpha}, E_{kH}, \theta_\alpha$).

3.   Excitation of Collective Modes in Ternary Fission

It is not difficult to imagine how collective modes occur in trinuclear molecules if the two heavier clusters (for example, the ^{96}Sr and ^{146}Ba) are connected via a smaller spherical nucleus (^{10}Be) [7]. The situation is illustrated in Fig. 1, where the main dynamical variables are indicated. In order to separate the center-of-mass motion, the Jacobi-coordinates are introduced

$$r = r_2 - r_1$$
$$\xi = \frac{m_1 r_1 + m_2 r_2}{m_1 + m_2} - r_3 \tag{6}$$

The classical expression of the kinetic energy of the three-body system, after removing the center of mass contribution, is expressed as a sum of

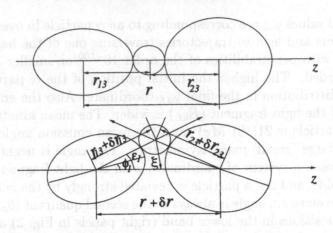

Figure 3 The main variables of the trinuclear molecule. The upper figure represents the linear equilibrium configuration.

translational and rotational degrees of freedom [8]:

$$T = \frac{1}{2}\mu_{12}\dot{r}^2 + \frac{1}{2}\mu_{(12)3}\dot{\xi}^2 + \frac{1}{2}{}^t\omega_1 J_1 \omega_1 + \frac{1}{2}{}^t\omega_2 J_2 \omega_2 + \frac{1}{2}{}^t\omega_3 J_3 \omega_3 \quad (7)$$

where $\mu_{12} = \frac{m_1 m_2}{2(m_1+m_2)}$ and $\mu_{(12)3} = \frac{m_3(m_1+m_2)}{2(m_1+m_2+m_3)}$. The vector ω_i denotes the angular velocity of the i'th cluster referred to the laboratory frame. Its relation to the components of the angular velocities of the molecular frame ω_i' and those of the clusters referred to the molecular frame, ω_i'' is written bellow

$$\omega_i = R(\Omega_i)\omega_i' + \omega_i'',$$

where $R(\Omega_i)$ denotes the transformation matrix which connects the axes of the molecular frame and the principal axes of each constituent nucleus.

Eventually we obtain an expression quadratic in the angular velocities components ω_i' and the butterfly angular velocities $\dot{\varepsilon}$ [9]:

$$T = \frac{1}{2}\Theta_{11}(\omega_1'^2 + \omega_2'^2) + \frac{1}{2}\Theta_{33}\omega_3'^2 - \Theta_{13}\varepsilon\omega_1'\omega_3' + \frac{1}{2}\Theta_\varepsilon\dot{\varepsilon}^2 + \Theta_{2\varepsilon}\dot{\varepsilon}\dot{\omega}_2'$$

The next step consists in quantizing the expression of the kinetic energy. Then this quantized form of the kinetic energy splits-up in a diagonal contribution and a non-diagonal part, which can be treated as a perturbation

$$\hat{T} \approx \frac{\hbar^2}{2(\Theta_{11} - \frac{\Theta_{13}^2}{\Theta_\varepsilon})}(\hat{I}^2 - \hat{I}_3'^2) - \frac{\hbar^2}{2(\Theta_\varepsilon - \frac{\Theta_{13}^2}{\Theta_{11}})}\left\{\frac{\partial^2}{\partial\varepsilon^2} + \frac{1}{\varepsilon^2}\left(\frac{1}{4} - \hat{I}_3'^2\right)\right\}$$

$$+ \frac{\hbar^2}{\frac{\Theta_{11}\Theta_\varepsilon}{\Theta_{13}} - \Theta_{13}}\left\{\frac{1}{\varepsilon}\hat{I}_1'\hat{I}_3' - \hat{I}_2'\left(\frac{1}{i}\frac{\partial}{\partial\varepsilon}\right)\right\} \quad (8)$$

provided the kinetic terms in R and ξ freezed.

The interaction between two clusters composing the giant molecule can be calculated according to Equ. 1.

For small non-axial fluctuations, the total potential can be expressed in a simplified form, provided we keep terms up to the second power in ε

$$V_{\text{trinucl}} = V_{12}(R_1 + R_2 + 2R_3) + V_{13}(R_1 + R_3) + V_{23}(R_2 + R_3) + \frac{1}{2}C_\varepsilon \varepsilon^2 \quad (9)$$

We take the following ansatz for the unsymmetrized collective wave function corresponding to the diagonal part of the Hamiltonian

$$\phi = D_{MK}^{I*}(\vartheta)\chi_{K,n_\varepsilon}(\varepsilon) \quad (10)$$

Due to this factorization the corresponding Schrödinger equation in the ε mode, acquire the form

$$\left[-\frac{\hbar^2}{2\Theta_\varepsilon}\frac{\partial}{\partial \varepsilon^2} + \frac{1}{\Theta_\varepsilon}\left(K^2 - \frac{1}{4}\right)\frac{\hbar^2}{2\varepsilon^2} + \frac{C_\varepsilon}{2}\varepsilon^2 + \frac{\hbar^2}{2\Theta_{11}}[I(I+1) - K^2] \right] \chi$$
$$= E\chi, \quad (11)$$

with eigenvalues

$$E = \frac{\hbar^2}{2\Theta_{11}}\left[I(I+1) - K^2\right] + \hbar\sqrt{\frac{C_\varepsilon}{\Theta_\varepsilon}}\left(\mid K \mid +2n_\varepsilon + \frac{3}{2}\right) \quad (12)$$

For all linear configuration, the typical values of $\hbar\sqrt{\frac{C_\varepsilon}{\Theta_\varepsilon}}$ range between 2.4 MeV for the splitting $^{96}Sr + ^{10}Be + ^{146}Ba$ and 2.9 MeV for the splitting $^{112}Ru + ^{10}Be + ^{130}Ba$ making thus possible the observation of molecular butterfly excitations in such a decay phenomenon [9].

4. Polarization of the Light Cluster in Trinuclear Molecules

The existence of a shift of the first 2_1^+ state in ^{10}Be (Fig.5) was very recently claimed [10]. Using the Gammasphere with 72 detectors, the γ-ray from the first 2^+ state in ^{10}Be, was measured in coincidence with the γ rays of the fission partners ^{146}Ba and ^{96}Sr. It was observed that the spectrum corresponding to these γ quanta is not Doppler broadened as if the quanta are emitted from a resting source, and secondly the energy value of the 2_1^+ state in such a quasi-bound configuration is lowered by 6 keV compared to the value compiled for the free configuration ($E_\gamma = 3368.03\ keV$).

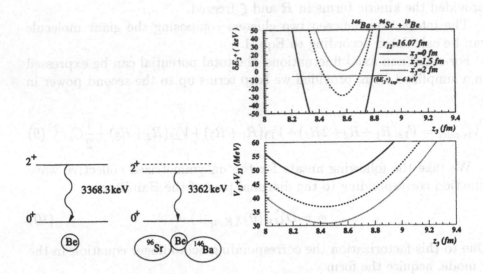

Figure 4 Polarization of ^{10}Be sandwiched between ^{96}Sr and ^{146}Ba

Figure 5 The energy shift of the 2_1^+ state (upper panel) and total potential energy $(V_{13} + V_{23})$ of ^{10}Be (lower panel) as a function of the z-component of the position of the third cluster.

Most probable, if the measurement was correct, the shift of the 2_1^+ level of ^{10}Be is a consequence of its polarization induced by the other two clusters in the quasi-molecular configuration. In order to evaluate this polarization we make recourse to the following assumptions:

a) For a a given configuration of the three clusters, which should correspond to a precise point on the fission path, the static energy of the quasi-molecule must have a minimum with respect to variations in the deformations, i.e.

$$\frac{\partial E_{\text{ternary}}}{\partial \beta_i} = 0, \quad (i = 1, 2, 3) \tag{13}$$

b) The deformation energy for the light cluster 3 resembles very much a vibrator with a small cubic anharmonicity for not too large departures from the spherical equilibrium position:

$$E_{\text{def}}(\beta_3) = \frac{1}{2}C_2\beta_3^2 + C_3\beta_3^3 \tag{14}$$

where the stiffness parameters have the numerical values: $C_2 = 7.688$ MeV and $C_3 = -2.855$ MeV, obtained upon interpolating the Hartree-Fock deformation curve.

c) The interaction between the light cluster 3 and the heavy clusters, 1 and 2, can be expanded with respect to the deformation of the third cluster β_3, keeping in the same time the deformations of the heavy clusters, β_1 and β_2, freezed.

Accordingly, discarding the constant terms, the vibrational Hamiltonian reads

$$H_3^{(\beta)} = -\frac{\hbar^2}{2B}\frac{\partial^2}{\partial\beta_3^2} + \frac{1}{2}C_2'\beta_3^2 + C_3'\beta_3^3 \tag{15}$$

where $C_{2,3}'$ are the modified stiffness parameters of the quadratic and cubic potential terms and B, the effective mass, is computed from the experimental values of $B(E2, 0_1^+ \to 2_1^+)$ and the energy of the 2_1^+ state. The corresponding spectrum can be deduced by applying the stationary perturbation theory in the second order approximation [11].

$$E_n' = \hbar\omega\left\{n + \frac{1}{2} - (\hbar\omega)^2\frac{15C_3''^2}{4C_2'^3}\left(n^2 + n + \frac{11}{30}\right)\right\} \tag{16}$$

where $\hbar\omega = \sqrt{\frac{C_2'}{B}}$. Then, the shift can be readily obtained by taking the difference between the energies of the 1-phonon state in the quasi-bound configuration (with primed $C_{2,3}$) and the free case (with unprimed $C_{2,3}$)

$$\delta E_{2^+} = E_1' - E_1 \tag{17}$$

In Fig. 5 we represent the energy shift of the 2_1^+ state, in the case of the splitting $^{146}Ba + {}^{96}Sr + {}^{10}Be$, as a function of the location of the light cluster on the molecular z-axis (upper panel). In the lower panel the sum $V_{13}+V_{23}$ of the interactions between clusters 1 and 3 and between clusters 2 and 3, is plotted. The heavy fragments were supposed to have their symmetry axes aligned. An interesting fact, which becomes apparent from the inspection of this figures, is the close proximity between the location of the minima of the two sets of curves, which may suggest that the lowest value of the energy shift is obtained in the case when the light cluster is located on, or very close from, the electro-nuclear saddle curve. The curves represented in Fig. 5 was obtained for a tip distances between the heavy clusters, $d = 2\ fm$. If we increase or we decrease this distance, the negative energy shifts will gradually disappear. If we make the assumption that the light cluster is located on the bottom of the potential pocket, or in quantum terms, is filling-up the first state of the quantum well produced by the interaction of the three fragments, then the experimental shift, $\delta E_{2^+} = -6\ keV$ can be reproduced if the tip distance between the heavy fragments is confined in the interval $1.5 - 2.5\ fm$, and the cluster ^{10}Be is located off the fission axis at a height x_3 not larger than $2\ fm$.

Acknowledgments

One of the authors (Ş.M.) is grateful to the Alexander von Humboldt-Stiftung for the financial support.

References

[1] A. Săndulescu et al., Phys. Part. Nucl. 30 (1999) 386

[2] F. Carstoiu et al., Phys. Rev. C61 (2000) 044606.

[3] P. Möller et al., At. Data Nucl. Data Tab. 59 (1995) 185

[4] A.V. Ramayya et al., Phys. Rev. C57 (1998) 2370

[5] P. Heeg, Dissertation T.H. Darmstadt (1990)

[6] Ş. Mişicu et al., Il Nuovo Cimento A112 (1999) 300

[7] P.O. Hess et al., J. Phys. G25 (1999) L139

[8] Ş. Mişicu et al., J. Phys. G25 (1999) L147

[9] P.O. Hess et al., J. Phys. G26 (1999) 957

[10] A.V. Ramayya et al., Phys. Rev. Lett. 81 (1998) 947

[11] Ş. Mişicu, A. Săndulescu and W. Greiner, Phys. Rev. C61 (2000) 041620(R)

New Aspects of Nuclear Structure and QE Electron Scattering

E. Moya de Guerra, J.A. Caballero, E. Garrido, P. Sarriguren and J.M. Udías

IEM, CSIC, Serrano 123, 28006 Madrid, Spain

imtem22@pinar2.csic.es

Keywords: Electron scattering, relativistic bound nucleons, halo nuclei

Abstract   We focus here on two aspects of Nuclear Structure that are of major interest currently. i) Relativistic dynamics of bound nucleons. ii) Halo nuclear structures. We discuss how these aspects can be probed with electron beams. In the context of point i) we discuss issues as the sensitivity to the effective mass in the nuclear interior, and to the dynamical enhancement of negative energy components. In the context of point ii), after a brief view on elastic electron scattering from a halo nucleus, we discuss inclusive and exclusive observables with particular attention to $^6He(e,e')n\alpha$ and $^6He(e,e'\alpha)nn$ reactions.

1.   Introduction

Electron scattering from nuclei provides reliable information on charge, spin, current, momentum distributions and, in general, on nuclear and subnuclear structure and dynamics [1, 2]. Given a nuclear target one may select to study the ground state, or the transition currents to specific fixed final states, choosing appropriately the kinematical conditions by detecting the scattered electron at selected energy and directions. For medium incoming electron energies ($E_0 \leq 1\ GeV$) the electron nucleus interaction is described in terms of elastic electron-nucleon scattering. The electromagnetic interaction is dominated by one photon exchange and the off-shell nucleon may change its "off-shellness" (depending on the selected final nuclear state) but not its internal structure. In particular it is well known that for a spin zero nuclear target the elastic electron nucleus cross section is proportional to the charge form factor

D. N. Poenaru et al. (eds.), Nuclei Far from Stability and Astrophysics, 221–232.
© 2001 *Kluwer Academic Publishers. Printed in the Netherlands.*

squared. Elastic scattering can therefore be used to map out the charge density distribution. For a non-zero spin target the elastic scattering cross section at forward angles is proportional to the squared Coulomb form factor, while at backward angles it is proportional to the transverse form factor. Forward and backward scattering can then be used to map out independently charge and magnetization distributions.

When the energy is such that one of the nuclear constituent particles can be knocked out from the nucleus quasielastic scattering (QE) takes place. QE becomes the dominant process when the energy transfer (ω) equals the ratio between the squared four-momentum transfer $(\mid Q^2 \mid)$ and twice the mass of the constituent (M_C). At this quasielastic kinematics the knock-out constituent can be detected in coincidence with the scattered electron, and the differential cross section for coincidence $(e, e'C)$ reactions becomes proportional to the momentum distribution of the constituent particle inside the nucleus [3, 4]. Thus, coincidence measurements at quasielastic kinematics allow to map out the momentum distributions of the nuclear constituents.

The subject of Relativistic mean field (RMF) theory with Walecka's model [1] is retrieving more and more weight with time. In spite of much criticism this model is becoming more and more popular as many features of nuclear structure can be described in a most elegant way with few free parameters. Although most of nuclear structure can be understood treating bound nucleons non-relativistically, it pays to study what are the implications of Dirac equation for bound particles. Such relativistic effects can be expected to be better seen in nucleon momentum distributions, which in turn are best seen in quasielastic electron scattering. In section 2 we discuss some implications that relativistic nucleon dynamics in the nuclear medium has for observational features in quasielastic electron scattering.

In section 3 we explore the possibility of using electron scattering to investigate one of the most intriguing objects recently discovered in nuclear physics: Halo Nuclei. Their small binding energies and large spatial extensions had no precedent and were totally unexpected in the realm of systems governed by the nuclear force. Borromean two-neutron halo systems show the additional peculiarity that the three-body (core+n+n) system completely falls apart when either of the three constituents drops out. Among the latter ^6He $(\alpha+n+n)$ is the one that is most stable and has been extensively studied. Since the initial discovery by Tanihata and collaborators [5], much progress has been made in theoretical [6] and experimental studies of halo nuclei. The available experimental information comes basically from nucleus-nucleus collisions, that has the inherent difficulty of disentangling nuclear reaction mechanisms from nuclear structure effects. This is why the idea of using nonhadronic probes to obtain complementary information on halo nuclei has been put forward.

2. QE Electron Scattering: Probe of Relativistic Bound Nucleons

When an electron interacts with a bound proton and knocks it out of the nucleus, the outgoing proton can be measured in coincidence with the electron and the bound nucleon wave function in momentum space can be mapped out. This process dominates the electron-nucleus cross section when the energy transfer $\omega \simeq |Q^2|/2M$ (i.e., when the scaling variable $y \simeq 0$ or the Bjorken $x \simeq 1$). Relativistic effects of bound nucleon dynamics are expected to be seen preferentially in their momentum distributions, and thus $(e, e'p)$ coincidence measurements at quasielastic kinematics can be thought of as the main source of information of the role of relativity in the nuclear interior.

Not very much is known phenomenologically about relativistic bound particles. On the one hand, in high energy physics, particles are always treated as plane waves and their internal structure is oversimplified. On the other hand in atoms and nuclei, electrons and nucleons can mainly be treated non-relativistically, even though it is well known that spin-orbit interaction originates from Dirac equation. In particular, the fact that central potentials (Coulomb and Woods-Saxon type potentials) are so different in atoms and nuclei explain the fact that spin-orbit is much stronger in nuclei. Except for spin-orbit interaction, there is very little knowledge about characteristic features of relativity for bound particles. An issue that may be crucial to understand *quark dynamics* in the hadron interior.

In a series of papers [7], we have investigated this issue comparing predictions on momentum distributions obtained with relativistic and non-relativistic nucleon wave functions. To make a long story short, we may summarize it as follows: 1) Relativistic (RDWIA) and non-relativistic (DWIA) approximations were used to analyze existing data on hole states close to Fermi level in ^{208}Pb and ^{40}Ca, with the conclusion that the spectroscopic factors S_α obtained in RDWIA were larger than those obtained in DWIA. For $3s_{1/2}$ shell in ^{208}Pb, $S_\alpha(\text{RDWIA}) \simeq 0.7$ and $S_\alpha(\text{DWIA}) \simeq 0.5$. A reduction from the HF value $S_\alpha = 1$ to $S_\alpha = 0.7$ at most is what can be expected from short range correlation effects [8], while long range correlations are expected to cause less than a 10 % effect, except in soft nuclei. Thus, on theoretical grounds, a spectroscopic factor of 0.7 for the $3s_{1/2}$ shell in ^{208}Pb is better justified than one of 0.5. 2) For high missing momentum P_m (i.e., for high bound nucleon momentum), RDWIA calculations agree with experiment much better than DWIA. The latter give too small cross sections when using wave functions and spectroscopic factors fitted to the lower P_m-sector. 3) The longitudinal-transverse response (R_{TL}) reaches much larger absolute values in RDWIA than in DWIA. In addition, for spin-orbit partner orbitals ($l_{j=l-1/2}$, $l_{j=l+1/2}$), RDWIA predicts larger $|R_{TL}|$ for the jack-knifed ($l_{j=l-1/2}$) than for the stretched state [9], a prediction that has been confirmed experimentally [10].

In what follows, we will describe why this is so. But first we explain what the terms RDWIA and DWIA mean: DWIA stands for distorted wave impulse approximation, the prefix R is used to denote relativistic approach. To avoid confusion, we have to make it clear that relativistic or non-relativistic refers only to the treatment of the nucleon wave functions and current operator. At the energies involved in ($e, e'p$) experiments, electrons are ultrarelativistic, and the expressions for differential cross sections are obtained from a relativistic invariant amplitude $\mathcal{M} = l^\mu J_\mu / |Q^2|$, with l^μ the leptonic current and J_μ the hadronic current. When final state interactions are taken into account and nucleon distorted waves are used for the outgoing nucleon, the four-vector hadronic current J_μ is written as $J_\mu(\omega, q) = \int d\vec{p}\,\bar{\Psi}_F(\vec{p}+\vec{q})\hat{J}_\mu(\omega, q)\Psi_B(\vec{p})$.

In RDWIA, Ψ_B and $\bar{\Psi}_F$ are relativistic bound [1, 11] and scattering [12] wave functions, respectively and $\hat{J}_\mu$ is the relativistic free nucleon current operator. In DWIA, the nucleon wave functions are obtained from Schrödinger equations with phenomenological (Woods-Saxon type) potentials and the 4×4 matrices $\hat{J}_\mu$ are replaced by non-relativistic 2×2 matrices $\hat{J}_\mu^{nr}$ for free bispinors, which are usually truncated to order $(P/M)^2$. Thus, there are three types of relativistic effects of different nature to be examined: Effects due to above mentioned truncations of the current operator which are kinematical, i.e., come from relativistic kinematics compared to the non-relativistic one. Effects due to the $S - V$ potentials, which are dynamical. Among the latter, we have to distinguish between complex $S - V$ optical potentials [12] used for the scattering nucleon wave and real $S - V$ potentials [11] used for the bound nuclear states. The effects of the $S - V$ potentials are best discussed transforming Dirac equations for Ψ

$$[\gamma_0 \bar{E} - \vec{\gamma} \cdot \vec{p} - \bar{M}]\Psi = 0 , \tag{1}$$

(with $\bar{E} = E - V$, $\bar{M} = M - S$) into Schrödinger like equations for the upper (Ψ_{up}) and lower (Ψ_{down}) components.

These equations show that Ψ_{up} and Ψ_{down} are subject to the same central energy dependent potential (U_C), but to different spin-orbit and

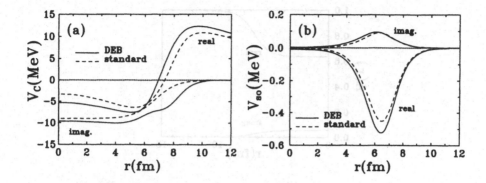

Figure 1 Real and imaginary parts of Central and spin-orbit standard and DEB optical potentials for 100 MeV protons on ^{208}Pb.

non-local potentials. Furthermore, one can remove the non-local terms to make the equations look like standard Schrödinger equations with central potentials. One can show that the function

$$K(r) = \sqrt{A}, \qquad A = \frac{A_+}{E+M} = 1 - \frac{V+S}{E+M} \qquad (2)$$

does this for us. The bispinor $\Phi_j^l(\vec{r}) = K(r)\Psi_{up}^\kappa(\vec{r})$, satisfies $\left(-\frac{\nabla^2}{2M} - U_{\text{DEB}}\right)\Phi_j^l(\vec{r}) = 0$. U_{DEB} is called the Dirac equivalent Schrödinger potential. It contains a new central potential ($V_C = U_C + V_D$) and the spin-orbit potential. In Figs. 1 and 2, we show by solid lines the various terms of the potentials and the function $K(r)$ for kinematical conditions that match those of $(e, e'p)$ experiments.

We recall that the label κ in four-spinors Ψ stands for j- and π-quantum numbers. The total angular momentum eigenvalue is $j = |\kappa| - 1/2$. We use the label l to characterize the l-eigenvalue of Ψ_{up}^κ, and the label $\bar{l}$ to characterize the l-eigenvalue of Ψ_{down}^κ:

$$\begin{aligned} \kappa > 0, \quad l = \kappa, \quad \bar{l} = \kappa - 1 \\ \kappa < 0, \quad l = |\kappa| - 1, \quad \bar{l} = |\kappa| \end{aligned} \qquad (3)$$

As we shall see, these relations are important to understand the third point above mentioned. They are also at the origin of pseudospin symmetry invoked by Ginochio [13]. Note that V includes Coulomb potential, and S is defined with opposite sign to the usual convention for convenience of notation (thus, in nuclear matter, $V_0 \simeq 330 \ MeV$ and $S_0 \simeq 430 \ MeV$).

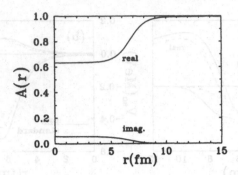

Figure 2   Real and imaginary parts of $A(r)$.

Let us then study point 1). The comparison of the spectroscopic factors $(S_\alpha(\text{RDWIA}) \simeq (7/5)S_\alpha(\text{DWIA}))$ indicates that RDWIA produces more absorption than DWIA. However, as seen in Figs. 1(a) and 1(b), neither real nor imaginary parts of standard and DEB potentials differ in a way that explains such a big effect. Why is then an increased absorption? The answer to this question is in Equ. 2 and in Fig. 2, which show that the relativistic component Ψ_{up} has an important depletion in the nuclear interior compared to $\Phi(r)$. This important effect of "increased absorption in the nuclear interior" is due to the Darwin or non-local potential $\left(V_{\text{non-loc}} = \frac{1}{2M} \frac{d\ln A_+}{dr} \frac{d}{dr}\right)$, which similarly to spin-orbit potential (V_{SO}), comes out automatically from Dirac equations with $S - V$ potentials. Such an effect can also be incorporated in phenomenological non-relativistic potentials. The advantage of using Dirac phenomenology is that both V_{SO} and $V_{\text{non-loc}}$ come from the same $S - V$ relativistic potentials. The increased suppression of the scattering wave function in the nuclear interior produced by $K(r)$ (non-locality factor) can also be interpreted as due to the reduced effective mass in the nuclear interior. *In this sense, we may consider that $(e, e'p)$ experiments are a probe of the effective mass* [14].

The answer to remaining questions 2, 3 (why in RDWIA we have larger high momentum components? and why increased asymmetry or R_{TL} response?), can be understood as due mainly to the role of the down (Ψ^κ_{down}) components of the bound nucleon wave function. More precisely, these effects are mainly due to the dynamical enhancement of down components of Ψ^κ_B. In non-relativistic approaches, the current operator $\hat{J}_\mu$ is expanded in a basis of free nucleon plane waves. This amounts to a truncation of the nucleon propagator that ignores negative energy solutions of the free Dirac equation, and thus negative energy projections of the bound nucleon wave functions. In RDWIA, this truncation is not made and non-zero overlaps of bound nucleon wave functions with Dirac sea (which are proportional to the dynamical enhancement of the down components) contribute to the cross section. This contributions enhance the high momentum components and the TL asymmetry of the reduced cross section [8, 9] in a way that makes theory compatible with experimental observations. This is illustrated in Fig. 3, where we show the TL response (R_{TL}) calculated in relativistic and non-relativistic approaches [9]. The RDWIA results seem to be confirmed by recent experimental data also shown in Fig. 3. Not only the enhancement of the asymmetry and $|R_{TL}|$, but also the fact that the latter reaches larger values for $p_{3/2}$ than for $p_{1/2}$ bound states seems to be confirmed by data and can be understood by inspection of the cross section in RPWIA [14].

3.   Electron Scattering on Halo Nuclei

The fact that electron scattering is one of the most powerful and cleanest tools for nuclear structure investigations lead us to consider these reactions to investigate the halo structure in a borromean nucleus [15]. To this purpose we used ^6He wave functions that give a good description of the known phenomenology. The three-body wave function of the nucleus is obtained by solving the Faddeev equations in coordinate space by means of the adiabatic hyperspherical expansion. As soon as the two-body interactions between the constituents of the nucleus are specified the three-body structure can be computed to the needed accuracy [16].

First we make a short incursion into **elastic electron scattering** where we see the variation of the charge form factor according to two limiting values for the charge r.m.s. radius, the core size and the ^6He size. The validity of the three-body picture requires that the charge form factor for ^6He and ^4He should be similar. For ^6He the elastic cross section in the nucleus frame takes the form

$$\frac{d\sigma}{d\Omega_e} = \frac{Z^2\sigma_M}{f_{rec}}\frac{Q^4}{q^4}|F_{ch}(q)|^2 \qquad (4)$$

where the ultrarelativistic limit for the electrons has been assumed, Ω_e defines the direction of the outgoing electron, and σ_M, f_{rec} are the Mott cross section and the recoil factor. Theoretical calculations of the charge form factor for ^6He are shown in Fig. 4. The solid line is computed assuming that protons in ^6He are distributed like in ^4He, while the dashed line is obtained assuming that the proton radius in ^6He is equal to the neutron radius. As seen from Fig. 4 and Equ. 4 elastic differential cross sections differ by orders of magnitude at moderate q values ($q \gtrsim 100\ MeV$) depending on the proton radius. Thus, experimental

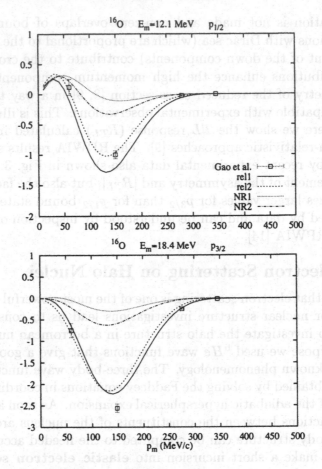

Figure 3 R_{TL} response for proton knockout from $p_{1/2}$ and $p_{3/2}$ orbitals in ^{16}O. Relativistic calculations with cc1 and cc2 current operators are labeled as rel1, rel2, respectively. NR1 and NR2 stand for nonrelativistic calculations. Experimental data are from [10].

measurements will provide an unambiguous test of the radial proton distributions and of the halo picture.

Next we study **coincidence** $(e, e'x)$ processes, with x either a halo neutron or the core. As shown in Fig. 5, a three-body system with energy and momentum (E_h, p_h) interacts with an electron with energy and momentum (E_0, p_0). The whole energy and momentum transfer (ω, q) is absorbed by one of the three constituents of the nucleus (constituent 3) that is ejected out from the three-body halo system. The energies and momenta in the final state are denoted by (E_i', p_i') where $i = 0$ for the electron and $i = 1, 2, 3$ for the three halo nucleus constituents.

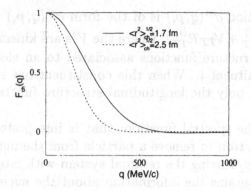

Figure 4 Charge form factor for elastic electron scattering from ^6He assuming that the charge density corresponds to the α charge density (solid line), and assuming that charge is spread out over the whole ^6He nucleus (dashed line).

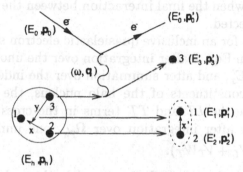

Figure 5 Scheme of the reaction and definition of the coordinates used to describe the exclusive $(e, e'x)$ process on a two-neutron halo nucleus. Only one of the constituents of the halo nucleus is considered to interact with the electron.

Working in the frame of the halo nucleus ($p_h = 0$) the differential cross section of the process shown in Fig. 5 is:

$$\frac{d^6\sigma^{(i)}}{dE_0'd\Omega_{p_0'}d\Omega_{p_i'}dE_x'} = (2\pi)^3 \frac{p_i'E_i'(m_j + m_k)}{M_h} f_{rec}\sigma^{ei}(q, p_i)S(E_x', p_i) \quad (5)$$

where the different momenta and energies are shown in the figure. p_x' is the relative momentum between particles j and k in the final state. m_j, m_k and M_h are the masses of the halo constituents j, k and the total mass of the halo nucleus, respectively. $E_x' = p_x'^2/2\mu_{jk}$ is the kinetic energy of the system made by particles j and k referred to its own center of mass and μ_{jk} its reduced mass. f_{rec} is the recoil factor, $\sigma^{ei}(q, p_i)$ is the cross section for elastic electron scattering on constituent i, and $S(E_x', p_i)$ is the spectral function.

The cross section $\sigma^{ei}(q, p_i)$ is of the form $\sigma^{ei}(q, p_i) = \sigma_M(V_L \mathcal{R}_L^{(i)} + V_T \mathcal{R}_T^{(i)} + V_{LT} \mathcal{R}_{LT}^{(i)} + V_{TT} \mathcal{R}_{TT}^{(i)})$, where the V's are kinematic factors and the $\mathcal{R}^{(i)}$'s are structure functions associated to an electron scattering process on constituent i. When this constituent has spin 0 (as the α particle in ^6He) only the longitudinal structure function appears (see [15] for details).

$S(E'_x, p_i)$ is the spectral function, that is interpreted as the probability for the electron to remove a particle from the nucleus with internal momentum p_i leaving the residual system with internal energy E'_x. This function contains the information about the nuclear structure, in particular it contains the whole dependence on the halo nucleus wave function and on the continuum wave function of the residual two-body system. It is defined as the square of the overlap between both wave functions, which reduces to the square of the Fourier transform of the halo wave function when the final interaction between the two surviving constituents is neglected.

The cross section for an inclusive quasielastic electron scattering process is obtained from Equ. 5 after integration over the unobserved quantities, i.e. $\Omega_{p'_i}$ and E'_x, and after summation over the index i, that runs over all the three constituents of the halo nucleus, the two neutrons and the core. Since the LT and TT terms in the cross section automatically disappear after integration over $\Omega_{p'_n}$, one immediately gets $\frac{d^3\sigma}{dE'_0 d\Omega_{p'_0}} = \sigma_M (V_L W_L + V_T W_T)$.

In Fig. 6 we show the two inclusive quasielastic peaks (top panels) and spectral functions (low panels) of the three-body wave function ($^6He = n + n + \alpha$). The large longitudinal peak W_L is due to the α-knockout by the electron, while the transverse peak W_T is due to the neutron knockout by the electron. At forward scattering angles, for low q and ω ($q \lesssim 200\ MeV/c$, $\omega \sim q^2/m_\alpha$) the behavior of the differential cross section and of the longitudinal structure function is dictated by the α-knockout process. In this domain the cross sections are large and dominated by the α-peak. It is therefore in this region where coincidence $(e, e'\alpha)$ measurements should be performed to determine the spectral function $S(E'_x, p_\alpha)$ that carries the information on the α-momentum distribution in the halo nucleus, or equivalently on the dineutron residual system. This spectral function is shown in the lower left panel of Fig. 6. Furthermore, in the $\omega \to 0$ limit elastic scattering from ^6He will provide the most stringent test of the charge distribution in ^6He. If the present halo picture is correct the charge distribution in ^6He must be dictated by that in the α-core. Backward scattering is more favorable to measure the halo-neutron peak. In particular, $\theta_e \gtrsim 160°$, $50 \lesssim E_0 \lesssim$

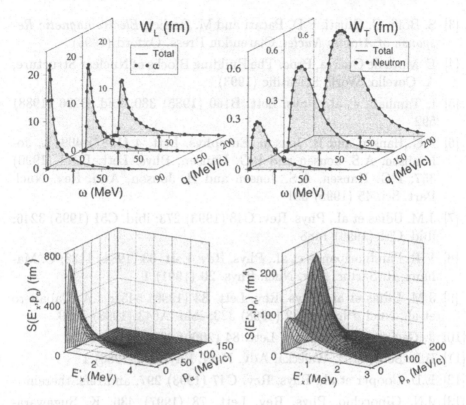

Figure 6 Upper part: W_L and W_T for momentum transfers $q=90$, 150 and 200 MeV/c as a function of ω. The arrows indicate the ω-values where core breakup contributions may show up. Lower part: Spectral functions for α-knockout (left) and halo neutron knockout (right).

100 MeV and $\omega < 40$ MeV define a region where the inclusive cross section is sizeable and dominated by the halo neutron peak. Thus coincidence $(e, e'n)$ measurements in this region will allow to determine the spectral function $S(E'_x, p_n)$ that carries the information on the halo neutron momentum distribution, or equivalently on the unbound ^5He residual system. This spectral function is shown in the lower right panel of Fig. 6.

References

[1] J.D. Walecka, *Theoretical Nuclear and Subnuclear Physics*, Oxford Univ. Pess. (1995)

[2] E. Moya de Guerra, Phys. Rep. 138 (6) (1986) 293

[3] S. Boffi, C. Giusti, F.D. Pacati and M. Radici, *Electromagnetic Response of Atomic Nuclei*, Clarendon Press, Oxford (1996)

[4] E. Moya de Guerra, Proc. The Building Blocks of Nuclear Structure, A. Covello, World Scientific (1993)

[5] I. Tanihata et al., Phys. Lett. B160 (1985) 380; ibid. B206 (1988) 592

[6] P.G.,Hansen and B. Jonson, Europhys. Lett. 4 (1987) 409; L. Johannsen, A.S. Jensen and P.G. Hansen, Phys. Lett. B244 (1990) 357; P.G. Hansen, A.S. Jensen and B. Jonson, Ann. Rev. Nucl. Part. Sci. 45 (1995) 591

[7] J.M. Udías et al., Phys. Rev. C48 (1993) 273; ibid. C51 (1995) 3246; ibid. C51 (1996) 1488

[8] V.R. Pandharipande et al., Phys. Rev. Lett. 53 (1984) 1133; C. Mahaux, R. Sartor, Adv. Nucl. Phys. 20 (1991) 1

[9] J.M. Udías et al., Phys. Rev. Lett. 83 (1999) 5451; J.A. Caballero et al., Nucl. Phys. A632 (1998) 323; ibid. A643 (1998) 1899.

[10] J. Gao et al., Phys. Rev. Lett. 84 (2000) 3265

[11] B.D. Serot, J.D. Walecka, Adv. Nucl. Phys. 16 (1986) 1

[12] E.D. Cooper et al., Phys. Rev. C47 (1993) 297; and refs. therein

[13] J.N. Ginocchio, Phys. Rev. Lett. 78 (1997) 436; K. Sugawara-Tanabe, A. Arima, Phys. Rev. C58 (1998) R3065.

[14] E. Moya de Guerra, in Proc. of VII Hispalensis Int. Summer School "Nuclear Physics 2000: Master's Lessons", J.M. Arias and M. Lozano, eds. (Springer-Verlag, in press)

[15] E. Garrido and E. Moya de Guerra, Nucl. Phys. A650 (1999) 387; Phys. Lett. B488 (2000) 68

[16] E. Garrido, D.V. Fedorov and A.S. Jensen, Nucl. Phys. A617 (1997) 153

New Results on the $2\nu\beta\beta$ Decay

A. A. Raduta

National Institute of Physics and Nuclear Engineering, Bucharest, MG6, Romania

raduta@theor1.theory.nipne.ro

Keywords: Ground state, double beta decay, canonical coordinates, equations of motion, phase space, phase transition, sum rule, isospin

Abstract Recent results for the ground state stability of proton neutron systems are reviewed. A schematic Hamiltonian is treated by a time dependent formalism. Harmonic motions static ground states are studied. Applications to the $2\nu\beta\beta$ decays are given. Also the isospin symmetry restoration is discussed.

1.   Introduction

In the last two decades the double beta decay process was extensively investigated from both experimental and theoretical sides. The process may take place through two independent modes, $0\nu\beta\beta$ and $2\nu\beta\beta$ [1, 2, 3]. The major interest comes from the $0\nu\beta\beta$ decay, since the results in this field may provide an answer to the question, whether neutrino is a Majorana or a Dirac particle. Theoretical investigations of this process need reliable matrix elements (m.e.) of the nuclear interaction. Due to the lack of experimental data, these m.e. cannot be tested. Fortunately, the same m.e. are involved in the study of the $2\nu\beta\beta$ process, for which many data are available. This is the reason why many theoreticians focused their attention on explaining the data concerning the life time of the $2\nu\beta\beta$ process. The formalism which produces results lying closest to the data, is the proton-neutron (*pn*) quasiparticle (qp) random phase approximation (pnQRPA) with inclusion of the particle-particle (*pp*) channel in the two body interaction[2]. The experimental value for the transition amplitude extracted from the $2\nu\beta\beta$ half-life, is obtained for a certain value of the *pp* interaction strength, g_{pp}. Indeed, the predicted amplitude, M_{GT}, plotted as function of g_{pp}, exhibits a plateau followed by a rapid decrease causing its vanishing at $g_{pp} \approx 1$. The value which agrees with the experimental result lies on this decreasing part.

D. N. Poenaru et al. (eds.), Nuclei Far from Stability and Astrophysics, 233–244.

Unfortunately this narrow interval is close to the critical value where the QRPA approach breaks down. Along the time many attempts have been made to construct a stable ground state (gs) for g_{pp} close to unity.

In the present talk I shall focus on some specific features of the formalisms devoted to the description of a many body pn system. First, I will show that considering the pp interaction in the RPA procedure is not consistent with the structure of the mean field of the single particle motion. Second, I address the question whether the violation of the gauge and isospin symmetry may influence the stability of the gs. In both cases we construct a stable ground state belonging to some new nuclear phases.

2.   New Approach for a pn System

The Gamow Teller (GT) double beta transition is assumed to take place via two consecutive single beta minus decays caused by a pn dipole interaction. The intermediate state describes an odd-odd nucleus. The main strength is absorbed by the GT giant resonance which lies around 15 MeV. By contrast, the Fermi (F) transition is determined by a pn monopole interaction and involves intermediate states centered around 3 MeV. The properties mentioned before are valid for both the GT and F transitions. In what follows I shall treat the problem of the ground state stability against the pn interaction in connection with the double beta Fermi transition. The reason is that for this case one can use an exactly solvable many body Hamiltonian. Indeed, for such a Hamiltonian all approximative methods could be tested by comparing their predictions with the exact results. Actually that is exactly what one wishes before applying a complex theory to realistic situations.

Therefore, we shall consider an heterogenous system of nucleons which move in a spherical shell model mean field and which interact among themselves in the following manner. Nucleons of similar charge interact through monopole pairing forces while protons and neutrons interact by a monopole particle-hole and a monopole particle-particle two body term. Such a system is described by a model Hamiltonian which, in a standard notation, reads [4, 5]:

$$H = \sum_{\tau,j,m} (\epsilon_\tau - \lambda_\tau) c^\dagger_{\tau jm} c_{\tau jm} \tag{1}$$

$$-\frac{G_p}{4} \sum_{j,m;j',m'} c^\dagger_{pjm} c^\dagger_{\widetilde{pjm}} c_{\widetilde{pj'm'}} c_{pj'm'} - \frac{G_n}{4} \sum_{j,m;j',m'} c^\dagger_{njm} c^\dagger_{\widetilde{njm}} c_{\widetilde{nj'm'}} c_{nj'm'}$$

$$+2\chi(c^\dagger_{pj} c_{nj})_0 (c^\dagger_{nj'} c_{pj'})_0 - 2\chi_1 (c^\dagger_{pj} c^\dagger_{nj})_0 (c_{nj'} c_{pj'})_0.$$

In next sub-sections, this Hamiltonian will be treated by two distinct approaches. In each case, a set of classical equations

$$\delta \int_0^t \langle \Psi | H - i\frac{\partial}{\partial t'} |\Psi \rangle dt' = 0. \tag{2}$$

is obtained for a specific trial function and model Hamiltonian.

2.1.   The pn qp Pairing Interaction

We treat first the pp and nn pairing interactions by the Bogoliubov-Valatin (BV) transformations defined by the coefficients U_p, V_p and U_n, V_n respectively. In the qp representation, the model Hamiltonian is a second degree polynomial in the operators:

$$A^\dagger = [a_p^\dagger a_n^\dagger]_{00}, A = (A^\dagger)^\dagger; \ B^\dagger = [a_p^\dagger a_n]_{00}, B = (B^\dagger)^\dagger, \tag{3}$$

where $a_p^\dagger$ and $a_n^\dagger$ denote the creation operators for p and n qp, respectively. The pnQRPA approach consists in solving the equations of motion for the operators $A^\dagger, A$ under the assumption that they satisfy bosonic commutation relation. Within the pnQRPA approach, the scattering quasiparticle operators $B^\dagger, B$ are ignored. Thus, in the qp representation, the effective Hamiltonian is [6]:

$$H = \epsilon(\hat{N}_p + \hat{N}_n) + \lambda_1 A^\dagger A + \lambda_2(A^\dagger A^\dagger + AA), \tag{4}$$

where $\hat{N}_p$ and $\hat{N}_n$ are the p and n qp number operators. For simplicity we consider $\epsilon_p = \epsilon_n$. The strengths λ_1 and λ_2 are:

$$\lambda_1 = 4\Omega[\chi(U_p^2V_n^2 + U_n^2V_p^2) - \chi_1(U_p^2U_n^2 + V_n^2V_p^2)],$$

$$\lambda_2 = 2\Omega(\chi + \chi_1)U_pV_pU_nV_n.$$

As variational states we use alternatively four wave-functions depending on $A^\dagger$ and A which satisfy certain commutation relation:

$$|\psi_1(z, z^*)\rangle = e^{[zA^\dagger - z^*A]}|0\rangle_q, \ [A, A^\dagger] = 1,$$

$$|\psi_2(z, z^*)\rangle = e^{[z\sum_m a_{pm}^\dagger a_{nm}^\dagger - z^* \sum_m a_{nm} a_{pm}]}|0\rangle_q,$$

$$[A, A^\dagger] = 1 - \frac{1}{2\Omega}(\hat{N}_p + \hat{N}_n),$$

$$|\psi_3(z, z^*)\rangle = e^{[zA^{\dagger 2} - z^*A^2]}|0\rangle_q, \ [A, A^\dagger] = 1,$$

$$|\psi_4(z, z^*)\rangle = \mathcal{N}e^{zA^{\dagger 2}}|0\rangle_q, \ [A, A^\dagger] = 1 - \frac{1}{2\Omega}(\hat{N}_p + \hat{N}_n). \tag{5}$$

The parameters z are complex functions of time, $z = \rho e^{i\phi}$, while $\mathcal{N}$ denotes a normalization factor. The pair (z, z^*) plays the role of canonical conjugate variables in the classical phase space. The formalism emerging from the use of $|\psi_k\rangle$ as variational state will be hereafter called VPk. The first three variational functions are coherent states with respect to Weyl's, $SU(2)$ and $SU(1,1)$ groups, respectively. For each of the four trial functions one finds a pair of variables (α_k, π_k), which bring the classical equations of motion to the canonical Hamilton form

$$-\dot{\pi}_k = \frac{\partial \mathcal{H}_k}{\partial \alpha_k}, \quad \dot{\alpha}_k = \frac{\partial \mathcal{H}_k}{\partial \pi_k}. \tag{6}$$

$\mathcal{H}_k$ denotes the expectation value of H with respect to the trial state $|\psi_k\rangle$. The expressions of (α_k, π_k) in terms of (z, z^*) are given in Tab. 1. Except for the case $k = 1$, the equations of motion (2.6) are highly non-linear. Analytical solutions are only possible if they are linearized around a minimum point of the energy surface. For $k = 2$, the minimum can be analytically expressed while for the remaining cases the minima were numerically calculated. The linearized equations of motion are satisfied by periodic orbits with the energies given in Tab. 1. To illustrate how energies depend on the strength of the pp interaction we plotted them in Figs. 1 and 2 for the case of $j = \frac{19}{2}$ and the scaled parameters $\chi' \equiv 2\Omega\chi = 0.5$ and $k' = 2\Omega\chi_1$ considered as free parameter. The number of nucleons are $N = 14$ and $Z = 6$. The energy for the mode constructed with the VP1 is identical with that obtained by a standard pnQRPA procedure. Such a mode describes the small oscillations around the static gs $|0\rangle_q$, where only nucleons with the same charge are paired with each other. The other mode, given by the VP2, is allowed for k' lying beyond the value where the above mentioned mode collapses. Therefore a phase transition takes place and one finds a Goldstone mode with energy zero between the two phases. In the static gs of this case, the p and n qp's are paired and therefore it is a second order BCS-like state. Energies for the second excited states are given in Fig. 2. Note that the energy of the mode determined by the VP3 vanishes already for a k' smaller than that where the energy of the VP1 mode vanishes. The $k = 4$ mode does not collapse at all. In contrast to VP3, the energy, after passing through a minimum value, increases again. The classical orbits are quantized and the results are used to evaluate the amplitudes for Fermi β^- and β^+ transitions, the Ikeda sum rule (ISR) [7] and the qp correlations in the ground state. The quantization is achieved by the algebra mapping: $(C, C^*; i\{,\}) \rightarrow (B, B^\dagger; [,])$, where (C, C^*) are canonical complex coordinates while $B^\dagger(B)$ denotes a boson creation (annihilation) operator. In Fig. 1 we plotted also the energy of the first excited state (fes) described by diagonalizing H(solid line) in the basis $\{(A^\dagger)^n |0\rangle_q\}$. Energies for the second excited states (ses) are plotted in Fig. 2. The pattern of the two harmonic modes, from Fig. 1, is reproduced by diagonalizing the Holstein-Primakoff (HP) second order boson expanded Hamiltonian $H_{B,2}^{(HP)}$. The difference between the energy of the VP2 harmonic mode and that of the *fes* suggests that for a realistic description of the *fes*, inclusion of the many boson states is necessary. It is important to note that the energy of this mode is increasing with k' and that such a behavior is caused by the part of this interaction included in the mean field of the qp motion. The agreement between the *ses* energies predicted by the VP4

Table 1   F and G depend on ρ, ρ_x'. $S = [2\overset{\circ}{x}(1 + 2\overset{\circ}{x})]^{\frac{1}{2}}$. For the four trial functions, energies achieve the minimum values in $(0,0), (\overset{\circ}{r}, \frac{\pi}{2}), (\overset{\circ}{x}, \pi), (\overset{\circ}{x}, \pi)$, respectively. The canonical coordinates are: $\xi = \sqrt{2}Re(z), \eta = \sqrt{2}Im(z); \; r = 2\Omega\sin^2\rho, \phi;$ $x = \frac{1}{2}\sinh^2(2\rho), \phi; \; x = -\rho\frac{\partial\ln N}{\partial\rho}, \phi.$

$|\psi\rangle$	Class. energy, $\mathcal{H}$	Mode energy
$|\psi_1\rangle$	$\frac{1}{2}(2\epsilon + \lambda_1)(\xi^2 + \eta^2)$ $+\lambda_2(\xi^2 - \eta^2)$	$\sqrt{(2\epsilon + \lambda_1)^2 - 4\lambda_2^2}$
$|\psi_2\rangle$	$2\epsilon r + \lambda_1[r(1 - \frac{r}{2\Omega}) + \frac{r^2}{4\Omega^2}]$ $+\lambda_2\frac{2\Omega-1}{\Omega}r(1 - \frac{r}{2\Omega})\cos(2\phi)$	$[\frac{4\lambda_2}{\lambda_1 - 2\lambda_2}(2\epsilon + \lambda_1 - 2\lambda_2 + \frac{\lambda_2}{\Omega})$ $\times(2\epsilon - \lambda_1 + 2\lambda_2 + \frac{\lambda_1 - \lambda_2}{\Omega})]^{\frac{1}{2}}$
$|\psi_3\rangle$	$\frac{\xi}{2}(\ln[1 + 4x \pm 2\sqrt{2x(1 + 2x)}])^2$ $+2\lambda_1 x + 2\lambda_2\sqrt{2x(1 + 2x)}\cos\phi$	$[\frac{4\lambda_2}{S^2}(2\epsilon +$ $\lambda_1(1 + 4\overset{\circ}{x}) - 4\lambda_2 S)]^{\frac{1}{2}}$
$|\psi_4\rangle$	$(2\epsilon + \lambda_1)2x + 2\lambda_2\frac{x}{\rho}\cos\phi$ $-\frac{\lambda_1}{2\Omega}(4x^2 - 2x + 2\rho\frac{dx}{d\rho})$	$[-2\lambda_2\frac{\overset{\circ}{x}}{\rho_0}(\frac{\lambda_1}{\Omega}G(\overset{\circ}{x})$ $+2\lambda_2 F(\overset{\circ}{x}))]^{\frac{1}{2}}$

and by diagonalizing H, is very good. In Ref. [4] a recipe for deriving an expression for the transition operators by using a quantization procedure, was described. The β^- and β^+ transition amplitudes predicted by the VP2 are very close to the exact results. Results for ISR are given in Fig. 3. For $k' \leq 1.5$, predictions deviate from the $N - Z$ value by less than one unit. A measure for the qp correlations in the static gs is the average of the half the total qp number operator. As shown in Fig. 4 the static gs obtained by VP2, contains a relatively large number of qp's. The exact result is quite close to that characterizing the new mode. Note that except for the $k = 1$ case, the remaining three static gs are characterized by non-vanishing average values for the total qp number operator. This shows, undoubtedly, that the situations presented here provides a more substantial renormalization for the pnQRPA, in comparison to the usual renormalized pnQRPA [8], where the static gs has vanishing expectation values for the total qp number operator.

To conclude, a new mode was defined in the region beyond the critical strength of the *pp* interaction where the standard RPA procedure breaks down.

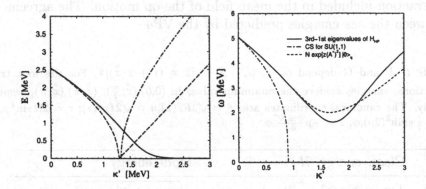

Figure 1 Energies obtained by the VP1 (dashed line, first branch) and VP2 (dash-dotted line) approaches, by diagonalizing $H_{B,2}^{(HP)}$ (dashed line, both branches) and H (solid line) are shown as function of k'.

Figure 2 Energies obtained with the VP3 (dash-dotted line) and VP4 (dashed-line) approaches and the exact energy of the second excited state of H (solid line) are shown as function of k'.

3.   The pn Pairing Interaction

Note that the *pp* interaction term is identical with a *pn* $T = 1$ pairing interaction. This suggests us to treat this interaction together with the *pp* and *nn* pairings through a generalized pairing formalism. To this aim, we consider the Hamiltonian (2.1) and a trial function which includes also the *pn* pairs [9].

$$|\Psi\rangle = \Psi(z_p, z_p^*; z_n, z_n^*; z_{pn}, z_{pn}^*) = e^{T_{pn}}e^{T_p}e^{T_n}|0\rangle. \tag{7}$$

where the transformations T are given by:

$$T_{pn} = \sum_{jm}(z_{pnj}c_{pjm}^\dagger c_{\widetilde{njm}}^\dagger - z_{pnj}^* c_{\widetilde{njm}}c_{pjm}),$$

$$T_\tau = \sum_{jm}(z_{\tau j}c_{\tau jm}^\dagger c_{\widetilde{\tau jm}}^\dagger - z_{\tau j}^* c_{\widetilde{\tau jm}}c_{\tau jm}), \tag{8}$$

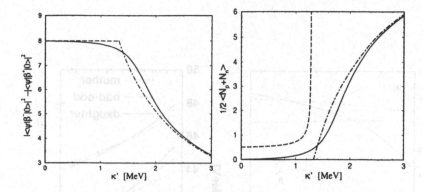

Figure 3 The ISR (Ikeda sum rule) obtained with the states provided by the VP1(dashed line), VP2 (dash-dotted line) and by the exact diagonalization of H (solid line).

Figure 4 The average value of $\frac{1}{2}(\hat{N}_p + \hat{N}_n)$ in the *fes* described by the VP1 (dashed line), the static gs of the VP2 (dash-dotted line) and the exact gs of H (solid line).

with $\tau = p, n$ and $|0\rangle$ standing for the particle vacuum state. The time depending parameters z are complex functions of time:

$$z_{pj} = \rho_{pj}e^{i\varphi_{pj}}, \; z_{nj} = \rho_{nj}e^{i\varphi_{nj}}, \; z_{pnj} = \rho_{pnj}e^{i\varphi_{pnj}}. \tag{9}$$

Denoting by Ω_j the shell degeneracy and changing the coordinates to:

$$r_{0j} = \Omega_j(2\sin^2\rho_{pnj} - 1)(1 - \sin^2 2\rho_{pj} - \sin^2 2\rho_{nj}), \; \varphi_{0j} = \varphi_{pnj},$$

$$r_{\mp j} = \frac{1}{2}\Omega_j(\sin^2 2\rho_{pj} \mp \sin^2 2\rho_{nj}), \; \varphi_{\mp j} = \varphi_{pj} \mp \varphi_{nj} - \begin{pmatrix} 0 \\ 2 \end{pmatrix}\varphi_{pnj}.$$

the corresponding equations of motion have the canonical form:

$$\frac{\partial\mathcal{H}}{\partial r_{kj}} = -\dot{\varphi}_{kj}, \; \frac{\partial\mathcal{H}}{\partial\varphi_{kj}} = \dot{r}_{kj}, \; k = 0 \pm . \tag{10}$$

Two of the chosen coordinates have a nice physical meaning. Indeed, $r_{0j}, 2r_{-j}$ are the classical variables associated to the z-components of the pn quasi-spin ($\hat{M}_{zj}$) and isospin ($\hat{T}_{zj}$), for each shell j, respectively. Using the explicit equations of motion (Equ. 9) one obtains:

$$\sum_j \dot{r}_{0j} = 0, \quad \sum_j \dot{r}_{-j} = 0. \tag{11}$$

which results in having two constants of motion:

$$\mathcal{M}_z = \sum_j \langle\Psi|\hat{M}_{zj}|\Psi\rangle, \quad \mathcal{T}_z = \sum_j \langle\Psi|\hat{T}_{zj}|\Psi\rangle. \tag{12}$$

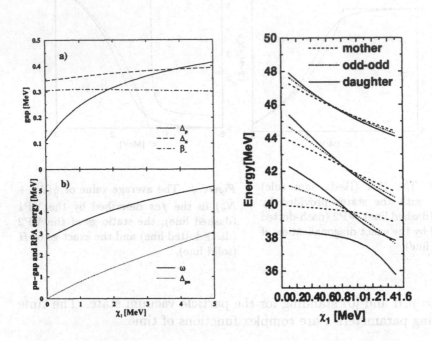

Figure 5 The gap parameters Δ_p, Δ_n, β_-, (a), Δ_{pn} (b) and the RPA energy (b) as functions of χ_1.

Figure 6 Exact energies for mother, daughter and intermediate odd-odd nuclei

Note that despite the fact that the trial function breaks the gauge and isospin symmetries, the classical trajectories conserve these symmetries. This is a reminiscence of the fact that the quantum mechanical operator H commutes with $\sum_j \hat{M}_{zj}$ and $\sum_j \hat{T}_{zj}$. As we shall see later on, the nice consequence of this property is that the spurious solutions of the RPA equations are fully separated from the physical ones. Static solutions are obtained by equating to zero the l.h.s. of Equ. 9. They are just the generalized BCS equations. Linearizing the equations of motion around the static solutions, one obtains the RPA equations. We solved them for the case of a single $j = \frac{19}{2}$ shell of energies $\epsilon_p = \epsilon_n = 3\ MeV$, occupied with 4 protons and 12 neutrons. The strengths of the two body interactions are: $G_p = 0.25\ MeV$, $G_n = 0.12\ MeV$, $\chi = 0.2\ MeV$. The strength for the pp interaction χ_1 is considered as a free parameter and varies in the interval from 0 to 5 MeV. Results are shown in Fig. 5, from where one sees that the RPA energy does not collapse at all. On the contrary it is an increasing function of the pp interaction strength.

For a single j shell, the model Hamiltonian is exactly solvable. The energies for mother, intermediate odd-odd and daughter states are given in Fig. 6. It is remarkable that for $\chi_1 \approx 0.9$ the gs of the odd-odd system is lower than that of the mother one. These energies are used in the calculations of the Fermi double beta transitions. The results are given in Fig. 7 where we also give the results of the pnQRPA treatment built on the top of the generalized BCS static ground state. The present approach predicts a transition amplitude which does not vanish at all and, moreover, agrees quite well with the exact result in the physical range of χ_1.

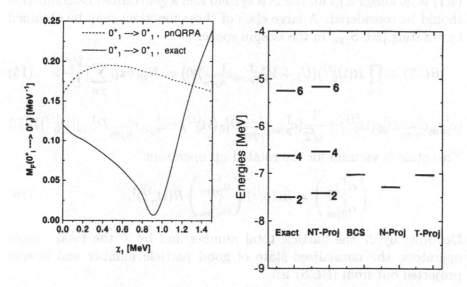

Figure 7 The Fermi $2\nu\beta\beta$ transition amplitude with exact description of the states involved (solid line) and the new QRPA approach (dashed line).

Figure 8 Energies for the exactly solvable Hamiltonian are given for the exact formula (4.6)(left column), *NT*-projection (second column), BCS, *N*-projection and *t*-projection. The total isospin of the states, 2, 4, 6, is also mentioned.

4.   Gauge and Isospin Symmetry

We consider a system of protons and neutrons moving in a mean field and interacting through a *pn* pairing force. Such a system has a gs described by a BCS type wave function [10]:

$$|BCS\rangle_{pn} = e^T|0\rangle \equiv \prod_{jm}(U_j + V_j^* c_{pjm}^\dagger c_{\widetilde{njm}}^\dagger)|0\rangle. \tag{13}$$

This is a vacuum state for the qp operators $\alpha_p^\dagger, \alpha_n^\dagger$ defined by:

$$e^T c_{\tau jm}^\dagger e^{-T} = U_j c_{\tau jm}^\dagger - V_j c_{\widetilde{\tau' jm}} \equiv a_{\tau jm}^\dagger, \tau = p, n; \ \tau' \neq \tau,$$

$$T = \sum_{jm} [z_{pn,j} c_{pjm}^\dagger c_{\widetilde{njm}}^\dagger - z_{pn,j}^* c_{\widetilde{njm}} c_{pjm}], z_{pn,j} = \rho_{pn} e^{i\phi_{pn}},$$

$$U_j = \cos(\rho_{pn}), V_j = e^{-i\phi_{pn}} \sin(\rho_{pn}). \tag{14}$$

If we add to the chosen Hamiltonian the pp and nn pairing, the state (4.1) is no longer a gs for the new system and a generalized BCS function should be considered. A large class of these functions may be obtained by rotating $|BCS\rangle_{pn}$ in the isospin space.

$$|BCS\rangle \equiv \prod_{jm} \hat{R}(\Omega_0^{(j)})(U_j + V_j^* c_{pjm}^\dagger c_{\widetilde{njm}}^\dagger)|0\rangle = N_{pn} \exp \sum_{jm} [\frac{V_j^*}{U_j} \times \tag{15}$$

$$(c_{pjm}^\dagger c_{\widetilde{njm}}^\dagger D_{00}^1(\Omega_0^{(j)}) + \frac{1}{\sqrt{2}}(c_{pjm}^\dagger c_{\widetilde{pjm}}^\dagger D_{10}^1(\Omega_0^{(j)}) + c_{njm}^\dagger c_{\widetilde{njm}}^\dagger D_{-10}^1(\Omega_0^{(j)})))]|0\rangle$$

This state is vacuum for the rotated qp operators:

$$\begin{pmatrix} \alpha_{1jm}^\dagger \\ \alpha_{2jm}^\dagger \end{pmatrix} = \hat{R}(\Omega_0^{(j)}) \begin{pmatrix} a_{pjm}^\dagger \\ a_{njm}^\dagger \end{pmatrix} \hat{R}(\Omega_0^{(j)})^\dagger. \tag{16}$$

Denoting by $\hat{N}$ the particle total number and by $\hat{T}$ the total isospin operators, the normalized state of good particle number and isospin projected out from $|BCS\rangle$ is:

$$|NTMK\rangle = \frac{2T+1}{16\pi^3} N_{NT} \int D_{MK}^{T}{}^* e^{i(\hat{N}-N)\phi} \hat{R}(\Omega) \hat{R}(\Omega_0)|BCS\rangle_{pn} d\Omega d\phi.$$

Due to the particular structure of $|BCS\rangle$, the norms and the m.e. between projected states can be analytically performed. Consider now the model Hamiltonian (2.1) for a single j-shell and a particular set of parameters:

$$\epsilon_\tau - \lambda_\tau = \epsilon, G_p = G_n = \frac{4\chi_1}{\Omega}, \chi = 0. \tag{17}$$

The new Hamiltonian is a scalar with respect to rotations in the space of isospin. Moreover it is exactly solvable and the eigenvalue is:

$$E = \epsilon N - \frac{G}{2} \left(\frac{N}{4}(4\Omega_j - N + 6) - T(T+1) \right). \tag{18}$$

Neglecting the m.e. between states of different K we approximate the eigenvalues of H, in the space of projected states by:

$$E_{NT} = \langle NTMK|H|NTMK\rangle \tag{19}$$

The results for energies are given in Fig. 8, for $j = \frac{23}{2}$, $\epsilon = 0$, $G = 0.125 \, MeV$, $Z = 4$, $N = 8$. One remarks that projection has an important effect on energies. Now let us investigate the influence of projection operation on the β^- and β^+ Fermi transitions. The transition operators are:

$$T_+ = \sum_m c^\dagger_{pjm} c_{njm}, \quad T_- = \sum_m c^\dagger_{njm} c_{pjm}. \tag{20}$$

For the unprojected state, $|0\rangle \equiv |BCS\rangle$ such a transition may take place to one of the three qp states:

$$|kk\rangle = \frac{1}{2\sqrt{\Omega}} \sum_m \alpha^\dagger_{kjm} \alpha^\dagger_{\widetilde{kjm}} |BCS\rangle, k = 1,2;$$

$$|12\rangle = \frac{1}{\sqrt{\Omega}} \sum_m \alpha^\dagger_{1jm} \alpha^\dagger_{\widetilde{2jm}} |BCS\rangle$$

In particular, for $\Omega_0 = (0, \frac{\pi}{3}, 0)$, the transition strengths are:

$$|\langle 0|T_\mp|11\rangle|^2 = \begin{pmatrix} 5.06 \\ 0.56 \end{pmatrix}, |\langle 0|T_\mp|22\rangle|^2 = \begin{pmatrix} 0.56 \\ 5.06 \end{pmatrix}, |\langle 0|T_\mp|12\rangle|^2 = 0.$$

The total strengths for β^- and β^+ transitions are both equal to 5.62. Concerning the projected states, the $\beta^\mp$ strength are:

$$\beta^{(-)} = 4\delta_{T,2} + 18\delta_{T,4} + 40\delta_{T,6}, \quad \beta^{(+)} = 14\delta_{T,4} + 36\delta_{T,6} \tag{21}$$

These results indicate that the ISR is exactly satisfied. The beta plus transition is forbidden for the $T = 2$ state while for the unprojected gs, the transition strength is 5.62. The beta minus strength for the unprojected gs is modified by projection from 5.62 to 4. In conclusion, the projection of particle number and total isospin is very important for a quantitative description of both energies and $\beta^\mp$ transitions.

5. Conclusions

The present lecture has the message that the gs instability suggested by the previous publications is due to an inconsistent treatment of the pp interaction. Indeed, except of our studies (2000), the pp interaction was treated at the QRPA level but not considered at all in the mean field. Here two methods to cure this deficiency are discussed, common to all formalisms used along the last 15 years. In the first method the p and n quasiparticles are paired with each other and a new ground state is defined by a second order BCS formalism. The second method treats the pp interaction on equal footing with the pp and nn pairing terms within a generalized BCS formalism. In both methods the resulting RPA energies are not decreasing any longer with the pp interaction strength. On the contrary they are increasing functions. The new formalism shows a quite good agreement with the exact results. Since the static gs is violating both the gauge and isospin rotation symmetries it is worth to investigate the effect of the NT-symmetry restoration on energies and single beta transitions. Within the space of states with definite total number of particles and definite total isospin, the results for energies and $\beta^{\mp}$ transitions differ substantially from those obtained with unprojected states.

Recently [11] it has been shown the renormalization of the pnQRPA formalism requires, for the sake of consistency, that the phonon operator is extended by adding the qp scattering terms. It seems [12] that the new collective state associated to the scattering terms is describing a wobbling motion for the isospin degrees of freedom.

The present results provide encouraging premises for extending the formalism to a large model space of single particle states and a realistic two body interaction. We hope that the ingredients pointed out in this lecture are sufficient for a realistic description of the $2\nu\beta\beta$ process.

Acknowledgments

The author wants to thank Drs. V. Baran, O. Haug, F. Simkovic, P. Sarriguren, Profs. E. Moya de Guerra, A. Faessler and Ms. L. Pacearescu for stimulating discussions and collaboration on this subject.

References

[1] W.C. Haxton and G.J. Stephenson, Jr., Prog. Part. Nucl. Phys. 12 (1984) 409

[2] A. Faessler and F. Šimkovic, J. Phys. G24 (1998) 2139

[3] J. Suhonen and O. Civitarese Phys. Rep. 300 (1998) 123

[4] A. A. Raduta et al., Nucl Phys. A671 (2000) 255

[5] A. A. Raduta et al., J. Phys. G26 (2000) L1

[6] M. Sambataro and J. Suhonen, Phys. Rev. C56 (1997) 782

[7] K. Ikeda, Prog. Theor. Phys. 31 (1964) 434

[8] J. Toivanen and J. Suhonen Phys. Rev. Lett. 75 (1995) 410

[9] A. A. Raduta et al., Nucl. Phys. A675 (2000) 503

[10] A. A. Raduta and E. Moya de Guerra, Ann. Phys. 284 (2000) 1.

[11] A. A. Raduta et al., Nucl. Phys. A634 (1998) 487

[12] A. A. Raduta, C. M. Raduta and B. Codirla, Nucl. Phys. in press.

II

Nuclear Astrophysics

The Past and Future of Coulomb Dissociation in Hadron- and Astrophysics

G. Baur[a], K. Hencken[b], D. Trautmann[b], S. Typel[c] and H. H. Wolter[c]

[a] Forschungszentrum Jülich, Institut für Kernphysik
D-52425 Jülich, Germany
g.baur@fz-juelich.de

[b] Institut für Theoretische Physik, Universität Basel
CH-4056 Basel, Switzerland

[c] Sektion Physik, Universität München
D-85748 Garching, Germany

Keywords: Electromagnetic (Coulomb) dissociation, nuclear structure, nuclear astrophysics, radiative capture

Abstract Breakup reactions are generally quite complicated, they involve nuclear and electromagnetic forces including interference effects. Coulomb dissociation is an especially simple and important mechanism since the perturbation due to the electric field of the nucleus is exactly known. Therefore firm conclusions can be drawn from such measurements. Electromagnetic matrixelements, radiative capture cross-sections and astrophysical S-actors can be extracted from experiments. We describe the basic theory, give analytical results for higher order effects in the dissociation of neutron halo nuclei and briefly review the experimental results obtained up to now. Some new applications of Coulomb dissociation for nuclear astrophysics and nuclear structure physics are discussed.

1. Introduction

One may regard the work of Oppenheimer and Phillips in 1935 [1, 2] as a starting point of the present subject. They tried to explain the preponderance of (d, p)-reactions over (d, n)-reactions by a virtual breakup of the deuteron in the Coulomb field of the nucleus before the actual nuclear interaction takes place. Because of the Coulomb repulsion of the proton this would explain the dominance of (d, p)-reactions. In this context, Oppenheimer [1] also treated the real breakup of the deuteron

247

in the Coulomb field of a nucleus. In the meantime, the subject has developed considerably. In addition to the deuteron, many different kinds of projectiles (ranging from light to heavy ions, including radioactive nuclei) have been used at incident energies ranging from below the Coulomb barrier to medium and up to relativistic energies.

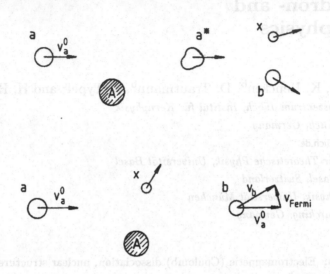

Figure 1 Two basic reaction mechanisms for breakup are shown schematically. In the upper figure (sequential breakup), the projectile a is excited to a continuum (resonant) state a^* which decays subsequently into the fragments b and x. In the lower part (spectator breakup) substructure x interacts (in all kinds of ways) with the target nucleus A, whereas $b = (a - x)$ misses the target nucleus ("spectator"). It keeps approximately the velocity which it had before the collision. [Fig. 1 of Ref. [3].]

In Fig. 1 we show two different kinds of reaction mechanisms. Since rigorous methods of reaction theory (like the Faddeev approach) cannot be applied in practice to such complicated nuclear reactions, we have to use different theoretical methods to treat the different cases as well as possible. In the spectator breakup mechanism the breakup occurs due to the interaction of one of the constituents with the target nucleus, while the spectator moves on essentially undisturbed. Another mechanism is the sequential breakup, where the projectile is excited to a continuum state which decays subsequently. Both mechanisms have been dealt with extensively in the past [3], see also Ref. [4], which provides a brief outline of the development over the last few decades. For low and medium energies (i.e., for energies not high enough for the Glauber theory to be applicable) it should be noted that the post-form DWBA is especially suited to treat the spectator process. For the sequential breakup mechanism the decomposition of the Hamiltonian in the initial and final channels is the same.

We start with a general discussion of Coulomb dissociation. Due to the time-dependent electromagnetic field the projectile is excited to a bound or continuum state, which can subsequently decay. We briefly mention the very large effects of electromagnetic excitation in relativistic heavy ion collisions. After a short review of results obtained for nuclear structure as well as nuclear astrophysics, we discuss new possibilities, like the experimental study of two-particle capture. We close with conclusions and an outlook.

2.   Electromagnetic and Nuclear Dissociation

Coulomb excitation is a very useful tool to determine nuclear electromagnetic matrixelements. The nuclei are assumed to interact with each other only electromagnetically. This can be achieved by either using bombarding energies below the Coulomb barrier or by choosing very forward scattering angles and high energy collisions. With increasing beam energy states at higher energies can be excited; this can lead, in addition to Coulomb excitation, also to Coulomb dissociation, for a review see, e.g., Ref. [5]. Such investigations are also well suited for secondary (radioactive) beams. The electromagnetic interaction, which causes the dissociation, is well known and therefore there can be a clean interpretation of the experimental data. This is of interest for nuclear structure and nuclear astrophysics [6, 7]. Multiple electromagnetic excitation can also be important. We especially mention two aspects: It is a way to excite new nuclear states, like the double phonon giant dipole resonance [7]; but it can also be a correction to the one-photon excitation [8, 9, 10].

In the equivalent photon approximation the cross section for an electromagnetic process is written as

$$\sigma = \int \frac{d\omega}{\omega} \, n(\omega)\sigma_\gamma(\omega) \tag{1}$$

where $\sigma_\gamma(\omega)$ denotes the appropriate cross section for the photo-induced process and $n(\omega)$ is the equivalent photon number. For sufficiently high beam energies it is well approximated by

$$n(\omega) = \frac{2}{\pi} Z^2 \alpha \ln \frac{\gamma v}{\omega R} \tag{2}$$

where R denotes some cut-off radius. More refined expressions, which take into account the dependence on multipolarity, beam velocity or Coulomb-deflection, are available in the literature [5, 9, 11]. The theory of electromagnetic excitation is well developed for nonrelativistic, as well as relativistic projectile velocities. In the latter case an analytical result for all multipolarities was obtained in Ref. [11]. The projectile motion was treated classically in a straight-line approximation. On the other hand, in the Glauber theory, the projectile motion can be treated quantally [5, 10]. This gives rise to characteristic diffraction effects. The main effect is due to the strong absorption at impact parameters less than the sum of the two nuclear radii.

If the above conditions are not met, nuclear excitation (or diffractive dissociation) also has to be taken into account. This is a broad subject and has been studied in great detail using Glauber theory, see, e.g., [5] for further references. Especially for light nuclei, Coulomb excitation tends to be less important in general than nuclear excitation. For heavy nuclei the situation reverses. The nuclear breakup of halo nuclei was more recently studied, e.g., in [12]. The nuclear interaction of course is less precisely known than the Coulomb interaction. In Ref. [12] the nuclear breakup was studied using the eikonal approximation as well as the Glauber multiple particle scattering theory. No Coulomb interaction was included in this approach, as the main focus was on the breakup on light targets. In Ref. [13] on the other hand, the combined effect of both nuclear and Coulomb excitation is studied. The nuclear contribution to the excitation is generally found to be small and has an angular dependence different from the electromagnetic one. This can be used to separate such effects from the electromagnetic excitation. We also mention the recent systematic study of ^8B breakup cross section in [14].

3.   Electromagnetic Excitation in Relativistic Heavy Ion Collisions

Electromagnetic excitation is also used at relativistic heavy ion accelerators to obtain nuclear structure information. Recent examples are the nuclear fission studies of radioactive nuclei [15] and photofission of ^{208}Pb [16]. Cross-sections for the excitation of the giant dipole resonance ("Weizsäcker-Williams process") at the forthcoming relativistic heavy ion colliders RHIC and LHC($Pb-Pb$) at CERN are huge [17, 18], of the order of 100 b for heavy systems ($Au-Au$ or $Pb-Pb$). In colliders, the effect is considered to be mainly a nuisance, the excited particles are lost from the beam. On the other hand, the effect will also be useful as a luminosity monitor by detecting the neutrons in the forward direction. Specifically one will measure the neutrons which will be produced after the decay of the giant dipole resonance which is excited in each of the ions (simultaneous excitation). Since this process has a steeper impact parameter dependence than the single excitation cross-section, there is more sensitivity to the cut-off radius and to nuclear effects. For details and further Refs., see [18].

4. Higher Order Effects and Postacceleration

Higher order effects can be taken into account in a coupled channels approach, or by using higher order perturbation theory. The latter involves a sum over all intermediate states n considered to be important. Another approach is to integrate the time-dependent Schrödinger equation directly for a given model Hamiltonian [19–22]. If the collision is sudden, one can neglect the time ordering in the usual perturbation approach. The interaction can be summed to infinite order. Intermediate states n do not appear explicitly.

Higher order effects were recently studied in [23], where further references also to related work can be found. Since full Coulomb wave functions in the initial and final channels are used there, the effects of higher order in $\eta_{\text{coul}} = \frac{ZZ_c e^2}{\hbar v}$ are taken into account to all orders. Expanding the T-matrixelement in this parameter η_{coul} one obtains the Born approximation for the dissociation of $a \to c + n$

$$T \propto \frac{\eta_{\text{coul}}}{(\vec{q}_n + \vec{q}_c - \vec{q}_a)^2} \left(\frac{1}{q_a^2 - (\vec{q}_n + \vec{q}_c)^2} + \frac{1}{q_c^2 - (\vec{q}_n - \vec{q}_a)^2} \right). \quad (3)$$

This expression is somehow related to the Bethe-Heitler formula for bremsstrahlung. The Bethe-Heitler formula has two terms which correspond to a Coulomb interaction of the electron and the target followed by the photon emission and another one, where the photon is emitted first and then the electron scatters from the nucleus. In the case of Coulomb dissociation we have a Coulomb scattering of the incoming particle followed by breakup $a = (c + n) \to c + n$ and another term, where the projectile a breaks up into $c + n$, and subsequently, c is scattered on the target. In the case of bremsstrahlung it is well known [24] that even for $\eta_{coul} \gg 1$ one obtains the Born approximation result as long as the scattering is into a narrow cone in the forward direction. This leads one to suspect that higher order effects are not very large in the case of high energy Coulomb dissociation, when the fragments are emitted into the forward direction.

We investigate higher order effects in the model of [8, 9, 10]. In a zero range model for the neutron-core interaction, analytical results were obtained for 1st and 2nd order electromagnetic excitation for small values of the adiabaticity parameter ξ. We are especially interested in collisions with small impact parameters. For these higher order effects tend to be larger than for the very distant ones. In this case, the adiabaticity parameter ξ is small. For $\xi = 0$ (sudden approximation) we have a closed form solution, where higher order effects are taken into account to all orders. In Equ. 37 of [8] the angle integrated breakup probability is given. We expand this expression in the strength parameter $\eta_{eff} = \frac{2ZZ_ce^2m_n}{\hbar v(m_n+m_c)}$. We define $x = \frac{q}{\eta}$ where the parameter η is related to the binding energy E_0 by $E_0 = \frac{\hbar^2\eta^2}{2m}$ and the wave number q is related to the energy E_{rel} of the continuum final state by $E_{rel} = \frac{\hbar^2 q^2}{2\mu}$. In leading order (LO) we obtain

$$\frac{dP_{LO}}{dq} = C \frac{x^4}{(1+x^2)^4} \tag{4}$$

where $C = \frac{128\pi^2\eta_{eff}^2}{3\eta^3 b^2}$. The next to leading order (NLO) expression is proportional to η_{eff}^4 and contains a piece from the 2nd order $E1$ amplitude and a piece from the interference of 1st and 3rd order. We find

$$\frac{dP_{NLO}}{dq} = C \left(\frac{\eta_{eff}}{b\eta}\right)^2 \frac{x^2(5 - 55x^2 + 28x^4)}{15(1+x^2)^6}. \tag{5}$$

The integration over x and the impact parameter b can also be performed analytically in good approximation. For details see [25]. We can easily insert the corresponding values for the Coulomb dissociation experiments on ^{11}Be and ^{19}C [26, 27] in the present formulae. We find that the ratio of the NLO contribution to the LO contribution in the case of Coulomb dissociation on ^{19}C [27] is given by -2 %. This is to be compared to the results of [23] where a value of about -35 % was found.

5.   Discussion of some Experimental Results for Nuclear Structure and Astrophysics

Coulomb dissociation of exotic nuclei is a valuable tool to determine electromagnetic matrix-elements between the ground state and the nuclear continuum. The excitation energy spectrum of the $^{10}Be + n$ system in the Coulomb dissociation of the one-neutron halo nucleus ^{11}Be on a Pb target at 72 $AMeV$ was measured [26]. Low lying $E1$-strength was found. The Coulomb dissociation of the extremely neutron-rich nucleus ^{19}C was recently studied in a similar way [27]. The neutron separation energy of ^{19}C could also be determined to be 530 ± 130 keV. Quite similarly, the Coulomb dissociation of the $2n$-halo nucleus ^{11}Li was studied in various laboratories [28, 29, 30]. In an experiment at MSU [31], the correlations of the outgoing neutrons were studied. Within the limits of experimental accuracy, no correlations were found.

In nuclear astrophysics, radiative capture reactions of the type $b+c \rightarrow a + \gamma$ play a very important role. They can also be studied in the time-reversed reaction $\gamma + a \to b + c$, at least in those cases where the nucleus a is in the ground state. As a photon beam, we use the equivalent photon spectrum which is provided in the fast peripheral collision. Reviews, both from an experimental as well as theoretical point of view have been given [6], so we want to concentrate here on a few points.

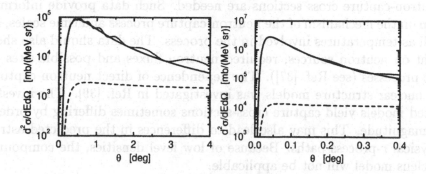

Figure 2 Coulomb dissociation cross section of ^8B scattered on ^{208}Pb as a function of the scattering angle for projectile energies of 46.5 $AMeV$ (left) and 250 $AMeV$ (right) and a $^7Be - p$ relative energy of 0.3 MeV. First order results $E1$ (solid line), $E2$ (dashed line) and $E1 + E2$ excitation including nuclear diffraction (dotted line). [From Figs. 4 and 5 of Ref. [10].]

The ^6Li Coulomb dissociation into $\alpha + d$ has been a test case of the method, see Ref. [6]. This is of importance since the $d(\alpha, \gamma)^6Li$ radiative capture is the only process by which ^6Li is produced in standard primordial nucleosynthesis models. There has been new interest in ^6Li as a cosmological probe in recent years, mainly because the sensitivity for searches for ^6Li has been increasing. It has been found in metal-poor halo stars at a level exceeding even optimistic estimates of how much ^6Li could have been made in standard big bang nucleosynthesis. For more discussion on this see [32].

The $^7Be(p, \gamma)^8B$ radiative capture reaction is relevant for the solar neutrino problem. It determines the production of ^8B which leads to the emission of high energy neutrinos. There are direct reaction measurements, for a recent one see Ref. [33]. Coulomb dissociation of ^8B has been studied at RIKEN [34], MSU [35] and GSI [36]. Theoretical calculations are shown in Fig. 2. It is seen that $E1$ excitation is large and peaked at very forward angles. $E2$ excitation is also present, with a characteristically different angular distribution. Nuclear diffraction effects are small. Altogether it is quite remarkable that completely different experimental methods with possibly different systematic errors lead to results that are quite consistent.

6.   Possible New Applications of Coulomb Dissociation for Nuclear Astrophysics

Nucleosynthesis beyond the iron peak proceeds mainly by the r- and s-processes (rapid and slow neutron capture) [37, 38]. To establish the quantitative details of these processes, accurate energy-averaged neutron-capture cross sections are needed. Such data provide information on the mechanism of the neutron-capture process and time scales, as well as temperatures involved in the process. The data should also shed light on neutron sources, required neutron fluxes and possible sites of the processes (see Ref. [37]). The dependence of direct neutron capture on nuclear structure models was investigated in Ref. [39]. The investigated models yield capture cross-sections sometimes differing by orders of magnitude. This may also lead to differences in the predicted astrophysical r-process paths. Because of low level densities, the compound nucleus model will not be applicable.

With the new radioactive beam facilities (either fragment separator or ISOL-type facilities) some of the nuclei far off the valley of stability, which are relevant for the r-process, can be produced. In order to assess the r-process path, it is important to know the nuclear properties like β-decay half-lifes and neutron binding energies. Sometimes, the waiting point approximation [37, 38] is introduced, which assumes an (n, γ)- and (γ, n)-equilibrium in an isotopic chain. It is generally believed that the waiting point approximation should be replaced by dynamic r-process flow calculations, taking into account (n, γ), (γ, n) and β-decay rates as well as time-varying temperature and neutron density. In slow freeze-out scenarios, the knowledge of (n, γ) cross sections is important.

In such a situation, the Coulomb dissociation can be a very useful tool to obtain information on (n, γ)-reaction cross sections on unstable nuclei, where direct measurements cannot be done. Of course, one cannot and need not study the capture cross section on all the nuclei involved; there will be some key reactions of nuclei close to magic numbers. It was proposed [40] to use the Coulomb dissociation method to obtain information about (n, γ) reaction cross sections, using nuclei like ^{124}Mo, ^{126}Ru, ^{128}Pd and ^{130}Cd as projectiles. The optimum choice of beam energy will depend on the actual neutron binding energy. Since the flux of equivalent photons has essentially an $\frac{1}{\omega}$ dependence, low neutron thresholds are favorable for the Coulomb dissociation method. Note that only information about the (n, γ) capture reaction to the ground state is possible with the Coulomb dissociation method. The situation is reminiscent of the loosely bound neutron-rich light nuclei, like ^{11}Be, ^{11}Li and ^{19}C.

In Ref. [8] the 1st and 2nd order Coulomb excitation amplitudes are given analytically in a zero range model for the neutron-core interaction (see section 4). We propose to use the handy formalism of Ref. [8] to assess, how far one can go down in beam energy and still obtain meaningful results with the Coulomb dissociation method, i.e., where the 1st order amplitude can still be extracted experimentally without being too much disturbed by corrections due to higher orders. For future radioactive beam facilities, like ISOL or SPIRAL, the maximum beam energy is an important issue. For Coulomb dissociation with two charged particles in the final state, like in the $^8B \to {}^7Be + p$ experiment with a 26 MeV ^8B beam [41] such simple formulae seem to be unavailable and one should resort to the more involved approaches mentioned in section 4.

A new field of application of the Coulomb dissociation method can be two nucleon capture reactions. Evidently, they cannot be studied in a direct way in the laboratory. Sometimes this is not necessary, when the relevant information about resonances involved can be obtained by other means (transfer reactions, etc.), like in the triple α-process.

Two-neutron capture reactions in supernovae neutrino bubbles are studied in Ref. [42]. In the case of a high neutron abundance, a sequence of two-neutron capture reactions, $^4He(2n, \gamma)^6He(2n, \gamma)^8He$ can bridge the $A = 5$ and 8 gaps. The ^6He and ^8He nuclei may be formed preferentially by two-step resonant processes through their broad 2^+ first excited states [42]. Dedicated Coulomb dissociation experiments can be useful, see [43]. Another key reaction can be the $^4He(\alpha n, \gamma)$ reaction [42]. The $^9Be(\gamma, n)$ reaction has been studied directly (see Ref. [44]) and the low energy $s_{1/2}$ resonance is clearly established.

In the rp-process, two-proton capture reactions can bridge the waiting points [45, 46, 47]. From the $^{15}O(2p, \gamma)^{17}Ne$, $^{18}Ne(2p, \gamma)^{20}Mg$ and $^{38}Ca(2p, \gamma)^{40}Ti$ reactions considered in Ref. [46], the latter can act as an efficient reaction link at conditions typical for X-ray bursts on neutron stars. A $^{40}Ti \to p + p + {}^{38}Ca$ Coulomb dissociation experiment should be feasible. The decay with two protons is expected to be sequential rather than correlated ("^2He"-emission). The relevant resonances are listed in Tab. XII of Ref. [46]. In Ref. [47] it is found that in X-ray bursts $2p$-capture reactions accelerate the reaction flow into the $Z \geq 36$ region considerably. In Tab. 1 of Ref. [47] nuclei, on which $2p$-capture reactions may occur, are listed; the final nuclei are ^{68}Se, ^{72}Kr, ^{76}Sr, ^{80}Zr, ^{84}Mo, ^{88}Ru, ^{92}Pd and ^{96}Cd (see also Fig. 8 of Ref. [45]). It is proposed to study the Coulomb dissociation of these nuclei in order to obtain more direct insight into the $2p$-capture process.

7.   Conclusions

Peripheral collisions of medium and high energy nuclei (stable or radioactive) passing each other at distances beyond nuclear contact and thus dominated by electromagnetic interactions are important tools of nuclear physics research. The intense source of quasi-real (or equivalent) photons has opened a wide horizon of related problems and new experimental possibilities especially for the present and forthcoming radioactive beam facilities to investigate efficiently photo-interactions with nuclei (single- and multiphoton excitations and electromagnetic dissociation).

Acknowledgments

We have enjoyed collaboration and discussions on the present topics with very many people. We are especially grateful to C. A. Bertulani, H. Rebel, F. Rösel, and R. Shyam.

References

[1] J.R. Oppenheimer, Phys. Rev. 47 (1935) 845

[2] J.R. Oppenheimer and M. Phillips, Phys. Rev. 48 (1935) 500

[3] G. Baur et al., Phys. Rep. 111 (1984) 333

[4] G. Baur et al., Mechanisms for direct breakup reactions, nucl-th/0001045, and to be published in the proceedings of the TMU-RCNP Symposium on Spins in Nuclear and Hadronic reactions, Tokyo, October 26-28 (1999)

[5] C.A. Bertulani and G. Baur, Phys. Rep. 163 (1988) 299

[6] G. Baur, C.A. Bertulani and H. Rebel, Nucl. Phys. A458 (1986) 188; G. Baur and H. Rebel, J. Phys. G: Nucl. Part. Phys. 20 (1994) 1; Ann. Rev. Nucl. Part. Sci. 46 (1996) 321

[7] G. Baur and C.A. Bertulani, Phys. Lett. B174 (1986) 23

[8] S. Typel and G. Baur, Nucl. Phys. A573 (1994) 486

[9] S. Typel and G. Baur, Phys. Rev. C50 (1994) 2104

[10] S. Typel, H.H. Wolter and G. Baur, Nucl. Phys. A613 (1997) 147

[11] A. Winther and K. Alder, Nucl. Phys. A319 (1979) 518

[12] K. Hencken, G. Bertsch and H. Esbensen, Phys. Rev. C54 (1996) 3043

[13] A. Muendel and G. Baur, Nucl. Phys. A609 (1996) 254

[14] H. Esbensen, K. Hencken, Phys. Rev. C61 (2000) 054606

[15] K.-H. Schmidt et al., Nucl. Phys. A665 (2000) 221

[16] M.C. Abreu et al., Phys. Rev. C59 (1999) 876

[17] G. Baur and C.A. Bertulani, Nucl. Phys. A505 (1989) 835

[18] G. Baur, K. Hencken and D. Trautmann, J. Phys. G: Nucl. Part. Phys. 24 (1998) 1657

[19] V.S. Melezhik and D. Baye, Phys. Rev. C59 (1999) 3232

[20] H. Esbensen, G.F. Bertsch and C.A. Bertulani, Nucl. Phys. A581 (1995) 107

[21] H. Utsunomiya et al., Nucl. Phys. A654 (1999) 928c

[22] S. Typel, H.H. Wolter, Z. Naturforsch. 54 a (1999) 63

[23] J.A. Tostevin, Paper presented at: 2nd *International Conference on Fission and Neutron Rich Nuclei, St. Andrews, Scotland, June 28 – July 2 1999*, ed. J.H. Hamilton et al., World Scientific, Singapore 2000

[24] L.D. Landau and E.M. Lifshitz, Quantenelektrodynamik, Lehrbuch der Theoretischen Physik, Band 4 (Berlin: Akademie 1986)

[25] S. Typel and G. Baur, Higher Order Effects in Electromagnetic Dissociation of Neutron Halo Nuclei, in preparation

[26] T. Nakamura et al., Phys. Lett. B331 (1994) 296

[27] T. Nakamura et al., Phys. Rev. Lett. 83 (1999) 1112

[28] T. Kobayashi et al., Phys. Lett. B232 (1989) 51

[29] S. Shimoura et al., Phys. Lett. B348 (1995) 29

[30] M. Zinser et al., Nucl. Phys. A619 (1997) 151

[31] K. Ieki, A. Galonski et al., Phys. Rev. C54 (1996) 1589

[32] K.M. Nollett, M. Lemoine, and D.N. Schramm, Phys. Rev. C56 (1997) 1144; K.M. Nollett et al., nucl-th/0006064

[33] F. Hammache et al., Phys. Rev. Lett. 80 (1998) 928

[34] T. Motobayashi et al., Phys. Rev. Lett. 73 (1994) 2680

[35] J.H. Kelley et al., Phys. Rev. Lett. 77 (1996) 5020

[36] N. Iwasa et al., Phys. Rev. Lett. 83 (1999) 2910

[37] C.E. Rolfs and W.S. Rodney, *Cauldrons in the Cosmos*, The University of Chicago Press (1988)

[38] J.J. Cowan, F.-K. Thielemann and J.W. Truran, Phys. Rep. 208 (1991) 267

[39] T. Rauscher et al., Phys. Rev. C57 (1998) 2031

[40] M. Gai, *ISOL workshop, Columbus/Ohio, July 30 – August 1, 1997*

[41] J. v. Schwarzenberg et al., Phys. Rev. C53 (1996) R2598

[42] J. Görres, H. Herndl, I.J. Thompson and M. Wiescher, Phys. Rev. C52 (1995) 2231

[43] T. Aumann et al., Phys. Rev. C59 (1999) 1252

[44] F. Ajzenberg-Selove, Nucl. Phys. A490 (1988) 1

[45] NuPECC Report, Nuclear and Particle Astrophysics, July 16, 1997, I. Baraffe et al., F.-K. Thielemann (convener)

[46] J. Görres, M. Wiescher and F.-K. Thielemann, Phys. Rev. C51 (1995) 392

[47] H. Schatz et al., Phys. Rep. 294 (1998) 167

Coulomb Dissociation as a Tool of Nuclear Astrophysics

H. Utsunomiya[a], S. Typel[b]

[a] *Department of Physics, Konan University, Okamoto 8-9-1, Higashinada*
Kobe 658-8501, Japan
hiro@konan-u.ac.jp

[b] *Sektion Physik, Universität München, Am Coulombwall 1*
D-85748 Garching, Germany

Keywords: Coulomb dissociation, nuclear astrophysics

Abstract   In the Coulomb dissociation method the electromagnetically induced breakup of light nuclei is studied in order to extract astrophysical S-factors. Possible pitfalls like higher order effects and a mixture of different multipolarities can be avoided by a careful study of the reaction mechanism. As an example a recent breakup experiment of ^7Li is discussed and it is shown that the Coulomb dissociation method is a useful tool for nuclear astrophysics.

1.   Introduction

Nuclear charged-particle reactions occur under various astrophysical conditions during stellar evolution or in the early universe. Ideally, cross sections are measured directly in laboratories. However, a direct measurement is very difficult or even impossible because the Coulomb barrier penetrability of two interacting particles is extremely small at energies over the Gamow peak. Often one has to rely on an extrapolation of the measured cross section to the Gamow peak region.

Coulomb dissociation is known to well simulate the inverse process to radiative capture [1, 2, 3]. It can provide independent information on the astrophysical S-factor for radiative capture reactions and is particularly useful whenever direct measurements are not feasible. Since its first applications to $^{7,6}Li$ [4, 5], the method has become vogue for unstable nuclei like ^8B [6] and aimed at a very challenging nucleus ^{16}O [7].

259

D. N. Poenaru et al. (eds.), Nuclei Far from Stability and Astrophysics, 259–270.
© 2001 *Kluwer Academic Publishers. Printed in the Netherlands.*

In the application of the Coulomb dissociation method, the reaction mechanism has to be understood with a reliable theoretical description. In this regard, the ^7Li nucleus has still carried such essential interest in the method as higher order effects and mixture of different electromagnetic multipoles in Coulomb excitation. Indeed, these two effects are typical complications which one encounters in the Coulomb dissociation method [8]. The ^7Li case can be a benchmark to ensure the validity of the method because substantial information on the astrophysical S-factor for $t(\alpha, \gamma)^7Li$ is available from direct measurements for the purpose of comparison.

2.   Experimental Technique

The double differential cross section for Coulomb excitation of the projectile a with electric multipole order λ is expressed in the first order perturbation theory by,

$$\frac{d^2\sigma}{d\Omega_a dE_\gamma} = \frac{1}{E_\gamma} \frac{dn_{E\lambda}}{d\Omega_a} \sigma_{E\lambda}^{photo}, \qquad (1)$$

where E_γ is the photon energy, $\frac{dn_{E\lambda}}{d\Omega}$ is virtual photon number per unit solid angle and $\sigma_{E\lambda}^{photo}$ is the photodisintegration cross section. The $\sigma_{E\lambda}^{photo}$ is related to the radiative capture cross section $\sigma_{E\lambda}^{capt}$ by

$$\sigma_{E\lambda}^{photo}(a + \gamma \to b + c) = \frac{(2j_b + 1)(2j_c + 1)}{(2j_a + 1)2} \frac{k^2}{k_\gamma^2} \sigma_{E\lambda}^{capt}(b + c \to a + \gamma), \qquad (2)$$

where j is the particle spin, k is the wave number of the $b + c$ relative motion, and k_γ is the photon wave number.

There are two amplification factors with which the capture cross section is boosted to the Coulomb excitation cross section; one is the ratio in wave number, $\frac{k^2}{k_\gamma^2}$, and the other is the virtual photon number, $\frac{dn_{E\lambda}}{d\Omega}$.

In a direct measurement, one usually measures the capture cross section by detecting γ rays emitted from the capture state. The count-rate of γ rays R_γ is given by $R_\gamma = I_B N_T \sigma_{E\lambda}^{capt} \epsilon$, where I_B is the flux of beam particles, N_T is the number density of target particles, and ϵ is the γ-ray detection efficiency. In contrast, the count-rate in Coulomb dissociation R_{bc} is given by $R_{bc} = I_B N_T \frac{d^3\sigma}{dE_{bc} d\Omega_{bc} d\Omega_a} \Delta\Omega^{coin}$, where the triple differential cross section is written by $\frac{d^3\sigma}{dE_{bc} d\Omega_{bc} d\Omega_a} = \frac{1}{4\pi} \frac{d^2\sigma}{d\Omega_a dE_\gamma}$ for an isotropic decay of a into $b + c$ and $\Delta\Omega^{coin}$ is an effective solid angle for particle-particle coincidences.

In the Coulomb dissociation method, care has to be taken to measure two breakup fragments at small relative energies of astrophysical interest. At the small relative velocities, the two fragments are emitted nearly at the same angle because of the high projectile velocity. In other words, the collinearity in the breakup channel, i.e. $a \rightarrow b + c$, is high where the two fragments, b and c, emerge nearly along the scattering axis of the projectile a. Therefore, good angular resolution is necessary in the particle-particle coincidence measurement. Further, measurements need to be made inside the grazing angle where nuclear breakup plays a minor role.

A collinear detection of two particles without the interference of the intense elastic scattering is possible by using a magnetic spectrograph. Two particles emitted within an entrance aperture of the magnetic spectrograph are momentum-analyzed and focused at different positions along the focal plane. Two particles with different charge-to-mass ratios can be measured even at zero relative velocity with position-sensitive focal-plane detectors. At low bombarding energies, the spectrograph is moved from one angle to another inside the grazing angle θ_{gr}. As the collinearity becomes high, the effective solid angle for particle-particle coincidences $\Delta\Omega^{coin}$ decreases. At sufficiently high bombarding energies, the spectrograph can cover the entire angular range inside θ_{gr} in the $0°$ setting of the spectrograph. Ray-trace capability is indispensable in this case.

The present experiment employed an Enge split pole magnetic spectrograph with $\rho_{max}/\rho_{min} \simeq 2.8$ and $\rho_{max} = 90\ cm$ to detect α-particles and tritons. A beam of $42\ MeV - {}^7Li$ was provided by the 12 UD Pelletron tandem accelerator of University of Tsukuba. The experiment covered seven targets and thirty-one detection angles inside θ_{gr}. More details are found in Ref. [9].

Let us compare experimental conditions of the present Coulomb breakup (CB) experiment with those of the direct measurement (DM) of Ref. [10]. The first amplification factor is ~ 90 at $200\ keV$, while the second one is $\sim 2 \times 10^4$ for $E2$ and ~ 50 for $E1$ at $200\ keV$ for ${}^{90}Zr$ at $15°$. The Coulomb field supplies more $E2$ photons than $E1$ photons. For $E1$ Coulomb excitation, the double differential cross section (Equ. 1) at $E_{at} = 200$ keV is enhanced by a factor of ~ 1000 (4500 divided by a spin factor $1/4$) from the radiative capture cross section.

In the present CB experiment, a target density $\sim 10^{19}/cm^2$ and a beam flux $\sim 10^{11}/s$ were used. In the DM [10], a low density of ${}^3H$ target $\sim 10^{17}/cm^2$ and a high intensity of α^+ beam $\sim 10^{13}/s$ were used. The product $I_B N_T$ is of the same order of magnitude in the two measurements. As shown in Fig. 3 of Ref. [11], the present effective solid angle $(\Delta\Omega^{coin})$ is $\sim 4 \times 10^{-4}[sr^2]$ at 200 keV. The photopeak efficiency of a large Ge detector in close proximity for $2.5-3.0$ MeV photons can be as large as $\sim 10^{-2}$. One can conclude that the count-rates of the DM and the CB experiment are similar though the quality of the experimental difficulties is very different. It is then noted that the CB experiment allows access to all E_{at} of interest in a single measurement. In addition, astrophysical S-factors for radiative capture of bare nuclei can, in principle, be deduced from the CB experiment.

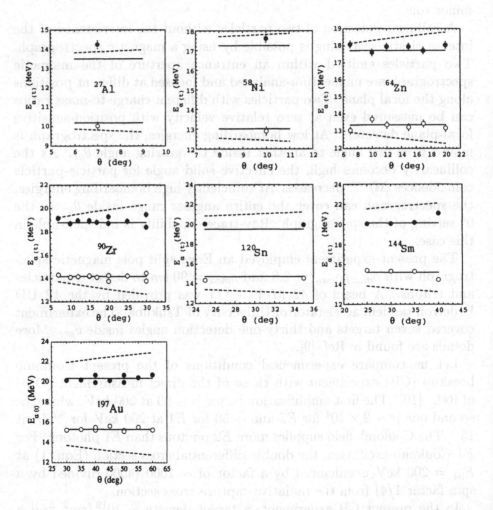

Figure 1 Location of the yield depletion in comparison with those expected for the asymptotic breakup (solid lines) and for the breakup at the distance of closest approach in the Rutherford trajectory (dashed lines).

3. Non-Resonant Breakup by Quantum Tunneling

Energy spectra of coincident α particles and tritons showed yield depletion corresponding to zero relative energy. The location of the yield depletion was closely examined and compared with the kinematics of asymptotic breakup and the breakup at the distance of closest approach [9]. In the asymptotic breakup at the distance of infinity compared to the size of the nucleus, no post-acceleration of the breakup fragments in the target Coulomb field can occur. In contrast, a strong post-acceleration is expected in the breakup at the closest distance. Fig. 1 shows α and triton energies at which the yield is depleted for all targets and angles presently measured in comparison with the two types of breakup reactions. The solid and open circles represent energies of α particles and tritons, respectively. Apparently, the non-resonant breakup is consistent with the asymptotic breakup.

It is tacitly believed that a ^7Li nucleus in continuum states above the α - t threshold behaves like kernels of an Indian corn that burst to *pop - corns* at the critical amount of heat. The term direct breakup well reflects this prompt nature with the threshold energy being the critical heat. However, the lack of post-Coulomb acceleration implies that the particle-unbound system may survive for a significant amount of time before decaying into an α particle and a triton. Although such system is no longer bound by a nuclear potential, it is bound by a Coulomb barrier between α and t. One may remember that non-resonant thermonuclear reactions between two charged particles in stars take place over the Gamow peak by tunneling through a Coulomb barrier [12, 13]. It is this continuum state that is populated after tunneling through the Coulomb barrier.

Fig. 2 schematically shows the nuclear and Coulomb potentials between α and t as a function of the distance r. The height of the Coulomb potential is $0.48\ MeV$ at $r = 6.0\ fm$. By definition $V(r) = 0$ at the α - t threshold. Assuming that a continuum state in ^7Li is an α - t cluster state, it is possible to evaluate the tunneling lifetime of the state. In an analogy to the α-decay theory [14], the particle decay rate (λ) of the continuum s-wave state can be defined by $\lambda = \omega P$, where ω is the number of particles that appear at the nuclear surface per second and P is the Coulomb barrier penetrability. The *frequency of s-wave vibration* can be estimated by $v/2R$ with the velocity of α - t relative motion v and the nuclear radius R [14]. For a resonant state, one may express ω by $\omega = \frac{3v}{R}\theta_0^2$ [12]. Here θ_0^2 is the dimensionless reduced width, representing a measure of to what degree the relevant state can be described as a cluster state of α and t. The simple estimate of the s-wave frequency corresponds to $\theta_0^2 = 1/6$. We evaluated P in the Wentzel-Kramers-Brillouin (WKB) approximation [12].

It was found [9] that the tunneling lifetime ($\tau = 1/\lambda$: reciprocal of the decay rate) of the continuum states can readily exceed the nuclear transit time of the order of 100 fm/c at energies of interest. Surprisingly it can be larger than the lifetime of resonance states in ^7Li at small energies: $\tau = 225 \ fm/c$ ($\Gamma = 875 \ keV$) for the $5/2^-$ state at 6.68 MeV and $\tau = 2122 \ fm/c$ ($\Gamma = 93 \ keV$) for the $7/2^-$ resonant state at 4.63 MeV.

Let us evaluate the Coulomb distortion of the α - t relative energy caused by the post-Coulomb acceleration. For simplicity, we assume that the Coulomb excitation takes place at the distance of closest approach. This assumption is reasonably justified by the first-order perturbation theory of Coulomb excitation [15]. After it keeps moving on a classical trajectory, the projectile breaks up at the time of its mean lifetime. We calculate the distance between projectile and target r_B when projectile breakup takes place. Assuming that the fragments are accelerated in the target Coulomb field without interacting each other, the Coulomb energy which the particle i ($i = \alpha$ or t) gains after breakup was calculated:

$$E_i - E_i^\ell = \frac{A_T}{A_a + A_T} \frac{Z_i Z_T e^2}{r_B}, \tag{3}$$

where A is the mass number, the final kinetic energy E_i corresponds to the experimental observable and E_i^ℓ is the local kinetic energy at the breakup point.

The α - t relative energy was calculated by using either the asymptotic energy (E_i) or the local energy at the breakup point (E_i^ℓ). Relative energy distributions were reconstructed. It was found [9] that the distribution in the collinear branch of $v_\alpha \geq v_t$ is not much affected by the post-Coulomb acceleration, whereas that in the branch of $v_\alpha \leq v_t$ are seriously distorted. Thus, the post-Coulomb acceleration after breakup is literally vanishing in the branch of $v_\alpha \geq v_t$, whereas it is surviving in the branch of $v_\alpha \leq v_t$.

4.   Quantum-Mechanical Treatment

We have quantum-mechanically investigated higher order effects and mixture of $E1$ and $E2$ multipoles in Coulomb excitation of ^7Li. The time evolution of the projectile system under the Coulomb perturbation from target was treated dynamically by solving the time-dependent Schrödinger equation. This framework has been used in order to study higher order effects in the Coulomb dissociation of nuclei like ^8B, ^{11}Li and ^{11}Be [16-21]. The numerical methods and technical details for the solution of the time-dependent Schrödinger equation are given in Ref. [22].

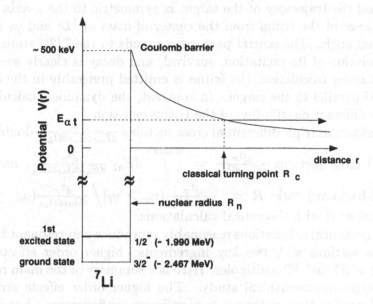

Figure 2 Nuclear and Coulomb potentials between α and t as a function of distance r.

We used a potential model for the ^7Li nucleus. The wave functions for the α - t relative motion were obtained by solving the stationary Schrödinger equation with central potentials of Woods-Saxon shape. The depths of the potential were adjusted in each partial wave in order to give the experimental energies of the bound states and resonances or the scattering phase shifts. Only s-, p-, d- and f-waves were considered. The center-of-mass of the ^7Li system was assumed to move on a hyperbolic trajectory in the target Coulomb field. The projectile experiences a time-dependent Coulomb perturbation

$$V_C(\vec{r}, t) = V_C^{\alpha T}(|\vec{r}_\alpha - \vec{R}(t)|) + V_C^{tT}(|\vec{r}_t - \vec{R}(t)|) - V_C^{LiT}(|\vec{R}(t)|) \quad (4)$$

where $\vec{r} = \vec{r}_\alpha - \vec{r}_t$ is the relative vector between the two clusters. The Coulomb interaction V_C^{LiT} between ^7Li and the target is subtracted since it is responsible for Coulomb scattering of the c.m. which is already included in the description. The perturbation potential was expanded into multipoles where we take into account only multipolarities up to $\lambda = 2$.

Fig. 3 shows the time-evolution of the triton angular distribution $\log P_{cont}(r_t, \varphi_t, t)$ in the rest frame of ^7Li within the scattering plane.

A coordinate system was employed such that the projectile is at rest; it is oriented in a way where the z-axis is perpendicular to the scattering plane and the trajectory of the target is symmetric to the x-axis. r_t is the distance of the triton from the center-of-mass of ^7Li and φ_t is the azimuthal angle. The central peak corresponds to the $7/2^-$ state. The time-evolution of its excitation, survival, and decay is clearly seen. In the first order calculation, the triton is emitted preferably in the direction anti-parallel to the target. In contrast, the dynamical calculation shows a different distribution of the triton emission.

Experimental triple differential cross sections $\frac{d^3\sigma}{dE_{\alpha t}d\Omega_{\alpha t}d\Omega_{Li}}$, double differential cross sections $\frac{d^2\sigma}{d\Omega_{\alpha t}d\Omega_{Li}} = \int\limits_0^{0.5\mathrm{MeV}} dE_{\alpha t}\, \frac{d^3\sigma}{dE_{\alpha t}d\Omega_{\alpha t}d\Omega_{Li}}$, and the forward-backward ratio $R = \frac{d^2\sigma}{d\Omega_{\alpha t}d\Omega_{Li}}(v_\alpha \geq v_t) \Big/ \frac{d^2\sigma}{d\Omega_{\alpha t}d\Omega_{Li}}(v_\alpha \leq v_t)$ were compared with theoretical calculations.

The dynamical calculations reasonably reproduce non-resonant break-up cross sections with two key ingredients: higher order effects and mixture of $E1$ and $E2$ multipoles. Here is a summary of the main results of the quantum-mechanical study. The higher order effects strongly reduced cross sections in the $v_\alpha \geq v_t$ collinear configuration. Admixture of $E2$ multipole was typically seen as the reduction of the $v_\alpha \leq v_t$ cross sections for heavy targets (^{197}Au, ^{144}Sm). The three results of the first order $E1$, the first order $E1 + E2$ and the dynamical calculations tend to merge at small angles for the medium-Z targets. This can be explained by the reduction of higher order effects and $E2$ multipole with decreasing scattering angle. Hence, the $v_\alpha \geq v_t$ branch for small scattering angles and medium-Z targets is affected least of all by higher order effects and $E2$ multipole.

5.   Astrophysical S-Factors

The nuclide ^7Li is produced in the early universe via the $t(\alpha, \gamma)^7Li$ reaction. The relevant energies are in the range of $0 - 500\ keV$ at $T_9 = 0.8$ [23]. Fig. 4 summarizes astrophysical S-factors for the $t(\alpha, \gamma)^7Li$ reaction. As of 1991, three direct measurements raised a question whether astrophysical S-factors are energy-dependent at small energies or not. The data of Griffiths et al. [24] showed constant S-factors ($S(E) = 0.064 \pm 0.016\ keV\ barn$) at energies down to $350\ keV$, while that of Schröder et al. [25] showed a marked rise to $S(0) = 0.14 \pm 0.02\ keV\ barn$ at small energies. The data of Burzyński [26] did not help to resolve the difference between the two data sets due to the limited energy range, i.e., $E \geq 297\ keV$. A new direct measurement was undertaken by Brune, Kavanagh, and Rolfs [10], providing $S(E)$ in the energy range

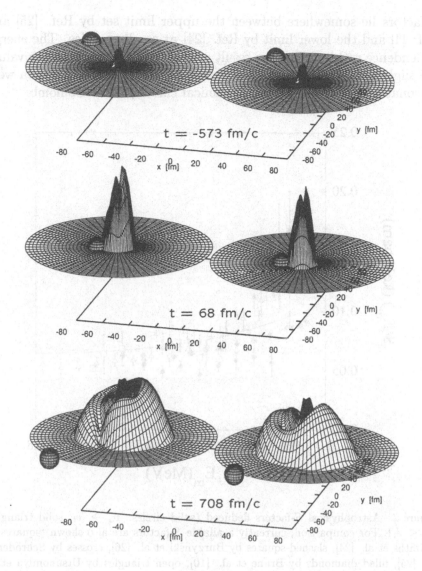

Figure 3 Probability $\log P_{cont}(r_t, \varphi_t, t)$ of finding the triton at a distance r_t and an azimuthal angle φ_t in the scattering plane for different time-steps in the evolution of the breakup reaction. Left: first order calculation, right: dynamical calculation. The position and size of the target in the scattering plane is indicated by a sphere.

$50-1200$ keV with a systematic uncertainty of 6 %. The data showed that $S(E)$ are indeed energy-dependent, but more moderate toward $S(0) = 0.1067(4)$ keV $barn$ than that of Schröder et al. [25].

The potential model of ^7Li employed in this work predicted astrophysical S-factors for $t(\alpha, \gamma)^7Li$ as shown by the solid line in Fig. 4. These

S-factors lie somewhere between the upper limit set by Ref. [25] and Ref. [4] and the lower limit by Ref. [24] at small energies. The energy dependence rather follow the result of Ref. [10] though absolute values are slightly different. With these S-factors, the experimental data were reasonably reproduced by the dynamical calculation of Coulomb.

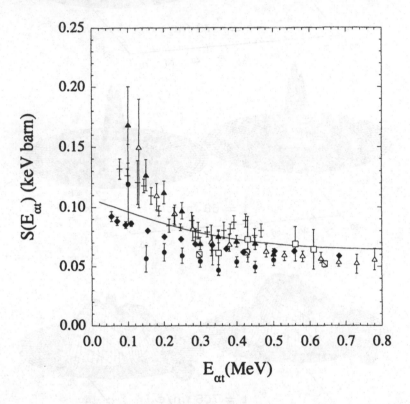

Figure 4 Astrophysical S-factors deduced (solid circles: $v_\alpha \geq v_t$, solid triangles: $v_\alpha \leq v_t$). For comparison, currently available S-factors are also shown (squares by Griffiths et al. [24], slashed-squares by Burzyński et al. [26], crosses by Schröder et al. [25], filled diamonds by Brune et al. [10], open triangles by Utsunomiya et al. [4]). The solid line stands for the result of our potential model calculation.

The method of Coulomb breakup [1] goes in the other way around, aiming at deducing S-factors from the data rather than reproducing the data with S-factors. Considering the dominance of the first-order $E1$ nature in adiabatic Coulomb breakup, we attempted to deduce astrophysical S-factors with the data for ^{90}Zr and ^{64}Zn at $7° - 15°$. Since the Coulomb breakup experiment gives $S(E_{\alpha t})$ for the ground state transition, the branching ratio $\gamma_1/\gamma_0 = 0.453$ was assumed [10]. As shown by solid circles in Fig. 4 the resultant $S(E_{\alpha t})$ show a very moderate energy dependence. They tend to be slightly smaller than those of the direct measurements. Possibly there are still higher order effects in the data which tend to reduce the cross section in the $v_\alpha \geq v_t$ branch. It should be pointed out that if cross sections in the $v_\alpha \leq v_t$ branch are used, the result of the previous Texas A & M experiment is reproduced (open and solid triangles). However, these are most likely Coulomb-distorted and should be replaced by the present $S(E_{\alpha t})$.

6. Summary

Non-resonant breakup of ^7Li was investigated for nuclear astrophysics with improved experimental technique and theoretical treatment. The close examination of energy spectra of coincident α particles and tritons have revealed the delayed nature of non-resonant breakup by quantum tunneling. Semi-classical discussions were made of the lifetime of continuum states in ^7Li and distortion of α - t relative kinetic energies by post-Coulomb acceleration. Dynamical calculations of Coulomb breakup were performed by solving a time-dependent Schrödinger equation. A simple potential model of ^7Li was employed. The dynamical calculations reasonably reproduced the non-resonant breakup cross sections with two key ingredients: higher order effects and mixture of $E1$ and $E2$ multipoles. Considering the dominant role of the first-order $E1$ nature in adiabatic Coulomb breakup, the forward-angle data in the $v_\alpha \geq v_t$ branch for the medium-Z targets were used to deduce astrophysical S-factors $S(E)$ for $t(\alpha, \gamma)^7Li$. They exhibit a moderate energy dependence at small energies. The strongly energy-dependent $S(E)$ resulted from the previous Coulomb breakup experiment based on cross sections with $v_\alpha \leq v_t$; they are most likely Coulomb-distorted and are revised in the present work.

Acknowledgments

This work was supported by the Japan Private School Promotion foundation.

References

[1] G. Baur, C.A. Bertulani and H. Rebel, Nucl. Phys. A458 (1986) 188

[2] G. Baur and H. Rebel, J. Phys. G: Nucl. Part. Phys. 20 (1994) 1

[3] G. Baur and H. Rebel, Annu. Rev. Nucl. Part. Sci. 46 (1996) 321

[4] H. Utsunomiya et al., Phys. Lett. B211 (1988) 24; Nucl. Phys. A511 (1990) 379; Phys. Rev. Lett. 65 (1990) 847; 69 (1992) 863(E)

[5] J. Kiener et al., Z. Phys. A332 (1989) 359; Z. Phys. A339 (1991) 489; Phys. Rev. C44 (1991) 2195; Nucl. Phys. A552 (1993) 66

[6] G. Baur, this proceedings

[7] J. Kiener, this proceedings

[8] S. Typel and G. Baur, Phys. Rev. C50 (1994) 2104

[9] H. Utsunomiya et al., Phys. Lett. B416 (1998) 43

[10] C.R. Brune, R. W. Kavanagh and C. Rolfs, Phys. Rev. C50 (1994) 2205

[11] Y. Tokimoto and H. Utsunomiya, Nucl. Instr. and Meth. A434 (1999) 449

[12] D.D. Clayton, *Principles of Stellar Evolution and Nucleosynthesis* The University of Chicago Press, Chicago and London, 1983

[13] C.E. Rolfs and W.S. Rodney, *Cauldrons in the Cosmos*, The University of Chicago Press, Chicago and London, 1988

[14] H.A. Bethe, Rev. Mod. Phys. 9 (1937) 69

[15] K. Alder et al., Rev. Mod. Phys. 28 (1956) 432

[16] G.F. Bertsch and C.A. Bertulani, Nucl. Phys. A556 (1993) 136; Phys. Rev. C49 (1994) 2839

[17] T. Kido, K. Yabana and Y. Suzuki, Phys. Rev. C50 (1994) R1276; Phys. Rev. C53 (1996) 2296

[18] H. Esbensen, G. F. Bertsch and C. A. Bertulani, Nucl. Phys. A581 (1995) 107

[19] C. A. Bertulani, Nucl. Phys. A587 (1995) 318

[20] H. Esbensen and G. F. Bertsch, Nucl. Phys. A600 (1996) 37

[21] V. S. Melezhik and D. Baye, Phys. Rev. C59 (1999) 3232

[22] S. Typel and H. H. Wolter, Z. Naturforsch. 54a (1999) 63

[23] M.S. Smith and L.H. Kawano, Astrophys. J. Supp. Ser. 85 (1993) 219

[24] G. M. Griffiths et al., Can. J. Phys. 39 (1961) 1397

[25] U. Schröder et al., Phys. Lett. B192 (1987) 55

[26] S. Burzyński et al., Nucl. Phys. A473 (1987) 179

Applications of Coulomb Breakup to Nuclear Astrophysics

J. Kiener

CSNSM Orsay, IN2P3-CNRS and Université Paris-Sud
F-91405 Orsay Campus, France
kiener@csnsm.in2p3.fr

Keywords: Coulomb breakup, fragment angular correlations, nuclear astrophysics

Abstract   The method of Coulomb breakup, applied to the determination of radiative capture cross sections at thermonuclear energies will be briefly introduced. In the following I will concentrate on the experimental requirements for a determination of the astrophysical S-factor using the Coulomb breakup reaction at intermediate energies for ^6Li and ^{16}O breakup. Interference of $E1$ and $E2$ contributions and of Coulomb and nuclear induced breakup are discussed and illustrated. Finally, some prospects concerning the future of Coulomb breakup studies are given.

1.   Introduction

Radiative capture of protons, neutrons and α-particles on nuclei play a major role in primordial nucleosynthesis and in quiescent and explosive stellar burning, largely responsible for the synthesis of the elements. Understanding of this chemical evolution of the universe and the explanation of the observed elemental abundance curve is one of the key motivations of nuclear astrophysics, and needs among others a good knowledge of the nuclear reactions governing the different thermonuclear burning modes, like e.g. the pp-chain, the $CNO/NeNa$-cycles and helium-burning in quiescent stellar burning or e.g. the rapid neutron capture process believed to operate in supernova explosions. The aim of laboratory nuclear astrophysics is to obtain the thermonuclear reaction rates for the implied reactions.

However, especially in quiescent stellar burning the energy region of thermonuclear burning, the so-called Gamow peak, is generally far below

D. N. Poenaru et al. (eds.), Nuclei Far from Stability and Astrophysics, 271–282.
© 2001 *Kluwer Academic Publishers. Printed in the Netherlands.*

the Coulomb barrier for radiative capture of charged particles. Therefore one has to deal with extremely small cross sections, very often in the picobarn and sub-picobarn domain. For most of the reactions experimental data exist only at energies above the Gamow peak and an extrapolation to astrophysical energies with the guide of theoretical models is necessary, which may enhance the uncertainties. A very comprehensive review of the field of nuclear astrophysics is given for example in Rolfs&Rodney [1].

The experimental situation today is even more difficult for the processes involved in explosive astrophysical scenarios where many proton, neutron and α-capture reactions on short-lived isotopes are essential pieces of the nuclear reaction network. Thermonuclear burning energies and thus cross sections here are generally higher than in quiescent burning, but the required radioactive beams suffer from moderate intensities, making direct cross section measurements with today's techniques practically impossible for a large majority of interesting cases. Neutron capture reactions on very short-lived nuclei would require dedicated sophisticated experimental set-ups and will probably be inaccessible to study in direct reactions still for quite some time.

Another way to obtain the required information is the use of indirect methods, like transfer reactions, decay studies and Coulomb breakup. Among these methods, Coulomb breakup is appealing by its straightforward relation to radiative capture reactions. This breakup of fast projectiles in the electric field of a heavy target nucleus has been proposed in 1986 by Baur, Bertulani and Rebel [2] as a new experimental approach to radiative capture cross sections of astrophysical interest. Another interesting fact are the relatively large cross sections compared to the direct reaction, which may yield experimental data directly at the thermonuclear burning energies or allow the use of low intensity beams of short-lived isotopes.

Since then, several Coulomb breakup experiments of astrophysical interest have been performed, investigating important radiative capture reactions for the primordial nucleosynthesis and quiescent and explosive stellar burning. It is the topic of this lecture to present some selected cases, and to give an impression of the achieved results and the problems inherent to the method which have still to be solved.

2.   The Coulomb Breakup Method

The passage of a fast nuclear projectile through the electric field of a target nucleus generates an electromagnetic impulsion $\vec{E}(t)$, $\vec{H}(t)$ seen by the projectile and the target, which can be interpreted as the interaction with a virtual photon field. The absorption of a virtual photon by one of the collision partners leads to so-called Coulomb excitation of this nucleus. Good knowledge of the electromagnetic interaction and generally comfortable cross sections made this method a widely used tool in nuclear spectroscopy, e.g. for the extraction of reduced transition probabilities.

If the projectile nucleus absorbs a virtual photon with energy above the threshold for particle emission, it may lead to dissociation of the latter, generally called Coulomb dissociation or Coulomb breakup. This reaction $a + \gamma \to b + c$ is the time reversed process of the radiative capture reaction $b + c \to a + \gamma$. In first order perturbation theory, which, for sake of simplicity, will be used throughout in this contribution, the Coulomb dissociation cross section can simply be factorized into the photoabsorption cross section $\sigma_{\pi\lambda}^{photo}$ and the virtual photon number density $\frac{dn_{\pi\lambda}}{d\Omega}$ at that specific energy [2] :

$$\frac{d^2\sigma}{d\Omega dE_\gamma} = \frac{1}{E_\gamma} \times \frac{dn_{\pi\lambda}}{d\Omega} \times \sigma_{\pi\lambda}^{photo} \tag{1}$$

$\pi\lambda$ stands for the electric ($\pi = E$) or magnetic ($\pi = M$) multipolarity λ. The photoabsorption cross section for the process $a + \gamma \to b + c$ is directly linked by the detailed balance theorem to the radiative capture cross section:

$$\sigma(a + \gamma \to b + c) = \frac{(2j_b + 1)(2j_c + 1)}{2(2j_a + 1)} \times \frac{k_{bc}^2}{k_\gamma^2} \times \sigma(b + c \to a + \gamma) \tag{2}$$

The wave number of the fragment relative momentum k_{bc}^2 is generally bigger than the photon wave number k_γ^2; for not too high breakup thresholds, the virtual photon numbers contribute another large factor in favor of Coulomb breakup, so that often the breakup cross sections are much larger than the corresponding radiative capture cross sections. The large photon numbers and the good theoretical understanding of electromagnetic excitation of nuclei (see e.g. [3, 4, 5]) is the principal interest of the method.

Practically, projectile energies of the order of several tens of MeV per nucleon seem to be best suited. For these energies the virtual photon spectrum extends to sufficiently high energies to reach the breakup threshold for most nuclei. Although for higher projectile energies (several hundred MeV per nucleon) differential cross sections may still increase for Coulomb breakup, the experimental requirements for a reasonable definition of the breakup reaction (center-of-mass fragment angular correlation and relative energy resolution) are hardly met by existing setups. For moderately high energies, nuclear interaction between target and projectile is to be taken into account. In first approximation nuclear breakup is minimized, when restricting the breakup reaction to scattering angles below the grazing angle.

A comparison between radiative capture and Coulomb breakup cross sections for different reactions of astrophysical interest is shown in Tab. 1. Relative energies E_{cm} in that list are chosen at the lowest energy reached in a radiative capture or a Coulomb breakup experiment. For the first two reactions the Coulomb breakup cross sections are several orders of magnitude larger than the capture cross sections, which can be explained by the relatively low Q-values — or breakup thresholds — which implies low gamma energies, where the virtual photon numbers are extremely high. Furthermore, a breakup experiment yields generally a whole relative energy spectrum over several hundred keV simultaneously, while in a direct measurement, only one data point per beam energy is obtained in the energy spectrum.

Table 1 Comparison between direct reaction and Coulomb dissociation cross sections for several radiative capture reactions of astrophysical interest. For Coulomb dissociation, the double differential cross section (Equ. 1) has been integrated from $0°$ to the grazing angle and for an E_{cm} interval of 10 keV. A projectile energy of 50 $AMeV$ was chosen.

$\pi\lambda$	REACTION E_{CM}	$\sigma^{capture}$	$\sigma^{Coul.diss.}$
$^7Be(p,\gamma)^8B$	$Q = 0.138\ MeV$		
$E1$	$100\ keV$	$1.7\ nb$	$145\ \mu b$
$d(\alpha,\gamma)^6Li$	$Q = 1.475\ MeV$		
$E2$	$100\ keV$	$29\ pb$	$200\ nb$
$t(\alpha,\gamma)^7Li$	$Q = 2.468 MeV$		
$E1$	$50\ keV$	$17\ nb$	$49\ nb$
$^{12}C(\alpha,\gamma)^{16}O$	$Q = 7.162\ MeV$		
$E1$	$1\ MeV$	$18\ pb$	$11\ pb$
$E2$	$1\ MeV$	$18\ pb$	$600\ pb$

An important fact in Coulomb breakup is seen for $^{12}C(\alpha,\gamma)^{16}O$, which proceeds by $E1$ and $E2$-capture with about the same magnitude. In Coulomb breakup at moderately high energies, the $E2$-amplitudes are largely favored over the $E1$-amplitudes. A way to disentangle these amplitudes, consists in measuring precisely fragment-angular correlations, which are extremely sensitive to $E1/E2$ interference. Apart from the interference between different multipolarities, interference between Coulomb and nuclear breakup amplitudes, even at angles below grazing, has to be taken care of. These items will be further developed in the example of ^6Li and ^{16}O breakup.

3. Applications

3.1. ^6Li Breakup

It is widely accepted that the origin of ^6Li is mainly from cosmic ray induced $\alpha + \alpha$ reactions and spallation of CNO nuclei and only a tiny part of its present abundance may be produced in primordial nucleosynthesis by the $d(\alpha, \gamma)^6Li$ reaction at c.m. energies from about 70 keV to 400 keV. Recent observations of ^6Li in two metal-poor halo stars [6] are consistent with this statement but one can hope that further progress in observational techniques may extend ^6Li abundance determinations to still metal-poorer stars, eventually approaching stars whose ^6Li-content results mainly from the primordial gas. This would bring another important constraint on the present-day baryonic density, which at the moment is derived by comparison of the inferred primordial ^2H, ^3He, ^4He and ^7Li abundances and the predicted production of these nuclei in Big-Bang nucleosynthesis models (see e.g. Schramm and Turner [7] for a review of Big-Bang nucleosynthesis). However, no direct measurement of the cross section below $E_{cm} = 0.6$ MeV is available, making the $d(\alpha, \gamma)^6Li$ reaction one of the most uncertain nuclear input parameters in those modelisations (see [8] for a recent discussion of nuclear physics uncertainties related to Big-Bang nucleosynthesis).

First ^6Li-breakup studies driven by the potential astrophysical interest of Coulomb breakup in general and of the $d + \alpha$ reaction in particular were done at Karlsruhe by the group of H. Rebel and the experimental aspects of the method were investigated in a pilot experiment at the Karlsruhe Isochronous Cyclotron. Several interesting features of the ^6Li nucleus make it a good test case. It has a low binding energy which results in comfortably large Coulomb breakup cross sections (see Tab. 1). The first excited state of ^6Li lies at 2.19 MeV, 710 keV above the $\alpha - d$ breakup threshold, whose $B(E2)$-value is well known and comparison with the value extracted from the breakup experiments permits a validation of the theoretical concept.

Sequential (resonant) breakup via the 2.19 MeV, 3^+ state of ^6Li has been measured at several angles below the grazing angle at a projectile energy of 156 MeV. This experimental angular distribution could be nicely reproduced by optical model calculations with the optical model parameters (slightly modified), obtained from fits of the elastic scattering cross section $^{208}Pb(^6Li,^6Li)^{208}Pb_{g.s.}$ at the same projectile energy and a Coulomb deformation parameter β_2, which corresponds to a $B(E2)$-value in agreement with the tabulated value [9].

Direct (nonresonant) breakup with center-of-mass fragment energies from about 50 keV up to the 710 keV resonance has also been observed in these experiments [10]. From the triple differential breakup cross section, obtained in a run at different reaction angles below the grazing angle, the radiative capture cross section $\sigma^{capt}(E)$ for $d(\alpha,\gamma)^6Li$ was extracted, assuming pure $E2$ Coulomb excitation. These data are shown in Fig. 1 in form of the astrophysical S-Factor $S(E) := E \times \sigma^{capt}(E) \times e^{2\pi\eta}$, η being the Sommerfeld parameter. The lowest energy data for the nonresonant radiative capture from direct measurements can be seen above 1 MeV. It gives an impression of the progress that could be obtained with Coulomb breakup, reminding that the cross section falls exponentially with decreasing energy and for example by about 2 orders of magnitude between 1 MeV and 100 keV.

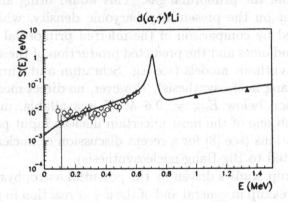

Figure 1   Astrophysical S-factor $S(E)$ for the reaction $d(\alpha,\gamma)^6Li$. Open circles represent the extracted values from the ^6Li breakup experiment, averaged over six different reaction angles from 2.5° to 6°. Filled triangles are data from a direct measurement [11]. The continuous line is the theoretical calculation of [12].

The absolute magnitude of the cross section is well reproduced, adding an experimental basis to the above mentioned astrophysical conclusions. However, the experimental data suggest an essentially constant S-factor below 400 keV, whereas the theoretical curve drops with decreasing energy. Anticipating an astronomical determination of the primordial

^6Li abundance, a good knowledge of the astrophysical S-factor below 400 keV is desirable, and it would be interesting to investigate this problem by additional breakup experiments with precise determination of the fragment angular correlation. A new experiment on the breakup of ^6Li at 150 $AMeV$, with possibility to obtain such correlations is planned to be performed in 2001 at GSI Darmstadt.

3.2. ^{16}O Breakup

The radiative capture reaction $^{12}C(\alpha, \gamma)^{16}O$ is one of the most important thermonuclear reactions in non-explosive astrophysical sites. Its thermonuclear reaction yield determines directly the ratio of carbon and oxygen abundances at the end of helium burning and indirectly the abundances of all elements between carbon and iron, which are produced in later burning stages like carbon-, oxygen- and silicon burning [13]. In spite of its importance in nuclear astrophysics, the cross section of this reaction is still very uncertain in the stellar energy domain ($E_{cm} \approx$ 300 keV for helium-burning at $T \approx 2 \times 10^8 K$).

The reason lies in the extremely small cross section at astrophysical energies and the superposition of $E1$ and $E2$ capture. Moreover, the high-energy tails of two sub-threshold resonances — a 1^- state at 7.12 MeV (45 keV below the $\alpha + ^{12}C$ threshold) and a 2^+ state at 6.92 MeV (245 keV below threshold) — are believed to influence strongly the cross section at 300 keV, interfering respectively with the low energy tail of a 1^--resonance at 2.4 MeV for the $E1$ component and with the direct $E2$ component. Several direct measurements of $^{12}C(\alpha, \gamma)$, elastic $\alpha - ^{12}C$ scattering and ^{16}N decay studies succeeded in determining reasonably well the $E1$-part. However, the $E2$-part of the capture cross section, which is thought to have about the same cross section at 300 keV as the $E1$-part, is still highly uncertain. A compilation of experimental data for $E2$ capture is presented in Fig. 3, showing e.g. the large dispersion and error bars of the $E2$-data.

The Coulomb breakup method is in this case an interesting alternative, because it favors greatly $E2$-excitations over $E1$-excitations, as explained in chapter 1 and as it can explicitly be seen in Tab. 1. Several factors, however, make this breakup reaction an extremely challenging case, as well from the experimental as from the theoretical point of view.

First, the relatively high breakup threshold of 7.162 MeV requires relatively high projectile energies around 100 $AMeV$ to produce a virtual photon spectrum which reasonable intensity above $E_\gamma \approx 8 MeV$. Even then, as seen in Tab. 1, only 600 pb cross section are expected for the breakup cross section at $E_{cm} = 1 MeV$ in an energy bin of 10 keV and integrated up to the grazing angle. This needs intense ^{16}O beams and good angular coverage of the detection system, requiring at the same time a good suppression of background events.

Second, the interesting $E2$ Coulomb induced breakup competes with strong contributions of nuclear induced $\Delta L = 2$ breakup, which is experimentally indistinguishable. Any extraction of $B(E2)$-values and thus capture cross sections must take into account the interference of Coulomb and nuclear induced breakup. The straightforward relation of the dissociation cross section (Equ. 1) with the photodissociation cross section is no longer valid. In principle, optical model calculations, including nuclear and Coulomb interaction could resolve the problem. However, there is no experience yet in optical model calculations of nuclear excitations to the continuum above the particle (or breakup) threshold, and they can therefore at the moment not be considered a-priori reliable.

Third, as well as the $E1$ Coulomb dissociation cross section is negligible compared to the $E2$ contribution, interference of $E1$ and $E2$ amplitudes may have a strong impact on the fragment angular correlations, as it is theoretically predicted by Baur and Weber [14]. For ^{16}O Coulomb breakup at 100 $AMeV$ on ^{208}Pb and a scattering angle below grazing, they predicted particularly strong asymmetries of the $\alpha - {}^{12}C$ velocity vector distribution in the coordinate system defined by their center-of-mass motion and the scattering plane.

Some optical model predictions of differential cross sections for the excitation of ^{16}O after collision with ^{208}Pb at 100 MeV per nucleon into the $\alpha - {}^{12}C$ continuum at $E_x = 8.66$ MeV — 1.5 MeV above the $\alpha + {}^{12}C$ threshold — are shown in Fig. 2. From Fig. 2 it is clear, that Coulomb induced breakup is far from predominant for the $E2$-part. An extraction of the $B(E2)$-values may still be possible by measuring precisely such angular distributions, especially in the angular region between 2 and 4 degrees, where the pattern is very sensitive to Coulomb/nuclear interference. It shows also, that the contribution of $E1$ excitations (nuclear $\Delta L = 1$ excitations are completely negligible) is more than a factor of ten smaller, although the astrophysical S-factor of $E1$-capture at $E_{cm} = 1.5$ MeV (≈ 20 $keVb$) exceeds the one of $E2$-capture (≈ 7 $keVb$).

It is on the other side precisely the sensitivity of these fragment angular correlations to the interference of $E1$ and $E2$ amplitudes and especially with $\Delta L = 2$ nuclear amplitudes that may help to extract the $E2$ Coulomb dissociation cross section. Predictions of optical model calculations including nuclear induced breakup show similar forward-backward asymmetries as predicted by Baur and Weber in semiclassical calculations of pure Coulomb induced breakup, although the effect is somehow

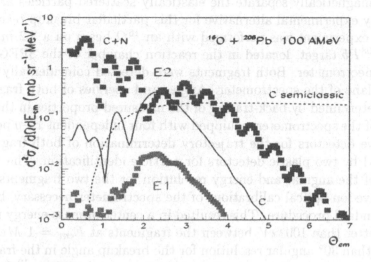

Figure 2 Results of optical model calculations for the reaction $^{208}Pb(^{16}O,^{16}O^*)^{208}Pb_{g.s.}$ at $E_{lab} = 1600$ MeV with the potential parameters of [15]. The excitations into the α-^{12}C continuum at $E_x = 8.66$ MeV were treated as collective vibrational excitations to a discrete 1^- or 2^+ state of ^{16}O. The curve labeled $C + N(E2)$ (dark grey squares) is calculated including nuclear $\Delta L = 2$ and $E2$-Coulomb excitation, while the solid line is only $E2$-Coulomb excitation, calculated by setting the nuclear deformation parameters $\beta_2^N = 0$. For comparison, a semiclassical calculation for pure $E2$-Coulomb excitation without nuclear absorption is shown by the broken line. The light gray squares show the calculation for $E1$-Coulomb excitation.

attenuated, probably due to the dominating influence of nuclear $\Delta L = 2$ amplitudes.

A first experiment to determine this radiative capture cross section by the method of breakup of a ^{16}O beam, with the aim to determine precisely the angular distributions of the fragments has been done at GANIL at a projectile energy of 95 MeV per nucleon. As seen from Fig. 2, most of the breakup cross section is concentrated in a narrow angular cone of $\approx 5°$ around the beam direction. Parally, the fragment angular distributions are squeezed in the laboratory system in an angular cone of $\Theta_{\alpha-^{12}C} \approx 3°$ around their center-of-mass diffusion angle $\Theta_{cm} \equiv \Theta_{^{16}O^*}$. This requires a laboratory angular resolution of the order of 0.2° for both fragments, to be measured close to the incoming beam direction. Furthermore, elastic scattering of the needed intense ^{16}O beam prevents any use solid-state detectors in this angular range. Magnetic spectrometers, which magnetically separate the elastically scattered particles are thus the only experimental alternative for this particular breakup reaction.

The experiment was performed with an ^{16}O beam on a 3.2 mg/cm^2 thick ^{208}Pb target, located in the reaction chamber of the SPEG magnetic spectrometer. Both fragments were detected coincidentally in the focal plane of the spectrometer. Angles and energies of both fragments were determined by back-tracing of the measured properties in the focal plane of the spectrometer, equipped with four independent $x-y$ position sensitive detectors for the trajectory determination of both fragments, followed by two plastic detectors for particle identification. The crucial point of the angular and energy resolution for the two fragments made extensive ion-optical calibrations of the spectrometer necessary, beyond the standard procedure. This resulted in a center-of-mass energy resolution better than $100\,keV$ between the fragments at $E_{cm} = 1\,MeV$ and better than 30° angular resolution for the breakup angle in the fragment center-of-mass rest frame. A total of 80000 α-α and 15500 α-^{12}C coincidences were accumulated in a coincidence run of about 13 h with 25 nA of beam current and a solid angle of the spectrometer of $44 \times 65\,mrad^2$.

Approximately 30 counts were found for the astrophysically interesting elastic breakup of ^{16}O into the α-^{12}C relative energy range E_{cm} = 0.9 – 1.8 MeV. Assuming only $E2$/nuclear $\Delta L = 2$ excitation into the continuum, the measured breakup cross sections were compared to optical model calculations. For the sake of simplicity, we assumed a constant S-factor in each of the defined relative energy bins, and used otherwise the same prescription for the model calculations as explained in the caption of Fig. 2. Our thus determined $E2$ capture cross sections are compared in Fig. 3 with a compilation of available experimental data from direct measurements.

Although our thus derived cross sections tend to be somewhat higher than the direct measurements, a reasonably good agreement with the absolute S-factor and the trend observed in the data points can be seen. The fact, that the $E1$ component in the breakup has been neglected and the relatively simple model for the coupling to the continuum could possibly explain the slight disagreement with the capture data. These results give evidence that the direct breakup into the α-^{12}C continuum has been observed for the first time in ^{16}O breakup and encourage new experiments which should be able to increase considerably statistics.

4.   Conclusion and Outlook

I see the future of Coulomb breakup in two different areas. The first domain concerns radiative capture reactions on short-lived nuclei. Here,

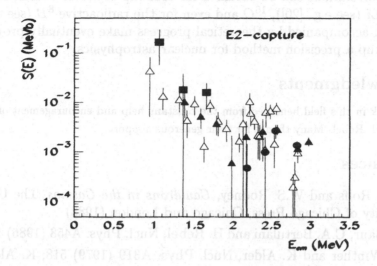

Figure 3 E2-component of the astrophysical S-factor for the $^{12}C(\alpha, \gamma)$ reaction. Data of direct measurements (triangles, filled circles), taken from the European compilation of thermonuclear reaction rates NACRE [16], and superposed (filled squares) are the values extracted from the ^{16}O experiment at GANIL.

the relatively high cross sections of Coulomb breakup are specially important, because the needed radioactive beams suffer generally from low intensities and direct measurements will in a majority of cases be unable to yield data, at least in the near future. Many of the thermonuclear reaction rates in explosive astrophysical scenarios involving those short-lived nuclei are dominated by one or several resonances. The extraction of the resonance parameters, particularly the reduced electromagnetic transition probabilities from Coulomb breakup appears to be well understood, and the "production era" has already begun with the breakup experiments of ^{14}O [17, 18] and ^{12}N [19].

The other interest lies in the better understanding of the breakup mechanism, especially for the direct, nonresonant breakup. The motivation seems to me twofold: First, progress in the understanding of the reaction mechanism is important for nuclear physics itself, thinking e.g. of the current interest in halo nuclei and the problem of post-acceleration of the fragments in the Coulomb field of the target nucleus, treated in the contribution of H. Utsunomiya. Second, progress in the modelisation of stellar burning, astronomical abundance determinations and solar neutrino determinations demands ever-increasing accuracy of thermonuclear reaction cross sections. Precise experimental data, especially fragment angular correlations can be obtained for breakup of stable beams, such as ^6Li, ^7Li (see e.g. [20]), ^{16}O and even for the radioactive ^8B (see e.g. [21]), and accompanied by theoretical progress make eventually projectile breakup a precision method for nuclear astrophysics.

Acknowledgments

My work in this field benefited from the constant help and encouragement of G. Baur and H. Rebel. Many thanks for their generous support.

References

[1] C.E. Rolfs and W.S. Rodney, *Cauldrons in the Cosmos*, The University of Chicago Press, Chicago and London (1988)

[2] G. Baur, C.A. Bertulani and H. Rebel, Nucl. Phys. A458 (1986) 188

[3] A. Winther and K. Alder, Nucl. Phys. A319 (1979) 518; K. Alder and A. Winther, *Electromagnetic Excitation*, North-Holland, Amsterdam (1975)

[4] C.A. Bertulani and G. Baur, Phys. Rep. 163 (1988) 299

[5] A.N.F. Aleixo and C.A. Bertulani, Nucl. Phys. A505 (1989) 448

[6] R. Cayrel et al., Astron. Astrophys. 343 (1999) 923

[7] D.N. Schramm and M.S. Turner, Rev. of Mod. Phys. 70 (1998) 303

[8] E. Vangioni-Flam, A. Coc and M. Cassé, Astron. Astrophys., in press

[9] J. Kiener et al., Z. Phys. A339 (1991) 489

[10] J. Kiener et al., Phys. Rev. C44 (1991) 2195

[11] R.G.H. Robertson et al., Phys. Rev. Lett. 47 (1981) 1867

[12] P. Mohr et al., Phys. Rev. C50 (1994) 1543

[13] T.A. Weaver and S.E. Woosley, Phys. Rep. 227 (1993) 65

[14] G. Baur and M. Weber, Nucl. Phys. A504 (1989) 352

[15] P. Roussel-Chomaz et al., Nucl. Phys. A477 (1988) 345

[16] C. Angulo et al., Nucl. Phys. A656 (1999) 3; astrophysical S-factor tables can be found in http://pntpm.ulb.ac.be/nacre.htm

[17] T. Motobayashi et al., Phys. Lett. B264 (1991) 259

[18] J. Kiener et al., Nucl. Phys. A552 (1993) 66

[19] A. Lefebvre et al., Nucl. Phys. A592 (1995) 69

[20] H. Utsunomiya et al., Phys. Lett. B416 (1998) 43

[21] T. Motobayashi et al., Phys. Rev. Lett. 73 (1994) 2680; T. Kikuchi et al., Phys. Lett. B391 (1997) 261

Nuclear Astrophysics Experiments with Radioactive Beams

Pierre Leleux

Université Catholique de Louvain
Louvain-la-Neuve, Belgium
Leleux@fynu.ucl.ac.be

Keywords: Nuclear astrophysics, radioactive elements

Abstract   In nuclear astrophysics, experiments with radioactive beams present particular problems to which specific solutions can be brought. Selected reactions measured in Louvain-la-Neuve are treated as practical examples.

1.   Introduction

As it was anticipated by Prof. W.A. Fowler in his Nobel lecture in 1983 [1], radioactive beams have brought exciting new developments to laboratory Nuclear Astrophysics in the last decade. This contribution presents an overview of this field without entering into details or astrophysical consequences of particular experiments. Section 2 is a general presentation of nuclear reactions of astrophysical interest. In Section 3, experimental problems related to experiments with radioactive beams will be described. Section 4 is devoted to a particular class of reactions, i.e. radiative capture on H. Section 5 emphasizes the importance of another type of experiments, namely the elastic scattering of radioactive ions.

2.   Nuclear Reactions of Astrophysical Interest

In the astrophysicist view, nuclear reactions can occur in two different stellar environments:

- In quiet situations, e.g. stars on the main sequence (like our sun), reaction happen in chains or cycles, in which radioactive elements

283

D. N. Poenaru et al. (eds.), Nuclei Far from Stability and Astrophysics, 283–293.
© 2001 *Kluwer Academic Publishers. Printed in the Netherlands.*

play no significant role, i.e. they decay as on earth. In this case, cycles are called "cold", e.g. cold CNO cycles in which four protons are burned to helium with C as a catalyst.

- In explosive situations, e.g. novae, supernovae, X-ray bursts, every time a radioactive element is produced in a cycle, a competition occurs between decay and reaction involving this element. Cycles are thus modified (they become "hot") and they even can be escaped.

In the nuclear physicist view, nuclear reactions of astrophysical interest involve either stable or radioactive elements in the entrance channel.

- In the first case, "classical" nuclear astrophysics [2] is concerned, in which very low energy beams are requested (mostly protons or alpha-particles) with very large intensity (several hundred of microamperes) in order to measure the (very low) cross sections down to very low energies, as close as possible to the Gamow window. The latter is usually located at energies lower than 100 keV.

- In the second case, reactions involving a radioactive element occur at large temperatures, i.e. large center-of-mass energies, implying larger cross sections. The Gamow window is located at energies well above 100 keV, even approaching 1 MeV. This second case can still be divided in two categories: depending on the lifetime (τ) of the radioactive element [3], this element will be a beam or a target; briefly speaking, for τ shorter than a few hours, a beam has to be preferred. This contribution deals with reactions involving radioactive beams.

3.   Experiments with Radioactive Beams: Specific Problems

Specific problems related to the beam, the target and the detectors are encountered by nuclear physicists aiming at measuring nuclear reactions induced by radioactive beams.

3.1.   Beams

The production of radioactive beams is a very particular matter, that will not be detailed here (see e.g.[4]). In brief, beams made by the ISOL or two-accelerator method are most appropriate to study reactions of astrophysical interest: a first accelerator induces a nuclear reaction on a stable target in which radioactive atoms are produced, which are then extracted from the target and ionized in an ion source; finally radioactive ions are accelerated in a second accelerator, up to the requested energy. The important point is that, despite many efforts in the last decade, the intensity of radioactive beams is presently less than a few 10^9 s^{-1}, this figure being obtained for "good" cases, like ^{19}Ne, while for more "difficult" cases, beams of $10^5 - 10^6$ s^{-1} are typical. The Cyclotron Research Center in Louvain-la-Neuve has accelerated a broad set of radioactive elements, with intensities that are quoted and updated in its website [5]. For comparison, remember that stable beams reach intensities of 10^{15} s^{-1}.

3.2.   Target

Reactions of astrophysical interest involve radioactive elements interacting with hydrogen and helium. Targets are either gaseous (with or without windows) [6] or solid; the latter are thin foils of polyethylene [7] or any substrate implanted with helium. Let us remark that the specific energy loss of a heavy ion beam in a CH_2 target is large, typically 10 20 $keV/(\mu g/cm^2)$. This means that while crossing the target (~ 200 $\mu g/cm^2$), the beam scans in one step a rather broad energy range in the c.m. system.

Deuterium (10^{-4} of natural Hydrogen) is a harmful poison in either CH_2 or H-gas targets: the (p, γ) and the (d, n) reactions lead indeed to the same final nucleus, the cross section for the latter being at least three orders of magnitude larger than the one for (p, γ).

3.3.   Detectors

Reactions are measured in the inverse kinematics mode (heavy beam on light target); this affects the geometry of the experimental set-up. Detectors will be placed at forward angles in order to detect particles recoiling from the target or scattered by the target.

Radioactive beams have two characteristics that have consequences to the detectors: i) beams have a low intensity and ii) beam particles are inducing background around any place where they are stopped. Accordingly detectors should: i) collect as many reaction products as possible, i.e. they should subtend a large solid angle and ii) have a low sensitivity to background. This last point deserves some comments: in the last decade, radioactive beams were mostly proton-rich nuclides decaying by positron emission. Detectors are thus subjected to background made of positrons and of 511 keV gamma-rays. Depending on the nature of the detectors (i.e. are they going to detect charged particles, gamma-rays or anything else produced in the reaction studied), different solutions have to be implemented. However there is a common principle applicable to all cases: detectors should be segmented (or pixellized) in order to have a total counting rate per pixel as low as possible.

In addition to the specific problems related to the beam, the target and the detectors mentioned above, some considerations about the observables of such reactions are worth doing.

The energy region of interest (the Gamow window) is broad and located well above the reaction threshold. In most cases, the cross section is dominated by the contribution of resonant states of the compound nucleus located in or above the Gamow window. Let us remind that the reaction yield Y from a state of resonant strength $(\omega \gamma)$ is given by:

$$Y = I\frac{\lambda^2}{2}(\frac{M+m}{m})\omega\gamma\frac{1}{\epsilon_{lab}} \qquad (1)$$

where I is the beam intensity; λ is the c.m. wave length; $M(m)$ is the mass of the beam (target); ϵ_{lab} is the beam stopping power through the target.

The resonance strength is thus determined from the measured yield. The contribution of a particular state to the reaction rate is obtained as follows:

$$< \sigma v >_i = \hbar^2(\frac{2\pi}{\mu kT})^{3/2}(\omega\gamma)_i exp(-E_i/kT) \qquad (2)$$

where $< \sigma v >$ is the reaction rate at temperature T; E_i is the resonance energy; μ is the reduced mass.

Summing the partial reaction rates defined in Equ. 2 one obtains the total reaction rate vs. the temperature. One can then compare this total reaction rate to the beta-decay rate of the radioactive element in the entrance channel (which is temperature-independent), in order to decide above which temperature the reaction rate dominates the decay rate.

4.   A Particular Class of Reactions: The Radiative Capture on Hydrogen

(p, γ) reactions are common in hot cycles involving nuclei of mass ≤ 40. In the following, these reactions are noted $A(p, \gamma)B$, where A is the radioactive beam and B the reaction product. If A is a positron emitter, then the same is true for B. Basically three types of detection set-ups can be implemented, to detect: i) gamma-rays; ii) the reaction products B; iii) the radioactive decay of the reaction products B. A common feature of ii) and iii) is the fact that, in inverse kinematics mode, reaction products are emitted at forward angle in the laboratory, in a very narrow cone around the beam direction; the opening of the cone is typically less than 1°.

In the last decade, experiments in Louvain-la-Neuve have used these methods, which are described and evaluated in this section.

4.1. Detection of Gamma-Rays

This method is the most straightforward as regard to the complexity of the set-up: one or more Ge detectors can do the job. However, not all the reactions are possibly studied that way: if the compound nucleus B has a large number of excited levels above and below the $(A+p)$ threshold, then complex gamma-ray spectra will result, that are not easily analyzed, in the presence of 511 keV background from beam particles. The reaction $^{19}Ne(p,\gamma)^{20}Na$ is such a poor case, as ^{20}Na has 7 levels below the reaction threshold of 2.195 MeV and 3 levels in the Gamow window, which is located between 300 keV and 800 keV above threshold for $T = 1 \cdot 10^9 K$. On contrary, the $^{13}N(p,\gamma)^{14}O$ reaction is a very nice case: ^{14}O has no level below the reaction threshold of 4.6 MeV, and only one level potentially important at 526 keV above threshold. Let us remark that, even in this favorable case, some actions had to be taken in order to improve the signal-to-background ratio: i) an active cosmic veto was protecting the Ge detector; ii) the time-of-flight w.r.t. the RF signal was recorded; iii) the Ge detector was shielded by a few cm of lead; iv) the beam was stopped several meters beyond the CH_2 target. Results of this measurements were reported in several publications [8].

Even when tractable, this method suffers from its low detection efficiency: even a 90 % Ge detector has a photopeak efficiency less than 0.1 % for 5 MeV gamma-rays.

4.2. Detection of Heavy Reaction Products

As mentioned above, in inverse kinematics, heavy products are contained in a narrow cone around the beam direction. In addition, as the momentum transferred to the γ-ray is very small, the beam and product ions have the same momentum in the laboratory system. A recoil spectrometer is a well adapted device to perform the separation between both species, taking into account the fact that the ratio of beam to product ions is larger than 10^9. Recoil spectrometers are planned or implemented in several places around the world to measure reactions induced by radioactive beams; let us mention Naples [9], Argonne [10], Louvain-la-Neuve [11], Tokyo [12], Vancouver [13]. The ARES project in Louvain-la-Neuve is described subsequently.

The separation is performed in three subsequent steps: i) the most abundant charge state of the product (and of the beam) is selected in a dipole magnet; ii) a velocity (or Wien) filter deflects the beam ions and transmits the product ions; iii) a $\Delta E - E$ detector consisting of a gas ΔE counter and a E-Silicon detector identifies the remaining particles. In-between these three devices, quadrupole doublets and triplet are placed to focus properly the product ions. The total length of the spectrometer is 10 m. First measurements with stable beams yielded rejection factors of the beam ions in excess of 10^9. In addition to kinematics, effects like the beam emittance on target and the multiple scattering of the beam in the target have to be considered, as they increase significantly the opening angle of the cone containing the product ions and the range of product momenta to be collected. The maximal detection efficiency is equal to the percentage of the product ions that are found in the most abundant charge state: for ions of mass $\leq$ 20, in the energy range of interest here (0.5 MeV/A to 1 MeV/A), this percentage amounts to 30 – 40 % [14], which is a significant improvement w.r.t. the detection of γ-rays. Fig. 1 shows a sketch of the ARES spectrometer.

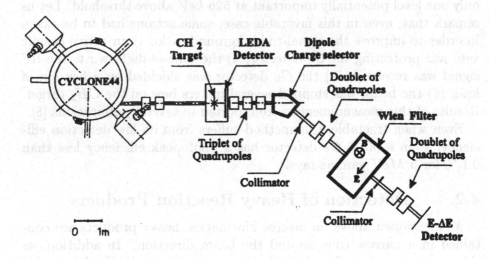

Figure 1 The ARES spectrometer in Louvain-la-Neuve

4.3.   Detection of the Radioactivity of the Product Ions

In a (p, γ) reaction induced by a proton-rich radioactive beam, the product nuclei are less bound than the beam nuclei. Consequently, the decay positron energy of the former is larger than the one of the latter (much larger even in some cases as $^{19}Ne(p,\gamma)^{20}Na$ in which the positron energies are 11.2 MeV and 2.2 MeV, respectively). This fact can be used to select positrons from product nuclei in a magnetic spectrometer: a solenoidal field transmits high-energy positrons up to a stack of detectors while low-energy positrons are stopped in an obstacle before reaching the detectors. Such a device was used to measure the cross section of the $^{19}Ne(p,\gamma)^{20}Na$ reaction [15], with a detection efficiency of 1-2 %.

While this difference in positron energy between product and beam nuclei is "universal", there exists an additional potential signature of product nuclei: in a very limited number of cases, product nuclei decay to unbound states of stable nuclei that subsequently emit prompt α-particles. Two cases in which the branching ratio for emission of α-particles is large are $^{7}Be(p,\gamma)^{8}B$ and $^{19}Ne(p,\gamma)^{20}Na$. The latter was measured in Louvain-la-Neuve. Two different set-up were used to detect α-particles from ^{20}Na decays: double-sided microstrip silicon detectors were developed by the Edinburgh group [16] while passive polycarbonate foils ("track detectors") were implemented by the Leuven group [17]. In both cases, a key point was the fact that detectors showed a very low sensitivity to positrons and 511 keV gamma-rays from the beam nuclei.

As a conclusion, it appears that a recoil spectrometer is definitely superior to the other methods as regard to the detection efficiency, at least by order of magnitude. However, the tuning of a spectrometer is a difficult task, particularly when it is set beyond an incident beam of poor emittance. In this case, more investment into the detection system is certainly needed, e.g. a timing signature from a microchannel plate.

A spectrometer presents another advantage: the kinematics of the (d,n) reaction is such that the opening angle of the product ions is much larger than in the (p,γ) case. As a consequence, most of the (d,n) contribution will not be collected.

5. The Detection of Elastic Recoils

Before measuring a reaction of astrophysical interest induced by a radioactive beam, it is mandatory that the level scheme of the compound nucleus is known to some extent. Usually, this was achieved by experiments involving stable beams, e.g. transfer reactions. Let us give an example: the $^{18}F(p,\alpha)^{15}O$ reaction is part of a hot CNO cycle of importance in novae, as it brings back the material to stable ^{15}N, in competition with the $^{18}F(p,\gamma)$ that leads eventually to an escape from the hot CNO cycle; this reaction was measured using a ^{18}F radioactive beam [18]. The level scheme of the ^{19}Ne compound nucleus was known from a measurement of the $^{19}F(^3He,t)^{19}Ne$ charge-exchange reaction [19] yielding the ratio of the proton-to-alpha-width for several levels above the $^{18}F + p$ threshold. The purpose of this section is to show that the elastic scattering of the radioactive beam (i.e. $^{18}F(p,p)$ in this example), can bring additional informations on the level scheme of the compound nucleus.

It was mentioned earlier in this contribution that the heavy ion beam scans a broad range of excitation energies above threshold, typically 50 – 100 keV in the c.m. system. Recoil protons detected at forward angle in the laboratory carry a precise picture of the levels present in the energy range covered: this is because protons experience a very small energy loss in the target and this small distortion of the spectrum can be corrected for. In the absence of any level of the compound nucleus, the proton spectrum is the square of the Coulomb amplitude. Any level will modify the spectrum, by adding two other terms, i.e. the square of a resonant (or Breit-Wigner) term, and an interference term between the Coulomb and resonant amplitudes. Fitting proton spectra over a broad angular range with the above expression yields informations on the following parameters: the level energy (E_R), the total width (Γ_{tot}), the partial proton width (Γ_p), and even the resonance strength $\omega\gamma$ where

$$\omega\gamma = \omega \frac{\Gamma_p \Gamma_{out}}{\Gamma_{tot}} \tag{3}$$

In Equ. 3, ω is the statistical factor; Γ_{out} is the partial width for decay to the reaction channel; in most cases, $\Gamma_{out} = 1 - \Gamma_p$.

The resonance strength from Equ. 3 should be in agreement with $\omega\gamma$ deduced from the reaction yield (Equ. 1).

It should be stressed that a broad coverage by detectors (in θ and ϕ) is requested in order to obtain good statistics in the proton spectra. LEDA (Louvain-Edinburgh-Detector Array) is a well-adapted answer to this problem: it is made of a total of 128 strips, covering about 20 % of the total proton solid angle when located at 25 cm from the target. This distance is needed to measure the time-of-flight of the particles and to identify them into two-dimensional spectra, i.e. time-of-flight vs. energy. Each strip is equipped with the requested electronics to provide with this information [20].

Several additional considerations have to be made:

i) A precise calibration of the proton spectra is requested in order to obtain absolute resonance energies. A brutal extrapolation from an Am alpha source is not a convenient solution [21].

ii) This method is limited as regard to the total width of the resonant

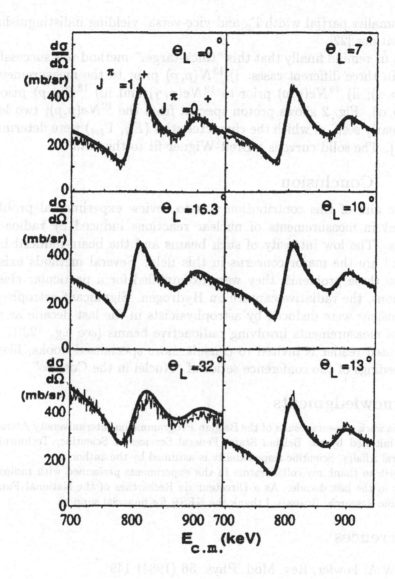

Figure 2 Proton spectra from the $^{19}Ne(p,p)$ scattering

state: a width of at least 1 *keV* is needed to distort in a significant way the Coulomb pattern [21].

iii) The *l*-value which is responsible for the transition is unambiguously determined by the proton spectra. Determining the J^{π}-value among the two possibilities, i.e. $l + 1/2$ and $l - 1/2$ is more difficult: a larger statistical factor (from the $l + 1/2$ alternative) can be compensated by a smaller partial width Γ_p and vice-versa, yielding indistinguishable calculations [22].

Let us remind finally that this "thick-target" method was successfully used in three different cases: i) $^{13}N(p,p)$ prior to the measurement of $^{13}N(p,\gamma)$; ii) $^{19}Ne(p,p)$ prior to $^{19}Ne(p,\gamma)$ and iii) $^{18}F(p,p)$ prior to $^{18}F(p,\alpha)$. Fig. 2 shows proton spectra from the $^{19}Ne(p,p)$; two levels are clearly seen, of which the characteristics (E_R, Γ_{tot}) were determined in [21]. The solid curve is a Breit-Wigner fit to the data.

6.   Conclusion

The aim of this contribution was to review experimental problems present in measurements of nuclear reactions induced by radioactive beams. The low intensity of such beams and the beam-induced background are the major concerns in this field. Several methods exist to bypass these problems; they were exemplified for a particular class of reactions, the radiative capture on Hydrogen. Significant astrophysical conclusions were deduced by astrophysicists in the last decade as a result of measurements involving radioactive beams (see e.g. [23]). The interested reader is invited to consult more specialized books, like the proceedings of the conference series of "Nuclei in the Cosmos".

Acknowledgments

This work presents results of the Belgian Programme on Interuniversity Attraction Poles initiated by the Belgian State, Federal Services of Scientific, Technical and Cultural Affairs. Scientific responsibility is assumed by the author.

I wish to thank my collaborators in the experiments performed with radioactive beams in the last decade. As a Directeur de Recherches of the National Fund for Scientific Research, Brussels, I thank the NFSR for financial support.

References

[1] W.A. Fowler, Rev. Mod. Phys. 56 (1984) 149

[2] C. Rolfs and W.R. Rodney, *Cauldrons in the Cosmos*, Chicago Univ. Press (1988)

[3] P. Leleux, Proceedings of the Joliot Curie Summer School, Maubuisson (1993)

[4] Proceedings of the Second (1991), Ed. Hilger and Third (1993), Ed. Frontières, Int. Conf. on Radioactive Nucl. Beams

[5] The URL address of the CRC web site is: www.cyc.ucl.ac.be

[6] J.W. Hammer, Proc. of the Int. Workshop Hirschegg (1998) p. 370

[7] W. Galster et al., Phys. Rev. C44 (1991) 2776

[8] P. Decrock et al., Phys. Rev. Lett. 67 (1991) 808; Phys. Lett. B304 (1993) 50; Phys. Rev. C48 (1993) 2057

[9] L. Gialanella et al., Nucl. Instr. Meth. A376 (1996) 174

[10] K.E. Rehm et al., Phys. Rev. C52 (1995) R460

[11] J.S. Graulich et al., Proc. of Nucl. in the Cosmos V(1998) Ed. Frontières, p. 471

[12] S. Kubono et al., INS-Report-1165 (1996)

[13] J. D'Auria et al., Proc. of Nucl. in the Cosmos V (1998), Ed. Frontières, p. 435

[14] K. Shima et al., At. Data Nucl. Tables 51 (1992) 173

[15] C. Michotte et al., Phys. Lett. B381 (1996) 402

[16] R.D. Page et al., Phys. Rev. Lett. 73 (1994) 3066

[17] G. Vancraeynest et al., Phys. Rev. C57 (1998) 2711

[18] J.S. Graulich et al., Nucl. Phys. A626 (1997) 751

[19] S. Utku et al., Phys. Rev. C57 (1998) 2731

[20] T. Davinson et al., accepted for publication in Nucl. Instr. Meth. A

[21] R. Coszach et al., Phys. Rev. C50 (1994) 1695

[22] J.S. Graulich, submitted to Phys. Rev. C.

[23] M. Arnould and K. Takahashi, Rep. Progr. Phys. 62 (1999) 395

Neutron Capture Nucleosynthesis: Probing Red Giants and Supernovae

F. Käppeler

Forschungszentrum Karlsruhe, Institut für Kernphysik, Postfach 3640,
D-76021 Karlsruhe, Germany
kaepp@ik3.fzk.de

Keywords: Nucleosynthesis, neutron capture reactions, stellar evolution

Abstract   Neutron reactions are responsible for the formation of all elements heavier than iron. The corresponding scenarios relate to helium burning in Red Giant stars (s process) and to supernova explosions (r and p processes). These processes are briefly sketched with respect to their characteristic time scales and their nuclear physics aspects. Examples for laboratory studies are discussed for the s process, which operates in or near the valley of β-stability. This process is mostly determined by the respective neutron capture cross sections along the reaction path. The resulting abundance patterns are reflecting the physical conditions at the stellar site and represent, therefore, sensitive tests for s-process models.

1.   Introduction

Our modern understanding of the origin of the chemical elements started in the 1930ies with the first tabulation of the solar (standard) abundance distribution derived from the composition of carbonaceous chondrites, a class of primitive meteorites which preserved the original composition of the protosolar nebula by Goldschmidt [1]. At about the same time, nuclear burning was identified as the stellar energy source, the pp-chain by Bethe and Critchfield [2] and the CNO cycle by von Weizsäcker and independently by Bethe [3, 4]. However, it was not before 1952, when Merrill [5] discovered Tc lines in the spectra of red giant stars — isotopes with much shorter half-lives than the stellar evolution time — that stellar nucleosynthesis was accepted as the origin of the chemical elements. The various aspects of this new field of Nuclear Astrophysics, the elemental composition of astronomical objects, the

D. N. Poenaru et al. (eds.), Nuclei Far from Stability and Astrophysics, 295–306.

standard abundance distribution, and the related nuclear physics, were eventually combined in the fundamental and seminal paper by Burbidge, Burbidge, Fowler, and Hoyle [6].

Since this presentation must necessarily be restricted to a coarse overview illustrated by a few examples, the interested reader may be referred to review articles dealing with the various topics of Nuclear Astrophysics: A comprehensive summary of the 40 years of progress in nucleosynthesis since B^2FH was published recently by Wallerstein [7] (more specific topics are reviewed in Refs. [8–15]).

2.   The Observed Abundances

Any nucleosynthesis model must be checked against observations. Naturally, the composition of the solar system was considered a standard which can be reliably derived by spectroscopy of the photosphere and by meteorite analyses [16, 17]. From this distribution (Fig. 1) the signatures of the dominant scenarios can be inferred, starting with the very large primordial H and He abundances from the Big Bang. The abundances of the rare elements Li, Be, and B, which are difficult to produce because of the stability gaps at $A = 5$ and 8, but are easily burnt in stars were mostly formed by spallation reactions induced by galactic cosmic rays.

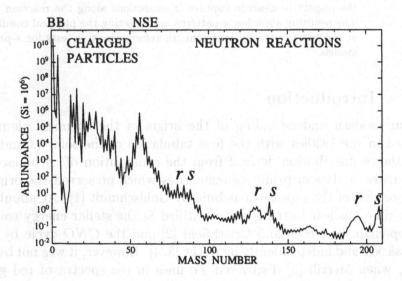

Figure 1 The isotopic abundance distribution in the solar system (from Ref. [17]).

Stellar nucleosynthesis starts with the ashes of He burning, ^{12}C and ^{16}O, which are partly converted to ^{14}N by the CNO cycle in later stellar generations. In subsequent stages of stellar evolution, the light elements up to the mass 40 to 50 region are produced by charged particle reactions during C, Ne, and O burning [7]. The corresponding yields show a strong preference for the most stable nuclei built from α-particles. This part of the distribution is strongly influenced by the Coulomb barrier, resulting in an exponential decrease with increasing atomic number Z. Ultimately, Si burning leads to such high temperatures and densities that nuclear statistical equilibrium is reached. Under these conditions matter is transformed into the most stable nuclei around Fe, giving rise to the pronounced maximum at $A = 56$. Due to the increasing Coulomb barriers the abundances of all heavier nuclei up to the actinides are essentially shaped by neutron capture nucleosynthesis, leading to a fairly flat distribution characterized by the pronounced r and s maxima. These twin peaks are the signatures of the slow (s) and rapid (r) neutron capture processes discussed below.

3. Neutron Capture Scenarios

When the concept of neutron capture nucleosynthesis was first formulated [6] the s and r processes were already identified as the mechanisms responsible for the sharp maxima in the abundance distribution. This is illustrated in Fig. 2, which shows the respective reaction paths in the chart of nuclides.

The s process being characterized by relatively low neutron densities implies neutron capture times much longer than typical β-decay half-lives. Therefore, the s-process reaction path follows the stability valley as indicated by the solid line in Fig. 2. The developing s abundances are determined by the respective (n, γ) cross sections averaged over the stellar neutron spectrum, such that isotopes with small cross sections are building up large abundances. This holds for nuclei with closed neutron shells giving rise to the sharp s-process maxima in the abundance distribution at $A = 88$, 140, and 208. By the way, this represents a good example for the intimate correlation between the relevant nuclear properties and the resulting abundances, a phenomenon that can be used for probing the physical conditions during nucleosynthesis.

The r-process counterparts of these maxima are caused by the effect of neutron shell closure on the β-decay half-lives. Since the r process occurs in regions of extremely high neutron density (presumably during stellar explosions in supernovae) neutron captures are much faster than β-decays. Therefore, the r-process path is driven off the stability valley until nuclei with neutron separation energies of $\approx 2\ MeV$ are reached. At these points, (n, γ) and (γ, n) reactions are in equilibrium, and the

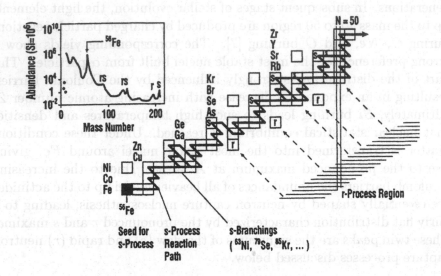

Figure 2   An illustration of the neutron capture processes responsible for the formation of the nuclei between iron and the actinides. The observed abundance distribution in the inset shows characteristic twin peaks. These peaks result from the nuclear
properties where the *s*- and *r*-reaction paths encounter magic neutron numbers. Note
that a *p* process has to be invoked for producing the proton rich nuclei that are not
reached by neutron capture reactions. (For details see discussion in text.)

reaction flow has to wait for β-decay to the next higher element. Accordingly, the *r* abundances are proportional to the half-lives of these
waiting point nuclei. This means that *r*-abundance peaks accumulate
also at magic neutron numbers, but at significantly lower Z compared
to the related *s*-process maxima, resulting in the typical twin peaks in
the abundance distribution.

While the observed abundances are dominated by the *s* and *r* components, which both account for approximately 50 % of the abundances
in the mass region $A > 60$, the rare proton-rich nuclei can not be produced by neutron capture reactions. This minor part of the abundance
distribution had to be ascribed to the *p* process that is assumed to occur
in explosively burning outer shells of supernovae [18].

Naturally, the *s* process is more easily accessible to laboratory experiments as well as to stellar models and astronomical observations
[9]. Attempts to describe the *r* and *p* processes are hampered by the
large uncertainties in the nuclear physics data far from stability [8, 19],
but also — and perhaps more severely — by the problems related to a
detailed modeling of the stellar explosion [18, 20, 21].

Obviously most isotopes received abundance contributions from the *s* and *r* processes. But as indicated in Fig. 2, there are neutron-rich stable isotopes (marked *r*) that are not reached by the *s* process because of their short-lived neighbors. Consequently, this species is of pure *r* process origin. In turn, these *r*-only nuclei terminate the β-decay chains from the *r*-process region, making their stable isobars an ensemble of *s*-only isotopes. The existence of these two subgroups is of vital importance for nucleosynthesis, since their abundances represent important model tests.

In the following, the origin of the heavy elements will be illustrated by some important features of the *s*-process.

4.   The Nuclear Physics of the *s*-Process

The main nuclear physics input for *s*-process studies are the (n, γ) cross sections of all nuclei along the reaction path from Fe to Bi as well as the β-decay rates at the branching points.

4.0.1   Laboratory Studies.   Neutrons in the energy range between 0.3 and 300 keV required for such measurements are produced in several ways: (i) At low-energy particle accelerators, nuclear reactions, such as $^7Li(p,n)^7Be$ offer the possibility of tailoring the neutron spectrum exactly to the energy range of interest. This has the advantage of low backgrounds, allowing for comparably short neutron flight paths to compensate limitations in the neutron source strength ([9, 22]).

(ii) Much higher intensities can be achieved at linear accelerators via (γ, n) reactions by bombarding heavy metal targets with electron beams of typically 50 MeV. The resulting spectrum contains all energies from thermal to near the initial electron energy. Since the astrophysically relevant energy range corresponds only to a small window in the entire spectrum, background conditions are more complicated and measurements need to be carried out at larger neutron flight paths. In turn, the longer flight paths are advantageous for high resolution measurements which are important in the resonance region. Refs. [23, 24] are recent examples of astrophysical measurements at such facilities.

(iii) Spallation reactions induced by energetic particle beams provide the most prolific sources of fast neutrons. An advanced spallation source suited for neutron time-of-flight (TOF) work is the LANSCE facility at Los Alamos, allowing for measurements on very small samples as well as on radioactive targets [25, 26]. Similar experiments are being prepared at CERN [27].

The experimental methods for measuring (n, γ) cross sections fall into two groups, TOF techniques and activations. In principle, TOF techniques can be applied to all stable nuclei and require a pulsed neutron source for determining the neutron energy via the flight time between neutron production target and capture sample. Capture events are identified by the prompt γ-ray cascade in the product nucleus.

The best signature for the identification of neutron capture events is the total energy of the emitted γ-cascade. Hence, accurate measurements of (n, γ) cross sections require a detector that operates as a calorimeter with good energy resolution such as the Karlsruhe 4π BaF_2 detector [28] (Fig. 3). In the γ-spectrum of a perfect calorimeter, all capture events would fall in a line at the neutron binding energy (typically between 5 and 10 MeV), well separated from backgrounds, which are inevitable in neutron experiments, and independent of the multiplicity of the γ-ray cascade. In this way, an efficiency for capture events of 96 to 98 % can be obtained, allowing for cross section uncertainties of ± 1 %.

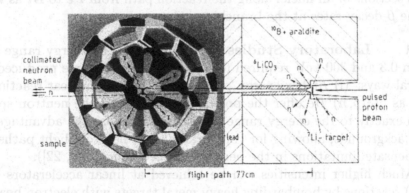

Figure 3 The Karlsruhe 4π BaF_2 detector.

Activation in a quasi-stellar neutron spectrum provides a completely different approach for the determination of stellar (n, γ) rates, but is restricted to those cases, where neutron capture produces an unstable nucleus. This method has superior sensitivity (very small samples sufficient) and is highly selective (enriched samples not required). Quasi-stellar neutron spectra can be produced via the $^7Li(p, n)^7Be$ [29, 30] by bombarding thick metallic lithium targets with protons of 1912 keV, only 31 keV above the reaction threshold. The resulting neutrons exhibit a continuous energy distribution with a high-energy cutoff at $E_n =$ 106 keV and a maximum emission angle of 60° as shown in Fig. 4. Since the angle-integrated spectrum corresponds closely to a Maxwell-Boltzmann distribution for $kT = 25$ keV, measurements yield immediately the proper stellar cross section.

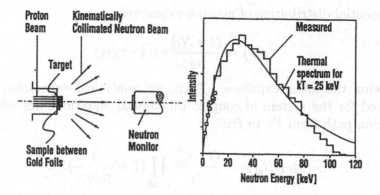

Figure 4 The activation technique: irradiation of a sample sandwich at the accelerator (left) and the angle-integrated neutron spectrum (right).

The possibility to use minute samples makes the activation technique an attractive tool for investigating unstable nuclei of relevance for *s*-process branchings [31]. For example, a measurement of the ^{155}Eu cross section ($t_{1/2} = 4.96 \ yr$) could be performed with a sample of only 88 ng corresponding to 3.4×10^{14} atoms, a must for minimizing the sample activity to a manageable level [32].

4.0.2 Theoretical Calculations.

Despite of the experimental progress, which is compiled in a collection of available data for *s*-process studies [33], cross section calculations remain indispensable for determining the (n,γ) rates of unstable nuclei with high specific γ-activity as well as the (possible) differences between the laboratory values and the actual stellar cross sections, which can be affected by thermally populated nuclear states with low excitation energies. Theoretical reaction rates are particularly important for explosive scenarios, where nuclei far from stability are involved and where experimental data are completely missing [34]. Another essential issue are weak interaction rates under astrophysical conditions, both for He burning [35] and for explosive scenarios [8].

4.1. *s*-Process Models

4.1.1 The Canonical *s*-Process.

This phenomenological model [9, 36] was suggested by the empirical assumptions that temperature and neutron density are constant during the *s*-process and that a certain fraction G of the observed ^{56}Fe abundance was irradiated by an exponential distribution of neutron exposures, τ,

$$\rho(\tau) = \frac{G \times N_{56}^{\odot}}{\tau_0} exp(-\tau/\tau_0)$$

τ_0 being the mean exposure. Then, an analytical expression can be derived for the system of coupled differential equations describing the reaction path from Fe to Bi:

$$\langle\sigma\rangle_{(A)} N_{s(A)} = \frac{G \cdot N_{56}^{\odot}}{\tau_0} \prod_{i=56}^{A} (1 + \frac{1}{\tau_0\langle\sigma\rangle_i})^{-1}.$$

This product of stellar cross section and resulting s abundance is the characteristic s-process quantity. Apart of the two parameters G and τ_0, which are adjusted to fit the abundances of the s-only nuclei, the stellar (n,γ) cross sections $\langle\sigma\rangle$ are the only input data required for determining the overall abundance distribution. This approach was later modified to include the particular patterns of the s-process branchings.

Given the very schematic nature of the classical approach, it was surprising to see that it provides an excellent description of the s-process abundances. Fig. 5 shows the calculated $\langle\sigma\rangle N_s$ values compared to the corresponding empirical products of the s-only nuclei (symbols) in the mass region between $A = 56$ and 209. The error bars of the empirical points reflect the uncertainties of the abundances and of the respective cross sections. One finds that equilibrium in the neutron capture flow was reached between magic neutron numbers, where the $\langle\sigma\rangle N_s$-curve is almost constant. The small cross sections of the neutron magic nuclei around A~88, 140, and 208 act as bottlenecks for the capture flow, resulting in the distinct steps of the σN-curve.

4.1.2   Stellar Models.

In terms of stellar sites, the *main* component can be attributed to helium shell burning in low mass stars, where neutron production and concordant s-processing occur in two steps, by the $^{13}C(\alpha,n)^{16}O$ reaction at relatively low temperatures of $T_8 \sim 1$ and by the $^{22}Ne(\alpha,n)^{25}Mg$ reaction at $T_8 \sim 3$ (see Refs. [10, 37] for details). The *weak* component indicated in Fig. 5 can be ascribed to core He burning in massive stars [38].

5. s-Process branchings

Branchings in the reaction chain of the s process occur at unstable nuclei with sufficiently long half-lives that neutron capture can compete with β-decay. The resulting abundance pattern can be used to derive information on the physical conditions during the s process. Fig. 6 shows the s-process branchings at ^{147}Nd and $^{147,148}Pm$, which are defined by the s-only nuclei ^{148}Sm and ^{150}Sm. Since ^{148}Sm is partly bypassed by the reaction flow, its $\langle\sigma\rangle N_s$ value will be smaller than that of ^{150}Sm, the ratio providing a measure for the combined strength of the branchings.

This strength can also be expressed in terms of the rates λ_β and $\lambda_n = n_n v_T \langle\sigma\rangle$,

$$f_\beta = \frac{\lambda_\beta}{\lambda_\beta + \lambda_n} \approx \frac{(\langle\sigma\rangle N_s)_{^{148}Sm}}{(\langle\sigma\rangle N_s)_{^{150}Sm}} \approx 0.9.$$

If — for simplicity — only the branching at ^{148}Pm is considered, the neutron density n_n is

$$n_n = \frac{1 - f_\beta}{f_\beta} \cdot \frac{1}{v_T\langle\sigma\rangle_{^{148}Pm}} \cdot \frac{ln2}{t^*_{1/2(^{148}Pm)}}.$$

This equation illustrates the input data that are important for reliable branching analyses:
(i) The first term depends on the cross sections for the s-only nuclei, which define the branching factor f_β. These data need to be known to about 1 %.

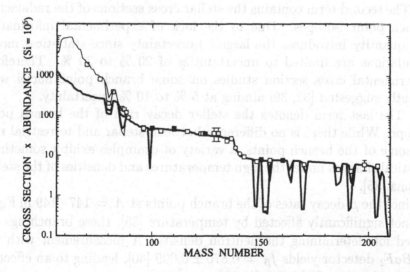

Figure 5 The characteristic product of cross section times s-process abundance plotted as a function of mass number. The solid line was obtained via the classical model, and the symbols denote the empirical products for the s-only nuclei. A complete representation of the empirical products requires at least two different processes, the **main** and **weak** components indicated by thick and thin solid lines, respectively. Some important branchings of the neutron capture chain are indicated as well.

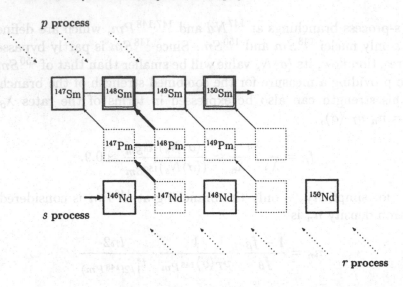

Figure 6 The *s*-process reaction path in the $Nd/Pm/Sm$ region with the branchings at $A = 147$, 148, and 149. Note that ^{148}Sm and ^{150}Sm are shielded against the *r* process. These two isotopes define the strength of the branching.

(ii) The second term contains the stellar cross sections of the radioactive branch point isotopes. Due to the lack of experimental information this quantity introduces the largest uncertainty since statistical model calculations are limited to uncertainties of 20 % to 30 %. Therefore, experimental cross section studies on some branch point nuclei were recently suggested [32, 39] aiming at 5 % to 10 % uncertainty.

(iii) The last term denotes the stellar decay rate of the branch point isotope. While there is no difference between stellar and terrestrial rate for some of the branch points, a variety of examples exhibit sometimes drastical changes under the high temperatures and densities of the stellar plasma [35].

Since the β-decay rates of the branch points at $A = 147-149$ in Fig. 6 are not significantly affected by temperature [35], these branchings are suited for determining the neutron density. A measurement with the $4\pi\, BaF_2$ detector yields $f_\beta = 0.870 \pm 0.009$ [40], leading to an effective neutron density of $(4.1 \pm 0.6) \cdot 10^8\ cm^{-3}$ [41].

6.   Summary

Neutron capture nucleosynthesis operates during the *He* burning stages of stellar evolution and (presumably) in the final supernova explosions of massive stars. The various scenarios can be identified by their typical abundance distributions as well as by the increasingly detailed astronomical observations. Comparison with model calculations can reveal the physical conditions of the respective astrophysical sites in quite some detail provided that the nuclear physics data are reliably known. This holds, in particular, for the *s* process, where quantitative analyses are beginning to yield a rather complete description.

References

[1] V. Goldschmidt, Norske Vidensk. Akad. Skr. Mat.-Naturv. Kl. IV (1937)

[2] H. Bethe and C. Critchfield, Phys. Rev. 54 (1938) 248

[3] C.F. von Weizsäcker, Physik. Zeitschrift 39 (1938) 639

[4] H. Bethe, Phys. Rev. 55 (1939) 103

[5] P. Merrill, Science 115 (1952) 484

[6] E. Burbidge et al., Rev. Mod. Phys. 29 (1957) 547

[7] G. Wallerstein et al., Rev. Mod. Phys. 69 (1997) 995

[8] F. Käppeler, F.-K. Thielemann, and M. Wiescher, Ann. Rev. Nucl. Part. Sci. 48 (1998) 175

[9] F. Käppeler, Prog. Nucl. Part. Phys. 43 (1999) 419

[10] M. Busso, R. Gallino, and G. Wasserburg, Ann. Rev. Astron. Astrophys. 37 (1999) 239

[11] R. Gallino, M. Busso, and M. Lugaro, in *Astrophysical Implications of the Laboratory Study of Presolar Material*, edited by T. Bernatowitz and E. Zinner (AIP, New York, 1997), p. 115

[12] D. Lambert, Astron. Astrophys. Rev. 3 (1992) 201

[13] J. Cowan, F.-K. Thielemann, and J. Truran, Phys. Rep. 208 (1991) 267

[14] E. Zinner, Ann. Rev. Earth Planet. Sci 26 (1998) 147

[15] H. Schatz et al., Physics Reports 294 (1998) 167

[16] E. Anders and N. Grevesse, Geochim. Cosmochim. Acta 53 (1989) 197

[17] H. Palme and H. Beer, in *Landolt-Börnstein New Series, Group VI, Vol. VI/3a*, edited by O. Madelung (Springer, Berlin, 1993), p. 196

[18] M. Rayet et al., Astron. Astrophys. 298 (1995) 517

[19] F.-K. Thielemann et al., in *Nuclear and Particle Astrophysics*, edited by J. Hirsch and D. Page (Cambridge University Press, Cambridge, 1998), p. 27

[20] W. Hillebrandt and P. Höflich, Rep. Prog. Phys. 52 (1989) 1421

[21] E. Müller, in *Nuclear Astrophysics*, edited by M. Buballa, W. Nörenberg, A. Wambach, and J. Wirzba (GSI, Darmstadt, 1998), p. 153

[22] Y. Nagai et al., Ap. J. 381 (1991) 444

[23] P. Koehler et al., Phys. Rev. C54 (1996) 1463

[24] H. Beer, F. Corvi, and P. Mutti, Ap. J. 474 (1997) 843

[25] P. Koehler and F. Käppeler, in *Nuclear Data for Science and Technology*, edited by J. Dickens (American Nuclear Society, La Grange Park, Illinois, 1994), p. 179

[26] R. Rundberg et al., Technical report, Los Alamos National Laboratory (unpublished)

[27] C. Rubbia et al., Technical report, CERN (unpublished).

[28] K. Wisshak et al., Nucl. Instr. Meth. A292 (1990) 595

[29] H. Beer and F. Käppeler, Phys. Rev. C21 (1980) 534

[30] W. Ratynski and F. Käppeler, Phys. Rev. C37 (1988) 595

[31] F. Käppeler, M. Wiescher, and P. Koehler, in *Workshop on the Production and Use of Intense Radioactive Beams at the Isospin Laboratory*, edited by J. D. Garrett (Joint Institute for Heavy Ion Research, Oak Ridge, 1992), p. 163.

[32] S. Jaag and F. Käppeler, Phys. Rev. C51 (1995) 3465

[33] Z. Bao et al., Atomic Data Nucl. Data Tables (2000), in print

[34] T. Rauscher and F.-K. Thielemann, Atomic Data Nucl. Data Tables 75 (2000) 1

[35] K. Takahashi and K. Yokoi, Atomic Data Nucl. Data Tables 36 (1987) 375

[36] P. Seeger, W. Fowler, and D. Clayton, Ap. J. Suppl. 97 (1965) 121

[37] R. Gallino et al., Ap. J. 497 (1998) 388

[38] C. Raiteri et al., Ap. J. 419 (1993) 207

[39] F. Käppeler, in *Radioactive Nuclear Beams*, edited by T. Delbar (Adam Hilger, Bristol, 1992), p. 305

[40] K. Wisshak et al., Phys. Rev. C48 (1993) 1401

[41] F. Käppeler et al., Phys. Rev. C53 (1996) 1397

Neutron Induced Reaction of Light Nuclei and its Role in Nuclear Astrophysics

Yasuki Nagai

Research Center for Nuclear Physics, Osaka University
Mihoga-oka 10-1, Ibaraki, Osaka 667-0047, Japan

nagai@rcnp.osaka-u.ac.jp

Keywords: Nuclear reactions and spectroscopy, nucleosynthesis of elements

Abstract   The abundance of the s-process isotopes for various metalicity stars has been considered to be used to construct models of the chemical evolution of the Galaxy. Hence efforts involving both observations and yield estimations of these isotopes are being made for a wide range of metalicities and stellar masses to compare the chemical evolution models with the observational data. In order to construct the models to predict the s isotope productions, it is necessary to know the neutron-capture cross sections of various nuclei at stellar neutron energy. The reaction cross section is also playing an important role in the inhomogeneous big bang models. The result of the measurement of the cross sections is discussed from both view points of nuclear astrophysics and nuclear physics.

1.   Introduction

Elemental abundance patterns are considered to give relevant information about the chemical evolution and the early history of the universe. In particular, light elements up to Li are believed to be synthesized in the early universe and their observed abundance is in good agreement with the calculated value of the standard big-bang model [1]. The agreement is one of the major triumphs of the standard model of cosmology. While heavier elements than Li are not produced in any significant abundance in the big-bang, since there are no stable elements of $A = 5$ and/or 8 [2]. Hence, a He can not capture a proton or another He and form a new stable isotope. The heavier elements have been synthesized by charged particle induced thermonuclear reactions in stars, up to the iron peak

D. N. Poenaru et al. (eds.), Nuclei Far from Stability and Astrophysics, 307–316.

region [3]. The elements heavier than iron, however, are not produced by the thermonuclear reactions, since the Coulomb energy between charged particles are quite large and at iron the binding energy of a nucleon is maximum. In the theory of stellar nucleosynthesis, neutron-capture reactions are considered to be necessary for the production of isotopes heavier than iron through both slow (s) and rapid (r) processes [3]. Because of the different timescales for these processes, they are assumed to occur during the He-burning phases of stellar evolution and during stellar explosions, respectively.

Recently, much interest has arisen in the abundance of the s-process isotopes for various metalicity stars to construct models of the chemical evolution of the Galaxy [4]. Efforts involving both observations and yield estimations of these isotopes are being made for a wide range of metalicities and stellar masses to compare the chemical evolution models with the observational data. Here, it is generally admitted that the s-isotopes with mass numbers of $A > 90$ are synthesized during He shell burning periods in low and/or intermediate asymptotic giant branch (AGB) stars [5]. While most of the s-process nuclei with $70 < A < 90$ are synthesized in massive AGB stars during their central He burning phase [6]. Two reactions of $^{13}C(\alpha, n)^{16}O$ and $^{22}Ne(\alpha, n)^{25}Mg$ are considered to be the neutron sources for producing the s-process isotopes. The neutrons from the $^{13}C(\alpha, n)^{16}O$ reaction are for producing s-isotopes with mass numbers of $A > 90$. Those from the $^{22}Ne(\alpha, n)^{25}Mg$ reaction are for producing s-isotopes with $A < 90$. Although the neutrons are used for the production of the s-process nuclei, they may be captured by abundant light nuclei, such as ^{12}C, ^{16}O, and ^{20}Ne, direct products of He burning. Namely, if the neutron capture cross sections of the above-mentioned light nuclei would be large, the yields of heavier s-isotopes would decrease; furthermore, the yields of p-process nuclei would decrease, since the s-process nuclei are the immediate predecessors of the p-nuclei [7].

In neutrino-induced nucleosynthesis energetic neutrinos emitted from a cooling neutron star can excite nuclei in ejected shells of a supernova via neutral and charged currents to particle-unbound states, leading to particle emission. The emitted particles can react with nuclei to produce heavy elements [8]. If the neutrino-induced r-process occurs in the He shell, where many carbons and oxygens exist, and if their neutron capture cross sections are large, they become neutron poisons and thus reduce the efficiency of r-process nucleosynthesis.

As discussed above a good agreement between the predicted primordial abundances of the light elements with $\Omega_b \approx 0.04$ with the observed values has been considered to be one of the major triumphs of the standard big bang model of cosmology, although dark matter problem remains to be open [9]. Because of the importance of the standard big bang nucleosynthesis theory, detailed studies of this theory are being carried out, motivated mainly by refined data for both the observed primordial abundances of the light elements and the nuclear reaction rates.

Recently, an inhomogeneous big-bang model, an alternative one of the standard one, has been proposed as discussed below. Namely, the possible first-order phase transition from a quark-gluon plasma to the hadronic phase could lead to the baryonic density fluctuations, which can form the inhomogeneous universe with two different regions of the neutron-poor and the neutron-rich [10]. Following the inhomogeneous models of nucleosynthesis many calculations have been carried out to test the possibility to reconcile observed abundances of primordial light elements with a flat $\Omega_b = 1$ universe. Another characteristic feature of the inhomogeneous models is the prediction of a sufficient production of heavy elements in low-density neutron-rich regions.

The main reaction sequence to synthesize the heavy elements in inhomogeneous big-bang models is [11]

$$^1H(n,\gamma)^2H(n,\gamma)^3H(d,n)^4He(t,\gamma)^7Li^7Li(n,\gamma)^8Li(\alpha,n)^{11}B(n,\gamma)^{12}B$$
$$(\beta^-)^{12}C(n,\gamma)^{13}C, \text{ etc.}$$

Although it has been claimed that the $\Omega_b = 1$ universe cannot be accepted even in inhomogeneous models, the possibility of the nucleosynthesis of heavy elements in a primordial rapid process cannot be ruled out because of the poor knowledge of the above reaction rates.

Therefore, in order to construct models to predict the s and p isotope productions as functions of the metalicity and stellar mass, and also to estimate the production yields of heavy elements in the r-process and in inhomogeneous models, it is necessary to know the neutron-capture cross section of light nuclei at stellar neutron energy. Here, the $p(n,\gamma)d$ reaction is one of the most important reactions in primordial nucleosynthesis; all deuterium is produced by the reaction, and the abundances of other primordial light elements (^3He, ^4He, and ^7Li) are affected appreciably by the cross section of this reaction. However, it has never been measured directly at astrophysically relevant energies between 10 and 80 keV and had therefore to be estimated on basis of theoretical calculation of deuteron photodisintegration. These calculations, however, have been plagued by the following two problems. First, although new calculations are in good agreement with each other, they are different from the old calculation. The difference is quite large, about 15 %, especially concerning the astrophysically important neutron energies of 10 and 100 keV and above 700 keV. Second, the calculated cross sections of the deuteron photodisintegration are about 15 % smaller than the experimental values. Hence, it is also necessary to directly measure the cross section of the p (n, γ) d reaction at the astrophysically relevant energy.

2.   Experimental Procedure and Results

The neutron capture (n, γ) reaction cross section of a nucleus has been measured by two methods: activation and prompt γ-ray detection. The former method is to detect a β- ray (and/or γ-ray) event from the β-activity produced by the reaction, which has following merits. First, the cross section can be determined reliably, since the β-decay half-life and/or the β-ray (or γ-ray) energy can be used to identify the final nucleus. Second, the measurement can be made using a small amount of sample, since a β-ray detector has a high detection efficiency with a large solid angle. However, it has the following demerits. First, it cannot be applied when the life time of the β-activity is longer than ~ 1 min, since activity can also be produced by thermalized neutrons. Second, the cross section can not be measured as a function of a small energy step of the neutron. The latter method is used to determine the cross section by detecting prompt γ-rays from a keV neutrons capture to its low-lying states, and it has the following strong points. First, it can be applied for all nuclei, irrespective of the β-decay half-life. Second, the cross section can be measured as a function of the neutron energy using a time-of-flight (TOF) method. Third, background γ-rays induced by scattered neutrons can be discriminated from those due to keV neutrons by the TOF method. Fourth, a prompt discrete γ-ray characterizing the final nucleus can be detected, which plays a critical role in determining the cross section unambiguously. Here, the discrete γ-rays, connecting particular nuclear states with definite spin-parities and having unique electromagnetic properties, are known to carry important information to study in detail both the reaction mechanism and nuclear structure. However, there are several weak points. First, in order to obtain a suffi- cient γ-ray yield from the keV neutron capture reaction it is necessary to use a large amount of sample material, a few moles for light nuclei. Second, the incident and/or scattered neutrons interact with the γ-ray detector, producing a large amount of background, and third they often damage the detector. These problems, however, could be overcome by constructing a spectrometer with high γ-ray detection efficiency and a powerful shield against neutrons. $NaI(Tl)$ detectors have been used for many years to detect a discrete γ-ray from the fusion reaction induced by charged particles. Yet, the $NaI(Tl)$ detector is known to be sensitive to neutrons; namely, the thermalized neutrons mentioned above are captured in the $NaI(Tl)$ detector via an $^{127}I(n,\gamma)^{128}I$ reaction, which has a large cross section of 6.1 b for thermal neutrons. However, if one could develop an $NaI(Tl)$ spectrometer system with powerful shield against neutrons, it would play an important role in studying the (n,γ) reaction. In the present work we have constructed an anti-Compton $NaI(Tl)$ spectrometer consisting of a central $NaI(Tl)$ detector surrounded by an annular one [12]. The spectrometer is well shielded against neutrons with use of ^6Li, H and B-doped polyethylene.

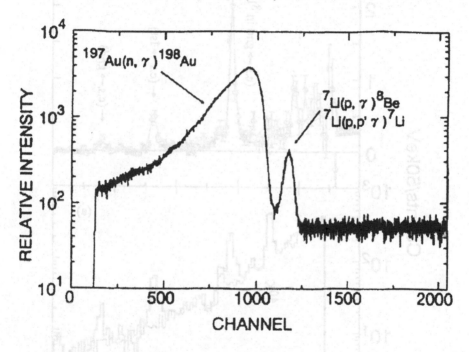

Figure 1 Time-of-flight spectrum measured by a $NaI(Tl)$ detector with a ^{197}Au sample.

The cross section has been measured by employing the prompt γ-ray detection method with use of the $NaI(Tl)$ detector, combined with pulsed keV neutrons. The neutrons are produced via the $^7Li(p,n)^7Be$ reaction using pulsed protons, which are provided from the 3.2 MV Pelletron Accelerator of the Research Laboratory for Nuclear Reactor at the Tokyo Institute of Technology. The neutron energy is determined by the TOF method. A typical TOF spectrum measured for Au by the spectrometer is shown in Fig. 1, where the sharp peak at about channel 1100 is due to a γ-ray event from the $^7Li(p,\gamma)^8Be$ reaction at the neutron target position, and the broad peak is due to events from the *keV* neutron-capture reaction by *Au*. The plateau region is due to background caused by scattered neutrons. Therefore, a true event can be obtained by subtracting the background (B) from the foreground (F); here, two spectra "B" and "F" were obtained by putting gates on the plateau and broad peak regions, respectively. Typical foreground, back-

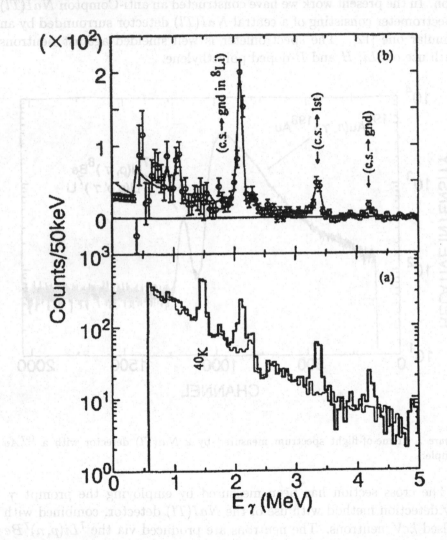

Figure 2 Foreground (*thickline*) and background (*thinline*) spectra (*a*), and background-subtracted γ-ray spectrum (*b*) at the average neutron energy (lab.) of 40 *keV*, respectively.

ground, and background-subtracted γ-ray spectra for the $^{16}O(n,\gamma)^{17}O$ reaction at an average neutron energy of 40 *keV* are shown in Figs.

2a and 2b. The 2.1 *MeV* peak was due to the γ-ray transition from the captured state, populated by the $^7Li(n,\gamma)^8Li$ reaction, to the 2^+ (ground) state in ^8Li; a weak 2.2 *MeV* γ-ray peaks of 3.3 and 4.2 *MeV* were assigned as being due to the $^{16}O(n,\gamma)^{17}O$ reaction. Namely, they correspond to the γ-ray transitions from a captured state to the $1/2^+$ (first) and $5/2^+$ (ground) states in ^{17}O, respectively, as shown in the partial level scheme of ^{17}O in Fig. 3. Here, both the $5/2^+$ and $1/2^+$

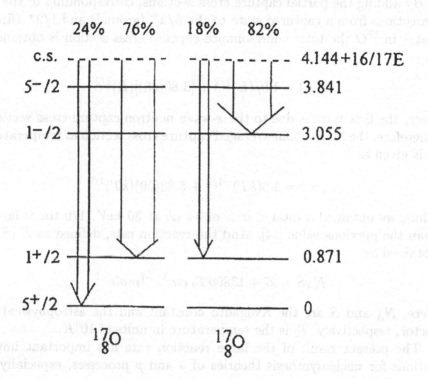

Figure 3 Partial level scheme of ^{17}O. Branching ratios from both the present experiment at $E = 40\ keV$ (right) and the thermal-neutron induced reaction (left) are shown, respectively.

states should be noted as having large neutron single-particle strengths of about 0.9. The γ-ray intensity in the background subtracted spectrum is analyzed by a stripping method using a response function of the γ-ray spectrometer. The absolute capture cross section $\{\sigma_\gamma(A)\}$ of sample A is obtained by comparing the γ-ray yield from the sample with that of Au as [13]

$$\sigma_\gamma(A) = C\frac{\phi_{Au}}{\phi_A} \cdot \frac{(\gamma^2 n)_{Au}}{(\gamma^2 n)_A} \cdot \frac{Y_\gamma(A)}{Y_\gamma(Au)} \cdot \sigma_\gamma(Au),$$

where Au is used, since the absolute capture cross section of Au is well known. Here, $\phi_{Au(A)}$ and $Y_\gamma\{Au(A)\}$ are the yields of the neutron and the γ-ray for $Au(A)$, respectively; $\sigma_\gamma\{Au(A)\}$ is the absolute capture cross section of $Au(A)$, γ and n are the radius and thickness of the sample, respectively. A correction factor (C) is introduced to correct for the multiple-scattering effect and the shielding of the incident neutrons in the sample.

By adding the partial capture cross sections, corresponding to the γ-transitions from a captured state to the $5/2^+$ (ground) and $1/2^+$ (first) states in ^{17}O the total nonresonance capture cross section is obtained, as

$$\sigma(E) = 1.0/(E)^{1/2} + [3.80(38)](E)^{1/2}$$

Here, the first term is due to the s-wave neutron-capture cross section. Therefore, the Maxwellian-averaged capture cross section at temperature T is given as

$$<\sigma> = 1.0(kT)^{-1/2} + 5.88(59)(kT)^{1/2}$$

Thus, we obtained a total $<\sigma>$ of 34 μb at 30 keV, 170 times larger than the previous value [14]. And the reaction rate, defined as $N_A S$, is obtained as

$$N_A S = 27 + 13800\, T_9 \; cm^3 s^{-1} mole^{-1}$$

Here, N_A and S are the Avogadro constant and the astrophysical S-factor, respectively. T_9 is the temperature in units of $10^9 K$.

The present result of the large reaction rate has important implications for nucleosynthesis theories of s and p processes, especially in metal-deficient massive stars, as a very important neutron poison. Actually, the s-process efficiency has been calculated for various metalicities of $Z = 1$, 0.1, 0.01 and 0.001. Thus, it is shown in particular that ^{16}O progressively emerges as a very effective neutron poison as metalicity decreases, the poisoning effect depending dramatically on σ_{16} [15].

The large capture cross section obtained here can be explained by nonresonant p-wave capture, which was demonstrated by two experimental facts: firstly, the drastic change in the γ-ray branching ratios from thermal to keV neutron capture; secondly, the $E^{1/2}$ energy dependence of the partial cross sections. The cross section has been calculated theoretically based on nonresonant p-wave capture; they are in good agreement with the present result. The nonresonant p-wave neutron capture process was also found in the $^{12}C(n,\gamma)^{13}C$ reaction [16]. It should be added that the nonresonant p-wave neutron capture process plays a crucial role for the neutron capture of light nuclei with $A \leq 18$, except for ^6Li and ^7Li, since there are s and/or d single-particle states with a considerably large spectroscopic factor in the residual nucleus, thereby allowing $E1$ γ-decay from a p-wave neutron-captured state to these states. Consequently, the neutron-capture cross section may deviate considerably from the extrapolated value of the measured thermal capture cross section by assuming a $1/v$ law, as found in the $^{16}O(n,\gamma)^{17}O$ and $^{12}C(n,\gamma)^{13}C$ reactions, and therefore the process significantly influences both the stellar and above mentioned primordial nucleosynthesis theories.

Similarly, the cross section of the $p(n,\gamma)d$ reaction was directly measured at the neutron energy between 10 and 550 keV, and the measured value is in good agreement with the theoretical value which includes the meson-exchange currents [17].

3.   Summary

In order to construct models predicting s and p isotope production as function of the metalicity and stellar mass and also estimating the production yields of heavy elements in the r-process and in the inhomogeneous big-bang, it appears quite important to measure the neutron capture cross section of light nuclei at stellar energy. The cross section, however, is expected to be very small, around 10 μb at 30 keV, and therefore one has to develop a high sensitive measuring system. We have developed a prompt γ-ray detection method with use of an $NaI(Tl)$ spectrometer, combined with pulsed keV neutrons. In the present work, we have succeeded to measure the cross sections for light nuclei, such as H, ^7Li, ^{12}C and ^{16}O, at the neutron energies of between 10 and 300 keV. The present results of the cross sections have important implications concerning nucleosynthesis from low-metalicity AGB stars, for the neutrino-induced nucleosynthesis as well as for the heavy-element nucleosynthesis yields of inhomogeneous big-bang models. The unexpected large cross sections for ^{12}C and ^{16}O can be explained quantitatively by a p-wave neutron capture.

References

[1] P.J.E. Peebles, Phys. Rev. Lett. 16 (1966) 410

[2] C.E. Rolfs and W.S. Rodney, *Cauldrons in the Cosmos* (1988)

[3] E.M. Burbidge et al., Rev. Mod. Phys. 29 (1957) 547

[4] J.C. Wheeler, C. Sneden and J.W. Truran. Jr., Ann. Rev. Astron. Astrophys. 27 (1989) 279

[5] I. Iben and A. Renzins, Astrophys. J. 259 (1982) L79

[6] S.A. Lamb et al., Astrophys. J. 217 (1977) 213

[7] D. Lambert, Astron. Astrophys. Rev. 3 (1992) 201

[8] S. E. Woosley et al., Astrophys. J. 356 (1990) 272

[9] D. N. Schramm and R. V. Wagoner, Ann. Rev. Nucl. Part. Sci. 27 (1977) 37

[10] J. Applegate, V. Hogan and R. J. Scherrer, Phys. Rev. D35 (1988) 1151

[11] R. A. Malaney and W. A. Fowler, Astrophys. J., 333 (1988) 14

[12] T. Ohsaki et al., Nucl. Inst. Math. A425 (1999) 302

[13] Y. Nagai et al., Astrophys. J. 372 (1991) 683

[14] M. Igashira et al., Astrophys. J. 441 (1995) L89

[15] M. Rayet and M. Hashimoto, Astron. Astrophys. 354 (2000) 740

[16] T. Ohsaki et al., Astrophys. J. 422(1994) 912

[17] T. S. Suzuki et al., Astrophys. J. 439 (1995) L59; Y. Nagai et al., Phys. Rev. C56 (1997) 3173

Astrophysical S-Factors from Asymptotic Normalization Coefficients

R.E. Tribble[a], A. Azhari[a], P. Bem[b], V. Burjan[b], F. Carstoiu[c],
J. Cejpek[b], H.L. Clark[a], C.A. Gagliardi[a], V. Kroha[b], Y.-W. Lui[a],
A.M. Mukhamedzhanov[a], J. Novak[b], S. Piskor[b], A. Sattarov[a],
E. Simeckova[b], X. Tang[a], L. Trache[a] and J. Vincour[b]

[a] *Cyclotron Institute, Texas A&M University, College Station, Texas 77843 USA* *

[b] *Institute for Nuclear Physics, Czech Academy of Sciences, Prague-Řež, Czech Republic*

[c] *Institute for Atomic Physics, Bucharest, Romania*

tribble@comp.tamu.edu

Keywords: Nuclear astrophysics, nuclear reactions

Abstract   Peripheral transfer reactions can be used to determine asymptotic normalization coefficients (ANC). These coefficients, which provide the normalization of the tail of the overlap function, determine S-factors for direct capture reactions at astrophysical energies. A variety of proton transfer reactions have been used to measure ANCs. As a test of the technique, the $^{16}O(^3He, d)^{17}F$ reaction has been used to determine ANCs for transitions to the ground and first excited states of ^{17}F. The S-factors for $^{16}O(p, \gamma)^{17}F$ calculated from these $^{17}F \rightarrow {}^{16}O + p$ ANCs are found to be in very good agreement with recent measurements. Following the same technique, the $^{10}B(^7Be, ^8B)^9Be$ and $^{14}N(^7Be, ^8B)^{13}C$ reactions have been used, along with optical model parameters for the radioactive beams that were obtained from a study of elastic scattering of loosely bound p-shell nuclei, to measure the ANC appropriate for determining $^7Be(p, \gamma)^8B$. The results from the two transfer reactions provide an indirect determination of $S_{17}(0)$. Recent measurements have been completed on the $^{14}N(^{11}C, ^{12}N)^{13}C$ reaction which will allow us to define the astrophysical S-factor for the $^{11}C(p, \gamma)^{12}N$ reaction.

1.   Introduction

The advent of space based telescopes has led to an explosion of new information about elemental abundances for a wide range of stars in the

---

*Supported in part by the U.S. Dept. of Energy and by the Robert A. Welch Foundation.

D. N. Poenaru et al. (eds.), Nuclei Far from Stability and Astrophysics, 317–328.
© 2001 *Kluwer Academic Publishers. Printed in the Netherlands.*

cosmos. Processing of nuclear fuel generates the power that fuels stars and produces these elements, often through explosive processes such as supernovae. Understanding the stellar evolution that leads to the observed abundances requires knowing reaction rates and half lives for a wide range of nuclei. Capture reactions involving stable nuclei have been studied for many years by measuring cross sections in the laboratory at low incident bombarding energies. However, very few reactions involving unstable nuclei have been measured by these techniques. To make further progress in understanding stellar evolution, new techniques for determining capture reaction rates must be found.

Direct capture reactions of astrophysical interest usually involve systems where the binding energy of the captured proton is low. Hence, at stellar energies, the capture proceeds through the tail of the nuclear overlap function. The shape of the overlap function in this tail region is completely determined by the Coulomb interaction, so the amplitude of this function alone dictates the rate of the capture reaction. The $^7Be(p, \gamma)^8B$ reaction is an excellent example of such a direct capture process. Indeed recent calculations of the normalization constant have been used to predict the capture rate [1, 2]. But new measurements, both direct and indirect, are still needed as was underscored in a recent review of stellar reaction rates [3] which includes a detailed discussion of the uncertainties in our present knowledge of $S_{17}(0)$ and its importance to the solar neutrino problem.

The asymptotic normalization coefficient (ANC) C for $A + p \leftrightarrow B$ specifies the amplitude of the tail of the overlap function for the system. In previous communications [1, 4], we have pointed out that astrophysical S-factors for peripheral direct radiative capture reactions can be determined through measurements of ANCs using traditional nuclear reactions such as peripheral nucleon transfer. Direct capture S-factors derived with this technique are most reliable at the lowest incident energies in the capture reaction, precisely where capture cross sections are smallest and most difficult to measure directly. Of course it is extremely important to test the reliability of the technique in order to know the precision with which it can be applied. Determining the S-factors for $^{16}O(p, \gamma)^{17}F$ from its ANCs has been recognized as a suitable test for this method [3] because the results can be compared to existing direct measurements of the cross sections [5, 6]. Furthermore, the $^{16}O(p, \gamma)^{17}F$ reaction has substantial similarities to the $^7Be(p, \gamma)^8B$ reaction. As part of an ongoing program to measure ANCs, we have used the proton exchange reactions $^9Be(^{10}B, ^9Be)^{10}B$ and $^{13}C(^{14}N, ^{13}C)^{14}N$ along with the $^{13}C(^3He, d)^{14}N$ reaction to measure the ANCs for $^{10}B \rightarrow ^9Be + p$ and $^{14}N \rightarrow ^{13}C + p$. Below we briefly summarize these results. This is followed by a discussion of a measurement of another proton transfer reaction, $^{16}O(^3He,d)^{17}F$, which is used to determine the ANCs for the ground and first excited states in ^{17}F. From these ANCs, we calculate S-factors for both $^9Be(p,\gamma)^{10}B$ and $^{16}O(p,\gamma)^{17}F$ and compare to experimental results. Then we discuss our measurements of the $^{10}B(^7Be,^8B)^9Be$ and $^{14}N(^7Be,^8B)^{13}C$ reactions, the extraction of the ANCs for $^8B \to {}^7Be + p$ and our determination of $S_{17}(0)$. Finally we present preliminary results on a recent measurement of the $^{14}N(^{11}C,^{12}N)^{13}C$ reaction which will allow us to determine the astrophysical S-factor for $^{11}C(p,\gamma)^{12}N$.

2.   ANCs from Proton Transfer Reactions

Traditionally spectroscopic factors have been obtained from proton transfer reactions by comparing experimental cross sections to DWBA predictions. For peripheral transfer, we show below that the ANC is better determined and is the more natural quantity to extract. Consider the proton transfer reaction $a + A \to c + B$, where $a = c + p$, $B = A + p$. As was previously shown [7] we can write the DWBA cross section in the form

$$\frac{d\sigma}{d\Omega} = \sum_{j_B j_a} \frac{(C^B_{Apl_Bj_B})^2}{b^2_{Apl_Bj_B}} \frac{(C^a_{cpl_aj_a})^2}{b^2_{cpl_aj_a}} \tilde{\sigma}^{DW}_{l_Bj_Bl_aj_a}, \qquad (1)$$

where $\sigma^{DW}_{l_Bj_Bl_aj_a}$ is the reduced DWBA cross section and j_i, l_i are the total and orbital angular momenta of the transferred proton in nucleus i. The factors $b_{cpl_aj_a}$ and $b_{Apl_Bj_B}$ are the ANCs of the bound state proton wave functions in nuclei a and B. If the reaction under consideration is peripheral, the ratio

$$R_{l_Bj_Bl_aj_a} = \frac{\tilde{\sigma}^{DW}_{l_Bj_Bl_aj_a}}{b^2_{Apl_Bj_B} b^2_{cpl_aj_a}} \qquad (2)$$

is independent of the single particle ANCs $b_{Apl_Bj_B}$ and $b_{cpl_aj_a}$. Thus for surface reactions the DWBA cross section is best parameterized in terms of the product of the square of the ANCs of the initial and final nuclei $(C^B)^2(C^a)^2$ rather than spectroscopic factors.

We have used this formulation to extract ANCs for $^{10}B \leftrightarrow {}^9Be + p$, $^{14}N \leftrightarrow {}^{13}C + p$, and $^{17}F \leftrightarrow {}^{16}O + p$ from the reactions $^9Be(^{10}B,^9Be)^{10}B$, $^{13}C(^{14}N,^{13}C)^{14}N$, $^{13}C(^3He,d)^{14}N$, and $^{16}O(^3He,d)^{17}F$. The first two reaction studies were carried out with beams from the K500 superconducting cyclotron at Texas A&M University. The $^{13}C(^3He,d)^{14}N$ reaction was investigated at the U-120M isochronous cyclotron of the Nuclear Physics Institute of the Czech Academy of Sciences. The $^{16}O(^3He,d)^{17}F$

reaction was studied at both laboratories. Details of the experiments can be found in [7–10].

In order to extract ANCs, DWBA calculations were carried out with the finite range code PTOLEMY [11], using the full transition operator. A check on the extracted ANCs versus Woods-Saxon well radial parameters indicated that the calculated DWBA cross sections are insensitive to assumptions about the wave functions in the nuclear interior. For the $(^3He, d)$ reactions, normalizing the DWBA calculations to the data and dividing by the ANCs for the single particle orbitals yields the product of the ANCs for $^3He \rightarrow d+p$ and either $^{14}N \rightarrow {}^{13}C+p$ or $^{17}F \rightarrow {}^{16}O+p$. Dividing this product by the known ANC for $^3He \rightarrow d+p$ [12] provides C^2 for the other vertex of the reaction. For proton transfer between ground states in $^9Be(^{10}B, {}^9Be)^{10}B$ and $^{13}C(^{14}N, {}^{13}C)^{14}N$, the two vertices are identical and each cross section depends on C^4.

The two independent measurements of the $^{14}N \leftrightarrow {}^{13}C + p$ ANCs provide a consistency check on our ability to extract ANCs from proton transfer reactions. We find the ANC for the dominant $p_{1/2}$ component of the ^{14}N ground state to be $C^2 = 18.6 \pm 1.2 \ fm^{-1}$ from the $^{13}C\,(^{14}N, {}^{13}C)\,^{14}N$ reaction [8] and $17.8 \pm 1.3 \ fm^{-1}$ from the $^{13}C(^3He, d)^{14}N$ reaction [9]. We adopted the weighted average of these two results when analyzing the $^{14}N(^7Be, {}^8B)^{13}C$ and $^{14}N(^{11}C, {}^{12}N)^{13}C$ reactions described below. This would not have been practical if we had utilized spectroscopic factors determined from the DWBA fits to $^{13}C(^{14}N, {}^{13}C)^{14}N$ and $^{13}C(^3He, d)^{14}N$ instead of ANCs, because the inferred spectroscopic factors are very sensitive to the choice of single-particle bound state orbitals in the DWBA calculations.

The ANCs for $^{10}B \leftrightarrow {}^9Be+p$ have been used to predict the direct capture contribution to the $^9Be(p, \gamma)^{10}B$ astrophysical S-factor and in the analysis of the $^{10}B(^7Be, {}^8B)^9Be$ reaction. The ANCs for $^{17}F \leftrightarrow {}^{16}O+p$ have been used to predict the astrophysical S-factor for $^{16}O(p, \gamma)^{17}F$. The results are discussed in the following sections.

3. Using ANCs to Predict Astrophysical S-Factors: Test Cases

The ANC's found from the proton transfer reactions can be used to determine direct capture rates at astrophysical energies. The relation of the ANC's to the direct capture rate at low energies is straightforward to obtain. The cross section for the direct capture reaction $A + p \rightarrow B + \gamma$ can be written as

$$\sigma = \lambda |< I_{Ap}^B(\mathbf{r}) \mid \hat{O}(\mathbf{r}) \mid \psi_i^{(+)}(\mathbf{r}) >|^2, \tag{3}$$

where λ contains kinematical factors, I_{Ap}^B is the overlap function for $B \to A+p$, $\hat{O}$ is the electromagnetic transition operator, and $\psi_i^{(+)}$ is the scattering wave in the incident channel. If the dominant contribution to the matrix element comes from outside the nuclear radius, the overlap function may be replaced by

$$I_{Ap}^B(r) \approx C \frac{W_{-\eta, l+1/2}(2\kappa r)}{r}, \tag{4}$$

where C defines the amplitude of the tail of the radial overlap function I_{Ap}^B, W is the Whittaker function, η is the Coulomb parameter for the bound state $B = A + p$, and κ is the bound state wave number. The required C's are just the ANCs found above from transfer reactions. Thus, the direct capture cross sections are directly proportional to the squares of these ANCs.

Using the results for the ANCs discussed above, the S-factors describing the capture to both the ground and first excited states for $^{16}O(p,\gamma)^{17}F$ were calculated, with no additional normalization constants. The results are shown in Fig. 1 compared to the two previous measurements of $^{16}O(p,\gamma)^{17}F$ [5, 6]. Both $E1$ and $E2$ contributions have been included in the calculations, but the $E1$ components dominate the results. The theoretical uncertainty in the S-factors is less than 2 % for energies below 1 MeV. Above 1 MeV the nuclear interaction begins to be important in the evaluation of the scattering wave function. The agreement between the measured S-factors and those calculated from our $^{17}F \to {}^{16}O + p$ ANCs is quite good, especially for energies below 1 MeV where the approximation of ignoring contributions from the nuclear interior should be very reliable and the optical potential uncertainties are negligible. Overall, the results verify that the technique is valid for determining S-factors to accuracies of at least 9 %.

The S-factor for $^9Be(p,\gamma)^{10}B$ has contributions from both resonance and direct capture at stellar energies. Thus the connection between the ANC and the capture cross section is more complicated and is discussed elsewhere [14]. Our measurement of the ANCs for $^{10}B \to {}^9Be + p$ fixes the nonresonant capture component. With this result, an R-matrix fit was made to the existing data [15] using the known locations of the resonance states and their widths. The fit [14], which is shown in Fig. 2, does an excellent job of reproducing the data. Prior to our determination of the direct capture contribution, similar attempts to fit the data required substantial changes in the known resonance positions and widths.

Recently, the ANC technique has been tested in a neutron capture reaction for the first time. The ANC for $^{12}C + n \leftrightarrow {}^{13}C(1/2^+, 3.09\ MeV)$,

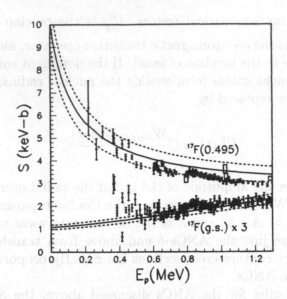

Figure 1 A comparison of the experimental S-factors to those determined from the ANCs found in $^{16}O(^3He,d)^{17}F$. The solid data points are from [5], and the open boxes are from [6]. The solid lines indicate our calculated S-factors, and the dashed lines indicate the $\pm 1\sigma$ error bands. Note that the experimental ground state S-factor may be contaminated by background [13] at energies below 500 keV.

measured in the $^{12}C(d,p)^{13}C$ reaction, has been found to predict the corresponding direct capture rate to better than 20 % [16].

4.   ANCs for Stellar Processes

We have measured the $(^7Be,^8B)$ reaction on a 1.7 mg/cm^2 ^{10}B target [17] and a 1.5 mg/cm^2 Melamine target [18] in order to extract the ANC for $^8B \rightarrow {}^7Be + p$. The radioactive ^7Be beam was produced at 12 MeV/A by filtering reaction products from the $^1H(^7Li,^7Be)n$ reaction in the recoil spectrometer MARS, starting with a primary ^7Li beam at 18.6 MeV/A from the TAMU K500 cyclotron. The beam was incident on an H_2 cryogenic gas target, cooled by LN_2, which was kept at 1 atm (absolute) pressure. Reaction products were measured by 5 $cm \times 5\ cm$ Si detector telescopes consisting of a 100 μm ΔE strip detector, with 16 position sensitive strips, followed by a 1000 μm E counter.

A single 1000 μm Si strip detector was used for initial beam tuning. This detector, which was inserted at the target location, allowed us to optimize the beam shape and to normalize the ^7Be flux relative to a Faraday cup that measured the intensity of the primary ^7Li beam. Following optimization, the approximate ^7Be beam size was 6 $mm \times 3\ mm$

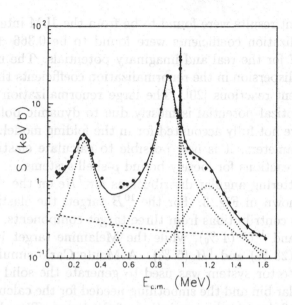

Figure 2 The result of an *R*-matrix fit to the direct plus resonance contributions to the *S*-factor for $^9Be(p,\gamma)^{10}B$. Standard resonance widths and positions were used in the fit. The direct capture contribution to the *S*-factor is the dashed line.

(FWHM), the energy spread was ≈ 1.5 *MeV*, the full angular spread was $\Delta\theta \approx 28$ *mrad* and $\Delta\phi \approx 62$ *mrad*, and the purity was $\geq 99.5\%$ ^7Be for the experiment with the ^{10}B target. The beam size and angular spread were improved for the experiment with the ^{14}N target to 4 *mm* $\times 3$ *mm* (FWHM), $\Delta\theta \approx 28$ *mrad* and $\Delta\phi \approx 49$ *mrad*. Periodically during the data acquisition, the beam detector was inserted to check the stability of the secondary beam tune. The system was found to be quite stable over the course of the experiment with maximum changes in intensity observed to be less than 5 %. The typical rate for ^7Be was ≈ 1.5 *kHz/particle nA* of primary beam on the production target. Primary beam intensities of up to 80 *particle nA* were obtained on the gas cell target during the experiments.

In order to extract ANCs from reactions involving radioactive-ion beams, it is necessary to have optical model parameters. Typically radioactive beam intensities are too small to measure elastic scattering and obtain good optical model parameters. Consequently we have carried out a series of elastic scattering measurements with stable beam and target combinations that are close to those for our radioactive beam measurements. We used the folding model prescription to calculate the potential parameters and then renormalized the real and imaginary parts of the potentials to fit the data. Several different interactions were tried but the most consistent results were found to be from the JLM interaction [19]. The renormalization coefficients were found to be 0.366 ± 0.014 and 1.000 ± 0.087 for the real and imaginary potentials. The uncertainties are from the dispersion in the renormalization coefficients that we found for the different reactions [20]. The large renormalization for the real part of the optical potential is mostly due to dynamic polarization effects which are not fully accounted for in the folding model. With these potential parameters, it is now possible to calculate elastic scattering and transfer reactions for loosely bound p-shell systems.

Elastic scattering angular distributions for ^7Be on the ^{10}B and ^{14}N targets are shown in Fig. 3. For the ^{10}B target, the elastic scattering yield includes contributions from three target components, ^{10}B (86 %), ^{12}C (10 %) and ^{16}O (4 %), while the Melamine target includes ^{14}N (67 %), ^{12}C (28 %) and ^1H (5 %). A Monte Carlo simulation of the beam and detector system was used to generate the solid angle factor for each angular bin and the smoothing needed for the calculation to account for the finite angular resolution of the beam. The absolute cross section is then fixed by the target thickness, number of incident ^7Be, the yield in each bin, and the solid angle. The curves shown with the elastic scattering were found from the optical model, with the parameters discussed above, by adding together the cross section predictions for the target components in the laboratory frame and then transforming the result to the center of mass assuming kinematics appropriate for either the ^{10}B or ^{14}N targets. In both cases, the optical model calculations are compared to the data without additional normalization coefficients. The detector resolution is not sufficient to distinguish inelastic excitations from elastic scattering. This likely explains why the data exceed the calculations in the minima. Overall, the agreement between the measured absolute cross sections and the optical model predictions is excellent thus providing confidence that our normalization procedure is correct.

Angular distributions for the $(^7Be, {}^8B)$ reactions populating the ground states of ^9Be and ^{13}C were extracted using the same procedure as for the elastic scattering. The results are compared to DWBA calculations in Fig. 4. The normalization factors between the data and calculations were obtained from the fits to the respective Q-value spectra.

The ANC for $^8B \rightarrow {}^7Be + p$ was extracted based on the fit to the angular distributions and the ANCs for $^{10}B \rightarrow {}^9Be + p$ and $^{14}N \rightarrow {}^{13}C + p$ following the procedure outlined above in our test case. Two ^8B orbitals, $1p_{1/2}$ and $1p_{3/2}$, contribute to the transfer reaction but the $1p_{3/2}$ dominates in both cases. In calculating the angular distributions, we used the ratio for the two orbitals as given by a microscopic description

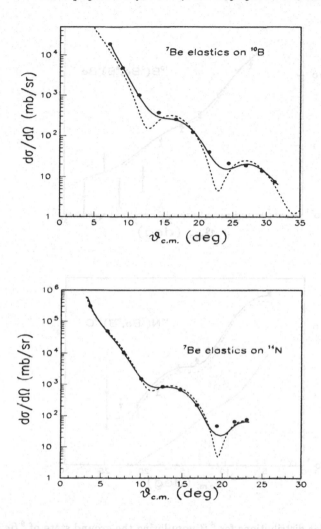

Figure 3 Angular distributions for elastic scattering from the ^{10}B and ^{14}N targets. The dashed curves are from optical model calculations of the target components and the solid curves are smoothed over the angular acceptance of each bin.

of the ^8B ground state [4]. The optical model parameters were obtained from renormalized microscopic folding potentials using the JLM effective NN interaction described above. The entrance channel parameters were the same as those used in calculating the elastic scattering angular distributions for ^7Be on ^{10}B and ^{14}N in Fig. 3. We have checked the sensitivity of the calculations by varying the normalization parameters. As in previous studies, the results are insensitive to bound state single particle well parameters in the DWBA calculations.

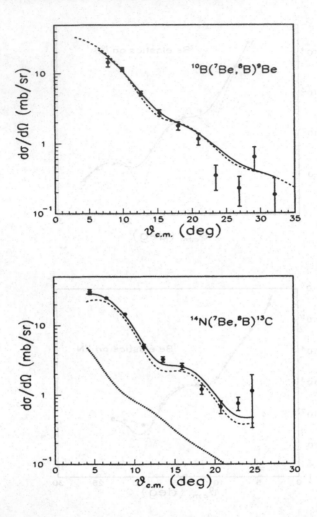

Figure 4 Angular distributions for ^8B populating the ground state of ^9Be from the ^{10}B target and ^{13}C from the Melamine target. In the top figure, the dashed curve shows the result of a DWBA calculation for the dominant component that contributes to the cross section. Two components are shown in the bottom figure. In both cases, the solid curve is smoothed over the angular acceptance of each bin.

The astrophysical S-factor for $^7Be(p, \gamma)^8B$ has been determined from the ANC for each target. The quoted uncertainties include 8.1 % for optical model parameters. Experimental uncertainties and the ANC of the other vertex contribute an additional 10.9 % for the ^{10}B target and 7.6 % for the ^{14}N target. The relative contribution of the two orbitals is straightforward to calculate and introduces a negligible additional uncertainty to the S-factor [1, 4]. The values that we find are $S_{17}(0) = 18.4 \pm 2.5 \; eV \; b$ for the ^{10}B target and $16.6 \pm 1.9 \; eV \; b$

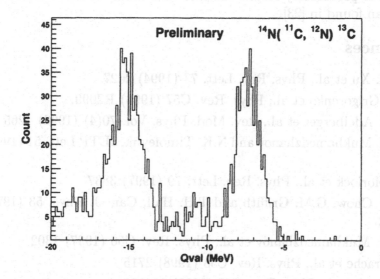

Figure 5 Preliminary result showing the Q-value spectrum for ^{12}N reaction products from the $^{14}N(^{11}C, {}^{12}N)^{13}C$ reaction. The right peak corresponds to excitation of the ground state of ^{13}C while the other peak is primarily from proton pickup to the ground state of ^{11}B from the ^{12}C in the target.

for the ^{14}N target, which are in good agreement with the recommended value [3] of 19^{+4}_{-2} eV b.

A primary source of uncertainty in the values quoted above for $S_{17}(0)$ is the optical model parameters that are used to predict the angular distribution. As indicated, we have developed a set of global optical model parameters for use with radioactive beams in this mass and energy region. Since the optical model parameters for the two different targets are derived by the same technique, this introduces correlations in the uncertainties between the two results. Accounting for this, we find the S-factor from the combined measurements to be 17.3 ± 1.8 eV b.

Recently we completed a measurement of the $^{14}N(^{11}C, {}^{12}N)^{13}C$ reaction, using procedures similar to those in the $(^7Be, {}^8B)$ measurements. This experiment will determine the ANC for $^{11}C + p \leftrightarrow {}^{12}N$ and, in turn, the direct capture rate for $^{11}C(p,\gamma)^{12}N$ at astrophysical energies. This reaction, which is a part of the hot pp chain, may provide a path for low-metalicity, super-massive stars to produce CNO nuclei while bypassing the triple-alpha reaction [22]. There have been two previous investigations of this reaction, both using Coulomb break-up of ^{12}N [23, 24], which have obtained conflicting results. A preliminary Q-value spectrum for ^{12}N reaction products from our measurement is shown in

Fig. 5. Our preliminary results indicate that the S-factor is somewhat larger than found in [23].

References

[1] H.M. Xu et al., Phys. Rev. Lett. 73 (1994) 2027

[2] L.V.Grigorenko et al., Phys. Rev. C57 (1998) R2099.

[3] E.G. Adelberger et al., Rev. Mod. Phys. Vol. 70(4) (1998) 1265

[4] A.M. Mukhamedzhanov and N.K. Timofeyuk, JETP Lett. 51 (1990) 282

[5] R. Morlock et al., Phys. Rev. Lett. 79 (1997) 3837

[6] H.C. Chow, G.M. Griffith and T.H. Hall, Can. J. Phys. 53 (1975) 1672

[7] A.M. Mukhamedzhanov et al., Phys. Rev. C56 (1997) 1302

[8] L. Trache et al., Phys. Rev. C58 (1998) 2715

[9] P. Bem et al., Phys. Rev. C62 (2000) 024320

[10] C.A. Gagliardi et al., Phys. Rev. C59 (1999) 1149

[11] M. Rhoades-Brown, M. McFarlane and S. Pieper, Phys. Rev. C21 (1980) 2417; Phys. Rev. C21 (1980) 2436

[12] A.M. Mukhamedzhanov, R.E. Tribble and N.K. Timofeyuk, Phys. Rev. C51 (1995) 3472

[13] R. Morlock, private communication.

[14] A. Sattarov et al., Phys. Rev. C60 (1999) 035801

[15] D. Zhanow et al., Nucl. Phys. A589 (1995) 95

[16] N. Imai et al., Proc. Conf. Nuclei in the Cosmos, 2000, in press

[17] A. Azhari et al., Phys. Rev. Lett. 82 (1999) 3960

[18] A. Azhari et al., Phys. Rev. C60 (1999) 035801

[19] J.P. Jeukenne, A. Lejeune and C. Mahaux, Phys.Rev. C16 (1977) 80

[20] L. Trache et al., Phys. Rev. C61 (2000) 024612-1

[21] S. Cohen and D. Kurath, Nucl. Phys. 73 (1965) 1

[22] M. Wiescher et al., Astrophys. J. 343 (1989) 352

[23] A. Lefebvre et al., Nucl. Phys. A592 (1995) 69

[24] T. Motobayashi, Proc. 2nd Int. Conf. Exotic Nucl. Atom. Masses, ed. by B. M. Sherrill, D. J. Morrissey, and C. N. Davids (AIP Conference Proceedings 455, Woodbury, NY, 1998) p. 882

III

Cosmic Rays

Massive Nuclei in the Cosmic Radiation

Arnold Wolfendale

Physics Department, University of Durham, Durham, DH1 3LE, U.K.

A.W.Wolfendale@durham.ac.uk

Keywords: Cosmic rays, energy spectrum, mass composition

Abstract   The most energetic nuclei in the contemporary universe are to be found in the Cosmic Radiation. The paper starts with comments about low energy nuclei but goes on to discuss the situation in the 'knee' of the cosmic ray spectrum at $\sim 3 \cdot 10^{15}$ eV. This is followed by an examination of the evidence for 'heavy' nuclei (probably iron) in the Galactic component, at higher energies, extending to somewhat above 10^{18} eV. Finally, the enigmatic extragalactic particles are considered, these particles being presumed to be mainly confined to energies above about 10^{19} eV. Arguments advanced by Wibig and Wolfendale [1] which claim evidence for heavy nuclei here, too, will be considered.

1.   Introduction

Although cosmic rays were discovered in 1912, by Viktor Hess in his perilous balloon ascents, there is still considerable debate about their origin, particularly above some 10^{11} eV — those below, at least, being almost certainly generated by supernova remnant (SNR) shocks, in the interstellar medium (ISM) (see Wolfendale and Zhang, [2]). It is likely, though by no means certain, that SNR continue to be a major mechanism to about 10^{16} eV but above this energy explanations are very uncertain (see Erlykin and Wolfendale [3] and [4]).

Of less uncertainty is the mass composition, although, as will be demonstrated, even here there are problems which develop above about 10^{14} eV. In what follows we examine the energy ranges in turn.

2.   The GeV Region

The discovery of nuclei heavier than those of hydrogen, was made a little over 50 years ago [5]. Since that date there has been considerable

D. N. Poenaru et al. (eds.), Nuclei Far from Stability and Astrophysics, 331–340.
© 2001 *Kluwer Academic Publishers. Printed in the Netherlands.*

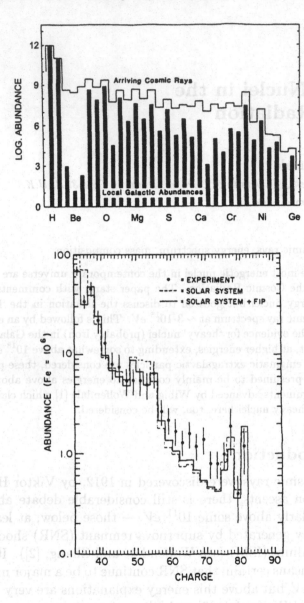

Figure 1 Relative abundance of CR nuclei in comparison with 'Local Galactic Abundances' (approx. equal to the 'solar system abundances') from the summary by [6]. The excess of *Li*, *Be* and *B* is due to fragmentation of heavier CR in the ISM. The excess in the range *Z*: 60 – 75 is indicative of *r*-process mechanisms.

work to determine the energy spectra of the mass components and their isotopic composition. In comparison with the composition of the solar system there are some remarkable differences, most dramatically in the case of Li, Be and B which are almost absent in the solar system abundances (SSA) but present to a significant degree in cosmic rays. The generally agreed reason is that these nuclei arise as fragments of nuclei which have struck nuclei in the ISM. Fig. 1 gives the CR and SS (or, equivalently, the Local Galactic) abundances. It will be noticed that H and He are seriously underabundant in CR, in comparison with the SSA. There are some interesting small CR excesses for high charges.

At the low energies, the measured CR intensities are known with good precision and it is often claimed that 'CR abundances are known more accurately than those of the solar system and the ISM'. However, this claim is not well-founded because of differences in spectral shape of the various masses and uncertainty in connection with what 'energy' to take: energy per nucleus, energy per nucleon (or rigidity: E/Z)?

There is, presumably, in the CR abundances information about the manner in which the CR gained their energy. If SNR are indeed responsible for CR acceleration then, presumably, it are nuclei from the ISM which are accelerated by the shocks. However, there is a standard problem about the early stages of acceleration ('the injection problem'). Early work pointed to a correlation of abundance with the 'first ionization potential' but more recent work has pointed to the likely influence of grains in the ISM playing a crucial role in the early stages. It is, of course, not unlikely that both factors are important.

Erlykin and Wolfendale [7] have very recently drawn attention to the likelihood of regions of the ISM which have recently been enriched by previous SN ejecta being important.

Contemporary research is devoted to all the various aspects: accurate spectra, isotopic composition and their synthesis to solve the origin problem: particular attention is being paid to unstable nuclei, both as regards to the overall lifetime of CR in the Galaxy (often estimated as ($\tau \simeq 2 \cdot 10^7$ y at GeV energies - but this may be an underestimate because of use of the, invalid, leaky box model) and a determination of the delay between nucleosynthesis and acceleration. The former uses data on ^{10}Be, ^{26}Al, ^{36}Cl and ^{54}Mn and the latter uses ^{59}Ni. It is interesting that the absence of primary ^{59}Ni in CR shows that the material from the source stayed in atomic form for at least 10^5 y, a time longer than the usual 10^4 y considered for the bulk of CR acceleration in the SNR shock model. The conclusion appears to be that the SNR providing the shock is not the one that synthesized the source material [8]. Such a result is consistent with the model advocated in Ref. [3].

Reference has been made to the spectra of the heavy nuclei being somewhat different, one from another. The differences are most marked at 'high' energies, above 10^{11} eV/nucleon and the situation is summa-

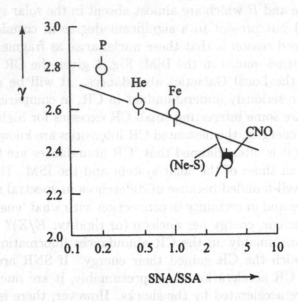

Figure 2 Spectral exponents for the differential spectrum (for $E > 10^{11}$ eV/nucleon) as a function of the ratio of the abundances of SN-enhanced to Solar System material. The fall of (with increasing ratio) lends support to the hypothesis that an SN-enhanced medium gives a big contribution to the intensity of high energy nuclei (after Erlykin and Wolfendale [7]).

rized in Fig. 2 which shows the exponent of the differential spectrum versus the ratio of SN abundance to SSA (after Erlykin and Wolfendale, [7])

The argument put forward by these workers is that previous SN ejecta play a significant part as 'sources' for a later SNR to accelerate them. In their model, which is not yet accepted, a nearby, recent SNR plays a significant part in accelerating CR in the energy range in question. At higher energies still, the role of this 'single source' is hypothesized to be even more important, as will be described in the next section.

3.   The 'Knee' Region

It has been known for over 40 years, following the pioneering extensive air shower work by the Moscow State University group (Kulikov and Khristiansen, [9]) that the primary spectrum is not featureless but has a pronounced "knee" at $E \sim 3 \cdot 10^{15}$ eV. This knee has long been attributed to propagation effects but Erlykin and Wolfendale ([3, 10] and later papers) have argued that it is too sharp to be explained in this

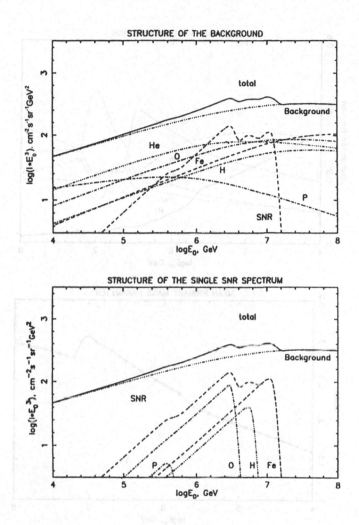

Figure 3 CR spectra from the Single Source Model of Erlykin and Wolfendale [3, 4]. Upper shows the total SNR intensity and the constituent 'background' contributions and lower shows the components from the single source itself.

way. Instead, they argue that it is due to the near - 'poking' - through of a component due to a local, recent SN. They go further and argue that there is a second 'knee' which they attributed to iron nuclei from the same SN. The situation is summarized in Figs. 3 and 4 which show the various spectra — background and single source — and the derived quantities; abundance fraction and mean logarithm of the atomic mass.

The Single Source Model has been shown by Erlykin and Wolfendale (see [3, 4], and references therein) to be consistent with many CR phe-

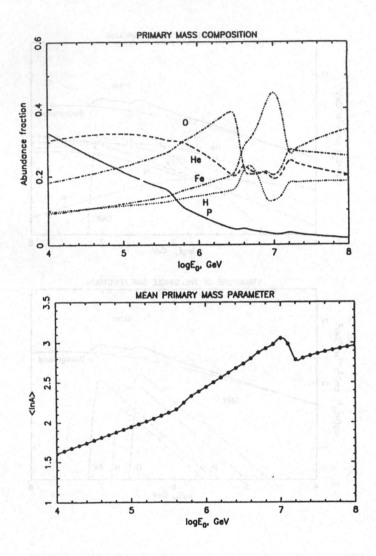

Figure 4 Primary mass composition and the average of the logarithm of the masses < ln A > in the Single Source Model of Erlykin and Wolfendale [3, 4].

nomena, including directional anisotropy measurements, but it is not yet accepted. One problem is that there is a wide spread in inferred < ln A > values in the knee region, from different experiments, the knee energy being some 2 orders of magnitude higher than the maximum energy at which direct measurements of any precision have been made.

4.　　The Range $10^{16} - 10^{18} \; eV$

It would be expected, and it seems to be the case in practice, that in this energy range the fraction of heavy nuclei would increase, the reason being that the Galactic trapping efficiency, caused by the Galactic magnetic field, is rigidity dependent. The result is that the Galactic protons escape more readily than do iron nuclei of the same energy. Fig. 5 shows some of the evidence for the contention that nuclei with $Z > 1$ predominate to at least $10^{19} \; eV$ and perhaps beyond (the EG situation at $E > 5 \cdot 10^{18} \; eV$ is considered in more detail in the next section).

The results shown in Fig. 5 relate to the depth of shower maximum from the important Fly's Eye experiment of Bird et al. [11]. The predictions for different A are from the calculations of Wibig [12]; admittedly they relate to a specific interaction model but it is one that enables a consistent picture to be drawn for a variety of CR phenomena.

5.　　The Extragalactic CR Component

It is generally regarded that the well known 'ankle' in the CR spectrum at an energy approaching $10^{19} \; eV$ signals the arrival of an abundance of EG particles. A contributory factor is the demonstration (Chi et al., [13]), Wibig and Wolfendale [1, 14]) that above a few times $10^{18} \; eV$ even iron nuclei would give unacceptably large anisotropies if they were generated in the Galaxy.

A crucial question now is the mass composition at the highest energies, say above $10^{19} \; eV$. The conventional view is that they are protons, in which case the top-down hypothesis could be valid. In this mechanism, very massive particles (from cosmic strings, 'cryptons' etc.) decay to produce protons and neutrons of energies to $10^{21} \; eV$ or so. In the opposite case, down-up, low energy particles are accelerated in exotic EG sources, such as Colliding Galaxies, just as in the case (presumably) for Galactic CR.

It is here that Wibig and Wolfendale [1] differ from the majority. They make the case for heavy nuclei continuing into the EG region. Fig. 5 (lower part) illustrates the argument; the flattening of $< \ln A >$ at $\sim 3 \cdot 10^{18} \; eV$, where EG particles are starting to become important, is rather exciting. Further evidence is provided by a study of the frequency distribution of the depth of shower maximum values (Wibig and Wolfendale [1]); the highest energy range — $(3 - 10)10^{18} \; eV$ shows a clear need for particles other than protons.

It is interesting to note that Tkaczyk et al. [18] had 'made the case' for a mixed composition of EG particles; they studied the (partial) frag-

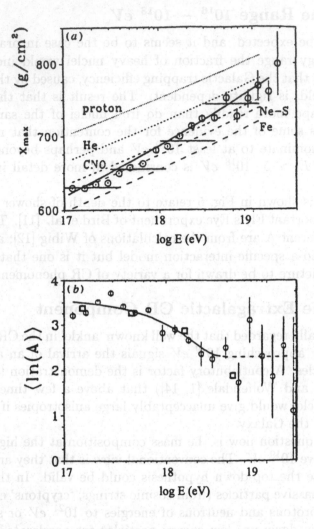

Figure 5 Depth of maximum of EAS development (X_{max}) versus primary energy, E_{prim} (after Wibig and Wolfendale, [15]). The experimental values are from Bird et al. (1993) and the predictions for the different elements are from Wibig (1999). The interest for the EG component is the apparent flattening of $< \ln A >$ above $\sim 3 \times 10^{18}$ eV, which is where the EG particles are starting to become significant. If the EG particles were all protons we would expect the value of $< \ln A >$ to fall towards zero continuously.

mentation of iron nuclei, which had been accelerated by EG sources, on the radiation fields in the Intergalactic Medium (IGM).

If presence of non-protons in the EG CR flux is confirmed there will be important consequences, as follows:

(i) The up-down mechanism will be disallowed.

(ii) The lack of an obvious grouping of EG arrival directions (with the exception of some hints, of low statistical significance, in the work of Chi et al. [20] and Al-Dargazelli et al. [19]) will be easier to understand. This fact comes from the greater random magnetic deflections of nuclei on the magnetic fields in the IGM.

(iii) The role of EG sources in accelerating nuclei preferentially, with respect to protons, will indicate a particular class of sources.

References

[1] T. Wibig and A.W. Wolfendale, J. Phys. G: Nucl. Part. Phys. 25 (1999) 1099

[2] A.W. Wolfendale and L. Zhang, J. Phys. G., 20 (1994) 935

[3] A.D. Erlykin and A.W. Wolfendale, Astropart. Phys. 8 (1998) 265

[4] A.D. Erlykin and A.W. Wolfendale, Nuovo Cim (2000), submitted

[5] P. Freier et al., Phys. Rev. 74 (1948) 1818

[6] N. Lund, 1986, *CR in Contemporary Astrophys*, M.M. Shapiro, Ed. D. Reidel Publ. Co. 1.

[7] A.D. Erlykin and A.W. Wolfendale, J. Phys. G: Nucl. Part. Phys. (2000), submitted

[8] J.J. Beatty, *Proc. 26th Int. Cosmic Ray Conf., (Salt Lake City)* 1999, Inv., Rapp. & Highlight Vol., p. 169

[9] G.V. Kulikov and G.B. Khristiansen, J.E.T.P., 35 (1958) 635

[10] A.D. Erlykin and A.W. Wolfendale, J. Phys. G: Nucl. Part. Phys. 23 (1997) 979

[11] D.J. Bird et al., *Proc. 23rd Int. Cosmic Ray Conf. (Calgary)* 2 (1993) 38

[12] T. Wibig, J. Phys. G: Nucl. Part. Phys. 25 (1999) 557

[13] X. Chi et al., J. Phys. G: Nucl. Part. Phys., 20 (1994) 673

[14] T. Wibig and A.W. Wolfendale, J. Phys. G: Nucl. Part. Phys. 25 (1999) 2001

[15] A.D. Erlykin and A.W. Wolfendale, J. Phys. G: Nucl. Part. Phys.(2000), submitted

[16] A.D. Erlykin and A.W. Wolfendale, J. Phys. G: Nucl. Part. Phys. (2000), submitted

[17] A.D. Erlykin and A.W. Wolfendale, J. Phys. G: Nucl. Part. Phys. (2000),submitted

[18] W. Tkaczyk, J. Wdowczyk, and A.W. Wolfendale, J. Phys. A., 8 (1975) 1518

[19] S.S. Al-Dargazelli et al., J. Phys. G: Nucl. Part. Phys. 22 (1996) 1825

[20] X. Chi et al., J. Phys. G: Nucl. Part. Phys. 18 (1992) 539

Particle Acceleration in Astrophysics

Luke O'C Drury

Astrophysics Section, School of Cosmic Physics, Dublin Institute for Advanced Studies
5 Merrion Square, Dublin 2, Ireland
ld@cp.dias.ie

Keywords: Particle acceleration, shock waves, Fermi mechanism

Abstract The general problem of particle acceleration is considered in the Astronomical and Astrophysical context. Recent developments in Fermi acceleration theory, usually called diffusive shock acceleration, are explained in a physically motivated and heuristic fashion.

1. Introduction

There is abundant evidence that non-thermal distributions of accelerated particles are common-place in the Universe. In our own solar system we see particles, typically of MeV energies, accelerated by Solar flares, associated with corotation interaction regions in the Solar wind, and produced at planetary and cometary bow shocks. Of particular interest are the so-called anomalous cosmic rays, interstellar neutral atoms which penetrate into the Solar system, are photo-ionized by the Solar ultraviolet radiation, picked up by the Solar wind magnetic field and swept out and accelerated at the Solar wind termination shock [4]. In the local interstellar medium we see the Galactic cosmic rays which at energies above about a GeV per nucleon can penetrate into the inner heliosphere. The field of radio astronomy exists largely because non-thermal electron populations occur in so many Astronomical objects producing strong radio synchrotron emission. Gamma-ray astronomy above about 100 MeV is dominated by inverse Compton upscattering of low-energy photons on energetic electrons, or photons produced by the decay of neutral pions produced in high-energy hadronic interactions. Even in X-ray astronomy, long dominated by thermal models, there is a growing recognition of the importance of non-thermal processes. It is clear that in many

341

D. N. Poenaru et al. (eds.), Nuclei Far from Stability and Astrophysics, 341–351.
© 2001 *Kluwer Academic Publishers. Printed in the Netherlands.*

cases these non-thermal populations carry a significant part of the total energy in the system and are not mere trace components. The problem of particle acceleration in Astrophysics is to understand why this should be the case and how these populations are produced [1].

It is remarkable that in very many cases the particles appear to have power-law differential energy distributions, $N(E) \propto E^{-\alpha}$, with exponent close to $\alpha \approx 2.0$ and this has lead to much speculation about the possible existence of a 'universal' acceleration process.

2.   Astrophysical Plasmas

The environments where these non-thermal populations are seen can be characterized as extremely low-density (relative to anything we are familiar with from terrestrial laboratory physics). As an order of magnitude guide the mean density of the interstellar medium in our Galaxy is only about one hydrogen atom cm^{-3}. The interplanetary medium near the Earth has a comparable but slightly higher mean density, of order 10 hydrogen atoms cm^{-3}. An important consequence of this very low density is that the two-body collision time-scales are long; for a relativistic proton in a medium of density 1 cm^{-3} both the nuclear collision time-scale and the ionization and coulomb energy loss time-scale are of order 10^8 years (a useful compilation of interaction cross-sections and rates is given in [2]). This clearly goes some way towards explaining the occurrence of non-thermal distributions. If they can be created, they can persist for (literally) astronomical lengths of time. However we still have the problem of creating them in the first place and here there are two major problems.

The first is a rather technical point, usually called the injection problem. Although the energy loss time-scales are very long for relativistic particles, they increase rapidly at lower energies becoming as short as 50 years for ionization losses of protons below about 45 keV energy at densities of 1 cm^{-3} and proportionally shorter at higher densities. Thus if the acceleration process starts at low energies close to thermal, the initial acceleration must be fast enough to overcome these rather high losses.

The second problem is of a more fundamental nature. The interstellar and interplanetary media are always at least partially ionized so that we are dealing with tenuous, collision-less, plasmas. These are extremely good conductors of electricity because of the very high electron mobility. Thus it is usually a good assumption to assume that we are operating in the ideal magneto-hydrodynamic (MHD) regime and that

$$\mathbf{E} + \mathbf{v} \times \mathbf{B} \equiv 0 \qquad (1)$$

where **E** is the electric field, **B** the magnetic field and **v** is the bulk plasma velocity. Physically all this means is that in a frame moving with the bulk plasma the local electric field vanishes. One way to think of this is that the high electron mobility shorts out any electric field along the magnetic field lines, and any residual field perpendicular to the magnetic field induces an **E** × **B** drift in the plasma such that the above condition is satisfied identically. The problem this poses for particle acceleration is that magnetic fields can only deflect, but not accelerate, particles. Thus if the electromagnetic field is purely magnetic, acceleration of charged particles by electromagnetic forces is strictly impossible. It is rather as if the designer of a terrestrial particle accelerator was told that he had to build the machine entirely out of copper with no insulating materials.

There are two ways out of this dilemma. One is to simply look for places where the ideal MHD approximation breaks down. Then one can have strong electric fields and charged particles moving through these fields can be accelerated. One obvious possibility is in the neighborhood of pulsars, rapidly rotating neutron stars with strong magnetic fields. Pulsar periods range from a few milliseconds (the shortest allowed) down to seconds and have surface magnetic fields of up to 10^7 T, thus considered as unipolar inductors the pole to equator potential drop can easily be as much as $10^{18}V$. The one thing that is generally agreed about pulsar magnetospheres (the region surrounding the pulsar where the magnetic field of the pulsar is dominant) is that they are not described by classical ideal MHD, but unfortunately there is little agreement beyond this. However it is clear from observations that the Crab pulsar, for example, is putting most of its spin-down luminosity into a relativistic outflow of what is probably an electron-positron pair plasma. Unfortunately (or fortunately for the observers) the strong magnetic field means that most of this energy is then radiated away as synchrotron emission.

Another possibility is to look at sites of magnetic reconnection, where regions of oppositely directed magnetic field are forced together and annihilate. Reconnection is thought to be the fundamental process behind solar flares and tearing events in the Earth's magnetotail, both of which are associated with accelerated particle populations. Unfortunately, again the detailed microphysics is not well understood and it is difficult to make any firm predictions beyond order of magnitude dimensional estimates.

The other, extremely interesting, possibility was spotted by Fermi and published in his classic paper [3]. Fermi's key realization was that within ideal MHD the electric field vanishes locally everywhere, but not in general globally. Although Fermi's original theory was clearly ruled out by subsequent observations, it remains of great intrinsic interest and beauty. The most successful modern theory of particle acceleration is also in a direct line of descent from Fermi's theory, so it is worth briefly discussing his ideas.

3.   Fermi's 1949 Mechanism

Fermi pointed out that the interstellar medium is not static and uniform, but contains many condensations (which he termed clouds) which are in random turbulent motion. In the frame of each cloud cosmic ray particles scatter elastically off the magnetic field frozen into the cloud, but in the observation frame, if the cloud velocity is V and the cosmic ray particle is relativistic and moving at the speed of light c, the particle's energy changes by an amount of order $E(V/c)$. The change will be positive if the collision is what Fermi called "head-on" and negative if the collision is of the "over-taking" type. Fermi then argued that "head-on" collisions are slightly more frequent than "over-taking" collisions, again by a factor of order V/c, and that therefore although the positive and negative changes nearly cancel out there remains a positive mean energy change of order $\Delta E \approx (V/c)^2 E$ per collision. Interestingly Fermi relegates this estimate to what is essentially an appendix of his paper and bases the main discussion on the deeper and more physical argument that in this situation there must be a net energy gain because "ultimately statistical equilibrium should be established between the degrees of freedom of the wandering fields and the degrees of freedom of the particle. Equipartition evidently corresponds to an unbelievably high energy. The essential limitation therefore, is not the ceiling of energy that can be attained, but rather the rate at which energy is acquired." In fact the process of energy gain is better pictured as one of diffusion in phase space, but this does not significantly change the results.

Fermi, with characteristic insight, recognized that the large mean free paths and gyroradii of cosmic ray particles cause them to couple through the magnetic field to macroscopic fluctuations in the velocity field, and that this must inevitably lead to an acceleration process. However, as he also recognized, it can be painfully slow. He used a random velocity of 30 $km\,s^{-1}$ corresponding to $\beta = V/c = 10^{-4}$, so that the acceleration time-scale corresponded to $\beta^{-2} = 10^8$ collision times. Optimistically assuming a mean time between collision of about one year a particle's energy could increase by a factor of e every $10^8\ yr$.

This is of the same order of magnitude as the loss time-scale for nuclear collisions. Fermi's second major point was that an extended power-law spectrum could naturally be produced under these circumstances. If the individual particle energy is increasing exponentially with time, but the surviving fraction of particles is decreasing exponentially, the steady-state spectrum is a power-law. More formally, in general we can write a particle conservation equation

$$\frac{\partial N}{\partial t} + \frac{\partial}{\partial E}\left(\frac{E}{\tau_{acc}}N\right) = Q - \frac{N}{\tau_{loss}} \tag{2}$$

for the differential energy spectrum $N(E)$ where τ_{acc} is the acceleration time-scale, τ_{loss} is the loss time-scale and Q is a source term. If we assume that the source supplies particles at some relatively low energy and look for a steady solution at higher energies it is easy to see that the logarithmic slope of the spectrum is simply

$$\frac{\partial \ln N}{\partial \ln E} = -\left(1 + \frac{\tau_{acc}}{\tau_{loss}} + \frac{\partial \ln \tau_{acc}}{\partial \ln E}\right). \tag{3}$$

Thus one can produce a power-law tail extending to high energies with exponent determined by the ratio of the acceleration rate to the loss rate.

This is a very beautiful theory, but unfortunately it simply does not work. Basically there are three major problems. It is much too slow, it requires an additional injection process (the Q term) and it is not obvious why the ratio of the acceleration time scale to the loss time scale should be constant over a large energy range. The big advance in astrophysical acceleration theory was the discovery about twenty years ago of a Fermi-type process associated with shock waves which appears to overcome all of these problems [5–8].

4. Diffusive Shock Acceleration

An interstellar shock is essentially a thin transition layer in which there is a sudden change in the bulk flow velocity; low density interstellar matter (a fully or partially ionized plasma with embedded magnetic fields) flows into the shock front at high speed from upstream, is decelerated and compressed, and emerges in the downstream region as lower velocity and higher density material. Much of the kinetic energy associated with the flow of the incoming plasma is converted irreversibly by dissipative processes within the shock front to random motion of the ions (and electrons) of the plasma, that is the plasma is "heated" in the shock. The full details of how this occurs in these "collisionless shocks" are not understood; however, just as in the much simpler "collisional" shocks of gas dynamics where the dissipation occurs through two-body collisions rather than collective plasma processes, for many purposes the internal details are not important. On scales larger than a few downstream ion gyroradii (which is the typical thickness of the shock) we can treat the shock as a simple discontinuity in the plasma properties, and conservation of mass momentum and energy through the shock then rather tightly constrain the allowed jumps in velocity, density etc. If the magnetic field, with its anisotropic pressure, is not dynamically significant, the incoming and outgoing plasma flow velocities can be taken to be parallel to the shock normal, that is perpendicular to the plane of the shock front.

Let us now consider [6] a cosmic ray particle near an interstellar shock front. The cosmic ray particle scatters off the magnetic field structures which are embedded in the upstream and downstream plasma and, at the microscopic level, makes a random walk (and thus, at the macroscopic level, the cosmic ray density obeys a diffusion equation). Let us further assume that the magnetic scatterers are frozen into the plasma and have no peculiar velocity of their own (this assumption can easily be relaxed).

It is certainly possible for a particle to be scattered from the upstream region, across the shock front into the downstream region, and then back again from the downstream region into the upstream region. Looked at from the frame of the upstream plasma the downstream scattering is a head-on collision with a scatter coming towards us with speed $U_1 - U_2$ where U_1 is the speed with which the upstream plasma is flowing into the shock and U_2 the speed with which the downstream plasma is flowing out. Thus the particle's energy, E, will be increased by an amount of order $E(U_1 - U_2)/c$. Furthermore the particle is now back where it started from and can repeat the process. Clearly we have an acceleration process of Fermi type, but one where, unlike Fermi's original picture, the energy changes associated with crossing the shock are always positive (because shocks are always compressive). Bell's contribution was to work out rigorously the statistics of this process.

It turns out to be more convenient to work with the scalar momentum, $p = |\mathbf{p}|$, than the energy E of the particle. The crucial point is that p is not a true scalar quantity; it is the modulus of the spatial part of the energy-momentum four vector. Thus its value depends on the choice of reference frame. By convention we measure p at any point with respect to a reference frame moving with the mean velocity of the scattering magnetic structures at that point. With our simplifying assumption, that the scatterers are simply advected with the plasma, this means that we measure p in a frame moving with the local plasma velocity. The great advantage of this choice is that all the scatterings are then at constant particle energy (even if we relax the assumption that the scatterers are frozen in, the mean energy change per scattering still vanishes although some residual classical Fermi acceleration will remain). However every time a particle crosses the shock front, although nothing happens to the particle itself, the reference frame we use to measure p changes, and consequently the value of p changes. An elementary application of relativistic transformation formulae shows that for a particle crossing the shock at an angle ϑ with respect to the shock normal from upstream (1) to downstream (2) the change, Δp in p is, to first order in the small quantity U/v,

$$\Delta p = p\frac{U_1 - U_2}{v}\cos\vartheta.$$

For a particle crossing the other way, from 2 to 1, the subscripts must be interchanged, but the sign of $\cos\vartheta$ also changes. Thus Δp is always positive. The scattering in the upstream region drives the angular distribution function there towards isotropy in the upstream frame. Similarly the downstream distributions are driven towards isotropy in the downstream frame. The distribution at the shock front itself cannot be exactly isotropic, but for particles whose velocity is large compared to the flow velocities, $v \gg U_1, U_2$, the difference is of order $U/v \ll 1$ and, to first order, we can obtain the mean change in p, $\langle\Delta p\rangle$, by averaging over the flux of particles coming from an isotropic distribution. This gives for the mean change on crossing the shock (in either direction)

$$\langle\Delta p\rangle = p\frac{U_1 - U_2}{v}\frac{\int_0^{\pi/2}\cos^2\vartheta\sin\vartheta d\vartheta}{\int_0^{\pi/2}\cos\vartheta\sin\vartheta d\vartheta} = \frac{2}{3}p\frac{U_1 - U_2}{v}.$$

An alternative way of looking at this result, which turns out to be very useful in applications [10], is to think of the shock as being associated with an upward flux of particles in momentum space of magnitude

$$\Phi(p) = \int f(p)\,v\cos\vartheta\,\Delta p\,2\pi p^2\sin\vartheta d\vartheta = \frac{4\pi p^3}{3}f(p)\,(U_1 - U_2) \quad (4)$$

where $f(p)$ is the (isotropic) part of the phase-space density of the accelerated particles. The particles interacting with the shock are those located within one diffusion length upstream, $L_1 = \kappa_1/U_1$ and one diffusion length downstream, $L_2 = \kappa_2/U_2$, where κ is the spatial diffusion coefficient of the particles. If we think of this region as the "box" within which particle acceleration is occurring and assume that the distribution is approximately uniform throughout the box we can write down a very simple conservation law for the number of particles in the box,

$$\frac{\partial}{\partial t}\left[(L_1 + L_2)4\pi p^2 f(p)\right] + \frac{\partial\Phi}{\partial p} = Q - 4\pi p^2 f(p)U_2. \quad (5)$$

This simply says that the rate of change of the number of particles interacting with the shock plus the divergence of the momentum flux associated with the shock is equal to the source term Q (either particles injected at low energy at the shock, or particles advected in from upstream) minus the advection of particles away from the shock by the downstream flow U_2. Note that f is being used in three subtly different ways in this equation. In the total number it represents a spatially averaged value over the interaction volume, in the momentum flux term it is the local value at the shock, and in the downstream advective loss term it represents the value about one diffusion length behind the shock. It is easy to show that all three values are equal if we have a steady state solution. The basis of the "box model" approximation is to assume that this is the case even when conditions are not stationary.

The above equation is easily seen to be equivalent to

$$(L_1 + L_2)\frac{\partial f}{\partial t} + (U_1 - U_2)\frac{p}{3}\frac{\partial f}{\partial p} = \frac{Q}{4\pi p^2} - U_1 f. \qquad (6)$$

Let us first look at the steady state solution at energies above injection. Then we get the remarkable result that

$$\frac{p}{f}\frac{\partial f}{\partial p} = -\frac{3U_1}{U_1 - U_2}, \qquad (7)$$

the steady spectrum is a power-law in momentum with an exponent determined purely by the compression ratio of the shock. For strong shocks in an ideal gas the compression ratio is always close to 4, $U_1 \approx 4U_2$, and thus $f(p) \propto p^{-4}$ or (for relativistic particles) $N(E) \propto E^{-2}$.

There are several ways of looking at the power-law nature of the spectrum. One is to note that the basic acceleration process has no associated momentum or energy scale, and thus any steady solution must be self-similar and a power-law. Another is to note that as in all Fermi acceleration processes the exponent is determined by the ratio of the acceleration time to the loss time; however whereas these are usually determined by different physical processes (in which case it is not very natural for the ratio to be energy independent) in the case of shock acceleration both are determined by the velocity structure of the shock, the acceleration being proportional to the velocity jump in the shock and the loss to the downstream advection velocity.

Where do the accelerated particles come from? Ambient energetic particles swept into the shock will certainly be accelerated. However there is a much more interesting source. The dissipation in the shock, which "heats" the incoming plasma occurs not through the two-body collision processes familiar in gas dynamics but through collective electromagnetic effects. There is no reason why the downstream ion distribution function should be exactly Maxwellian, and indeed detailed simulations as well as theoretical arguments show that there should be a non-thermal tail of high-energy ions. This provides a natural injection mechanism for the further Fermi acceleration. In fact one can regard diffusive shock acceleration as a simple asymptotic theory for the high-energy part of the particle distribution function at the shock [14]

How fast is the acceleration? This is a crucial question because it determines whether the acceleration can win out in competition with the general energy loss processes and also sets a limit on the maximum energies which can be reached by acceleration at a shock of finite age. Looking again at Equ. 6 we can write it in characteristic form as

$$\frac{\partial f}{\partial t} + \dot{p}\left(\frac{\partial f}{\partial p} + \frac{3U_1}{U_1 - U_2}\frac{f}{p}\right) = \frac{Q}{4\pi p^2(L_1 + L_2)} \tag{8}$$

with the momentum gain rate

$$\dot{p} = \frac{p}{3}\frac{U_1 - U_2}{L_1 + L_2}. \tag{9}$$

Thus the acceleration time scale is

$$\tau_{acc} = \frac{p}{\dot{p}} = \frac{3}{U_1 - U_2}\left(\frac{\kappa_1}{U_1} + \frac{\kappa_2}{U_2}\right), \tag{10}$$

a result confirmed by more exact calculations [9], and for $U_1/U_2 = 4$

$$\tau_{acc} = \frac{4\kappa_1}{U_1^2}\left(1 + 4\frac{\kappa_2}{\kappa_1}\right). \tag{11}$$

This, as one would expect simply on dimensional grounds, is proportional to the upstream diffusion time scale κ_1/U_1^2, however this is one of those cases where the numerical factor is quite large, between 8 and 20 depending on what one assumes for the ratio κ_2/κ_1.

This is still very substantially faster than other Fermi acceleration processes because the shock velocities U are higher than other typical interstellar medium velocities, and, crucially, because the diffusion coefficient near the shock can be very much smaller than in the general interstellar medium. This last is a very important point. The accelerated particles diffusing into the upstream region can resonantly excite Alfvén waves (transverse perturbations of the magnetic field) which produce enhanced scattering of the particles. This in turn allows the efficient acceleration which is needed to produce the particles in the first place. This nonlinear "boot-strapping" is not easy to model which in turn means that the magnitude of the diffusion coefficient at the shock is not well known. However it is clear that for a given characteristic magnetic field strength, the mean free path for scattering cannot be reduced below the particle gyro-radius. If we assume that the scattering is so efficient that the mean free path is in fact reduced to something close to the gyro-radius we get the so-called Bohm diffusion limit,

$$\kappa_{\text{Bohm}} = \frac{r_g v}{3} = \frac{pv}{3ZeB},$$

for particles of charge Ze. At strong shocks where there is plenty of energy available the Bohm limit may in fact be attained. Certainly it is hard to see why the nonlinear bootstrap process would saturate at any lower value. In fact magnetic field amplification may even lead to local diffusion coefficients which are smaller than the Bohm estimate based on the upstream field [11]

The important thing to note is that the acceleration time-scale increases with particle energy. Particles of higher energy are harder to deflect with magnetic fields and in consequence are expected to have larger mean free paths and longer acceleration times. This means that in a system of finite size and age particles can be accelerated easily from low energies until either the acceleration time becomes comparable to the age of the system, or the diffusion length scale becomes larger than the size of the system. At this point we expect a rather abrupt cut-off in the spectrum.

In concrete terms, for a shock with characteristic size R and velocity U, the maximum attainable energy is of order $0.3RUBZe$ for a particle of charge Ze. Applied to supernova remnant shocks this gives cut-off energies of order a few $10^{14} ZeV$ [12]. If we adopt the Bohm value for the diffusion coefficient and $\tau_{\text{acc}} \approx 10\kappa/U^2$, then particles gain kinetic energy at rate $0.3ZeBU^2$. For a field of $0.3 nT$ and shock velocity of $10^3 km\,s^{-1}$ this gives an acceleration of $100 eV\,s^{-1}$ per charge, easily enough to overcome ionization and Coulomb losses even at energies close to thermal.

5.   Nonlinear Effects

The major problem in the theory of diffusive shock acceleration is the question of what happens when the pressure (or energy density) of the accelerated particles becomes dynamically important. This must happen if the process is to operate efficiently, and using the test-particle spectrum it is easy to see that for typical interstellar shocks the pressure of the accelerated particles will become significant if more than about 10^{-4} of the protons flowing into the shock are injected. Estimates of the injection rate from computer simulations [13], theory [14] and observations of heliospheric shocks [15] all suggest strongly that this will in fact be the case. Unfortunately the reaction of the accelerated particles on the shock structure then greatly complicates the theory and only recently has much progress been made. In particular mention should be made of Malkov's courageous attempt at an approximate analytic theory [16, 17] and Ellison and Berezhko's simple numerical approximation [18].

Acknowledgments

This article was prepared while visiting the Service d'Astrophysique of CEA Saclay with the support of the TMR programme of the EU under contract FMRX-CT98-0168. The hospitality of M Tagger and L Koch is gratefully acknowledged.

References

[1] R. D. Blandford, Astrophys. J. Suppl. 90 (1994) 515

[2] K. Mannheim and R. Schlickeiser, Astr. Astrophys. 286 (1994) 983

[3] E. Fermi, Phys. Rev. 75 (1949) 1169

[4] J. R. Jokipii and J. Giacolone, Space Science Rev., 83 (1998) 123

[5] W. I. Axford, E. Leer and G. Skadron, Proc. 15th Int. Cosmic Ray Conf. (Plovdiv) 11 (1978) 132

[6] A. R. Bell, MNRAS 182 (1978) 147

[7] R. D. Blandford and J. P. Ostriker, Astrophys. J. 221 (1978) L29

[8] G. F. Krymsky, Doklady Akademiia Nauk SSSR, 234 (1977) 1306

[9] L. O'C. Drury, MNRAS 251 (1991) 340

[10] L. O'C. Drury, P. Duffy, D. Eichler and A. Mastichiadis Astr. Astrophys. 347 (1999) 370

[11] S. G. Lucek and A. R. Bell, MNRAS 314 (2000) 65

[12] E. G. Berezhko, Astropart. Phys. 5 (1996) 367

[13] L. Bennett and D. C. Ellison, J. Geophys. Res. 100 (1995) 3439

[14] M. A. Malkov and H. J. Völk, Astr. Astrophys. 300 (1995) 605

[15] M. A. Lee, J. Geophys. Res. 87 (1982) 5063

[16] M. A. Malkov, Astrophys. J. 485 (1997) 638

[17] M. A. Malkov, Astrophys. J. 511 (1999) L53

[18] E. G. Berezhko and D. C. Ellison, Astrophys. J. 526 (1999) 385

Cosmic Rays in the Galaxy

Luke O'C Drury

Astrophysics Section, School of Cosmic Physics, Dublin Institute for Advanced Studies
5 Merrion Square, Dublin 2, Ireland
ld@cp.dias.ie

Keywords: Cosmic rays, Galaxy

Abstract   This article is a short introductory survey of some aspects of Galactic
Cosmic Ray Astrophysics.

1.   Introduction

Nearly a century ago Victor Hess published a remarkable and extraordinary result [1], the discovery of an extremely penetrating ionizing radiation of extra-terrestrial origin. We now know a great deal more about this phenomenon, the so-called Cosmic Rays, but much remains mysterious. A good general reference though somewhat uneven and with a rather heavy emphasis on Russian theoretical contributions is [2]. For specialist reviews of the current state of the field the rapporteur articles in the Proceedings of the biennial International Cosmic Ray Conferences should be consulted (the last was in Salt Lake City in 1999, the next will be in Hamburg in 2001).

2.   Observational Aspects

The Cosmic Rays observed at and near the Earth consist mainly of fully stripped atomic nuclei together with some electrons, positrons and anti-protons. They are strongly affected at magnetic rigidity (momentum per charge) below about 1 GV by the Solar wind. The Solar magnetic field, carried out by the Solar wind in the well-known Parker spiral configuration, pushes these lower energy particles out of the inner heliosphere, an effect known as the Solar modulation. The strength of the modulation varies with the Solar cycle and general level of activity on the Sun making observations of interstellar cosmic rays below about 100 MeV per nucleon very difficult and uncertain.

D. N. Poenaru et al. (eds.), Nuclei Far from Stability and Astrophysics, 353–360.
© 2001 *Kluwer Academic Publishers. Printed in the Netherlands.*

At higher energies, in the region up to a few hundred GeV or so, measurements are relatively straightforward. The fluxes are reasonably high, of order 1 $cm^{-2} s^{-1}$ at 1 GeV per nucleon, and charge identification is technically straightforward. In consequence this region of the energy spectrum is now rather well known, with good composition and spectral measurements available from a wide variety of experiments. At still higher energies it becomes technically harder to make direct charge-resolved measurements and the fluxes are so rapidly falling that experiments begin to require large collecting areas. However some direct information on the composition and energy spectrum is available up to about 10^{14} eV.

Beyond about 10^{15} eV there is little or no hope of making direct observations of the primary cosmic rays in the foreseeable future. However it is fortunate that at just this energy the cascade of secondary particles produced when a high-energy cosmic ray hits the upper atmosphere is sufficiently intense to reach sea level. These large air showers, first demonstrated to exist by Auger [3], enable one to study the higher-energy part of the cosmic ray spectrum with ground based installations. Unfortunately, although one can get a good idea of the total energy involved in the shower, it is very much harder to say anything about the nature of the particle initiating the shower. However using this technique the all-particle energy spectrum has been extended all the way up to the incredible value of 3×10^{20} eV. The ambitious Auger project aims to considerably improve our knowledge of this end of the spectrum over the next few years.

3.   Energy Spectrum

As indicated above the interstellar spectrum of sub-relativistic cosmic rays is inaccessible to us because of Solar modulation. In the region from about 1 GeV per nucleon to a few times 10^{14} eV the spectra of all species appear to be very well described by a single power-law $N(E) \propto E^{-2.65 \pm 0.5}$ [4, 5]. At a few PeV (10^{15} eV) there is the one well-established feature in the entire energy spectrum, the so-called "knee" above which the spectrum clearly steepens slightly. The extent to which the "knee" is smooth [6] or bumpy [7] is under active debate at the moment and this is one area which the beautiful data coming from the KASCADE collaboration should help resolve. At still higher energies the spectrum appears to flatten again (the so-called "ankle" at about $10^{18.5}$ eV) and continues on to at least a few 10^{20} eV with no sign of a cut-off in flat contradiction to theoretical expectations (the famous GZK cut-off does not appear to exist!). This end of the spectrum, however, is almost certainly of extra-Galactic origin and thus not properly the subject of this talk.

For a spectrum falling as $E^{-2.6}$ the energy density is contributed mainly by the low-energy particles, which unfortunately are in the region where Solar modulation complicates the measurements. However the total energy density in the observed cosmic ray spectrum is certainly of order 1 $eV\,cm^{-3}$ and according to a recent estimate of Webber possibly as high as 1.8 $eV\,cm^{-3}$ [8]. This is comparable, as has often been pointed out, to the energy density of the interstellar magnetic field and of the thermal gas (this is probably not a coincidence). It is also, remarkably, comparable to the energy density of starlight and of the cosmic microwave background (this does appear to be a genuine coincidence).

4.   Composition

The chemical composition, by which is normally meant the relative fluxes of different elements measured at fixed energy per nucleon, is well determined in the energy range from about 1 to 100 GeV per nucleon. The most striking feature is the clear evidence for interstellar spallation. The light elements, $Li\ Be$ and B, which are extremely rare in the rest of the Universe, have abundances in the low energy cosmic rays of order 10 % of those of the abundant CNO elements. The universally accepted explanation for this is that some of the CNO nuclei in the cosmic rays have collided with the nuclei of atoms of the interstellar medium during their propagation from their sources to us, and that these collisions have knocked a few nucleons out of the nucleus (a spallation reaction) and converted a CNO nucleus to a $LiBeB$ nucleus. In the same way the sub-iron region has been partially filled in from the abundant iron peak. The amount of spallation corresponds to propagation through about 5 $g\,cm^{-2}$ of interstellar matter (the so-called grammage) and is clearly energy dependent. The spallation secondaries are *less* abundant at higher energies and in fact the secondary to primary ratios appear to fall off at high energies as $E^{-0.6}$.

This is enormously valuable information and it teaches us an important but unpleasant fact. The spectrum we observe is clearly not the spectrum as it emerges from the accelerator (the source spectrum) but one which has been significantly modified by propagation effects. Not only has the composition been changed, but the energy spectrum has been steepened. We know this because, for the secondary spallation nuclei, we can calculate the true production spectrum from the observed flux of interstellar primaries and the known spallation cross-sections. To a good approximation above threshold the cross sections are geometrical and have little energy dependence, thus the source spectrum of the secondaries should have the same $E^{-2.65}$ energy dependence as the primary flux. The observed flux of secondaries is however considerably steeper than this and more like $E^{-3.3}$. Thus interstellar propagation effects have steepened the secondary spectrum from its production form of $E^{-2.65}$ to the observed $E^{-3.3}$ and we may infer that the primary spectrum has also been altered, perhaps from a production value as hard as $E^{-2.0}$ to the observed $E^{-2.65}$.

One can, with slight model dependencies, correct the chemical composition for the effects of secondary production and deduce the original source composition. The resulting cosmic ray source composition is discussed in detail in [9]. It is basically quite close to the standard local galactic abundance pattern, but with significant differences. These appear to have found a satisfactory theoretical interpretation in terms of standard shock acceleration operating in a dusty interstellar medium [11, 10].

Unfortunately, correcting the spectral shape for propagation effects is much more model dependent than correcting the composition. Advocates of propagation models which include significant reacceleration, eg. [12], favor source spectra which are softer and similar to $E^{-2.4}$ whereas models based on energy dependent escape with negligible reacceleration require harder source spectra similar to $E^{-2.1}$ [13]. A major part of the problem is that we only have good secondary to primary ratios over a relatively small and low energy part of the spectrum; the propagation at energies above a TeV is essentially unconstrained by observations (except, to a limited extent, by the isotropy data; see below).

The compositional data contains two further important pieces of information relating to time-scales. There are a few secondary spallation nuclei which are radioactive with relatively long half-lives, most notably ^{10}Be [19, 20, 21]. This is observed to have partially decayed in the arriving cosmic rays indicating a mean age for the secondary nuclei of order a few times 10^7 yr between production and observation. The other interesting recent result relates to K-shell capture nuclei. These are nuclei which decay by capturing an electron, and are therefore unstable while neutral atoms but stable when fully stripped (as in the cosmic rays). Recent observations from the ACE spacecraft [22] show that the nuclide ^{59}Ni has almost certainly completely decayed by electron capture in the cosmic ray source indicating that the source material for the accelerator is *not* fresh supernova ejecta, but at least 10^5 years old.

5.   Isotropy

The arriving cosmic rays are astonishingly isotropic. There is evidence for a genuine but very small anisotropy with amplitude of order a few 10^{-4} at about 10^{12} eV [14, 15] and there appears to be a slight Galactic plane effect in the 10^{17} to 10^{18} eV region [17, 18] but apart from these rather small effects the arrival directions, even at the highest energies [16], are consistent with an isotropic distribution. This is a serious constraint on cosmic ray propagation models; almost all models have difficulty explaining the high levels of isotropy observed. In particular propagation models based on strong energy dependent escape have a major problem with the lack of a significant anisotropy at 10^{14} eV.

6.   Galactic or Extra-Galactic?

The cosmic rays at low to moderate energy are almost certainly of Galactic origin. The strongest evidence comes from two arguments involving gamma-ray astronomy. The diffuse gamma-ray emission of the Galaxy above about 100 MeV is thought to be dominated by π^0 decay, the neutral pions in turn being produced by hadronic interactions of cosmic rays with the nuclei of the interstellar medium. Thus this emission traces the product of the cosmic ray intensity and the interstellar gas density and, if one has a model for the gas distribution in the Galaxy, one may infer the Galactic cosmic ray distribution. In practice the gas distribution is not as well known as one would like, but the best recent studies indicate that the cosmic ray intensity is higher in the Galactic center region and falls off radially as one goes outwards clearly indicating a Galactic origin, at least for the few GeV particles which produce this emission [24].

The other argument is based on a famous test proposed by Ginzburg many years ago. If the cosmic rays were a universal and constant radiation field (like the cosmic microwave background) the gamma-ray emissivity of the Magellanic clouds, per unit mass, would be exactly the same as that of our Galaxy. The EGRET instrument on the Compton gamma ray observatory succeeded in placing an upper limit on the gamma ray emission from the small Magellanic cloud which was clearly below the signal which would have been seen if the cosmic ray flux in the Magellanic clouds was comparable to that in the Solar neighborhood [23].

This, together with the general similarity of the composition to standard Galactic abundances, appears to settle the question for the low energy nuclear component of the cosmic rays. And, as there is no feature in the spectrum until the knee, nor any evidence for a change in composition, it appears certain that at least up to 10^{15} eV the cosmic rays must be of Galactic origin. In addition, for the high energy cosmic ray electrons, the synchrotron life times are so short that these particle must be produced locally, within about one kPc.

For the particles between the "knee" at 10^{15} eV and the "ankle" at $10^{18.5}$ eV the situation is unclear. The lack of any obvious discontinuity at the "knee" does rather suggest that these particles also must have a Galactic origin, but the evidence is hardly conclusive. Finally the ultra-high energy cosmic rays with energies above 10^{19} eV have gyroradii comparable to the size of the Galaxy and can certainly not be confined to the Galaxy. This, combined with the lack of any obvious directional anisotropy, suggests an extra-Galactic origin for these particles (and this would offer a natural explanation for the "ankle" as has often been pointed out).

7.   Power

On the basis of the observed energy density and the amount of inter-stellar material traversed it is possible to estimate the power required to maintain the cosmic ray population of the Galaxy. Clearly this is also a propagation model dependent result, but unfortunately nobody has (to my knowledge) done a proper analysis of this. However an order of magnitude estimate is easy to obtain and shows that the Galactic cosmic ray luminosity is quite substantial, of order several times 10^{33} W. This should be compared with the mechanical energy input to the Galaxy from supernovae, which, if one adopts the canonical figures of 10^{51} erg and one event per 30 yr is 10^{35} W. Thus the Galactic supernovae could power the Galactic cosmic ray accelerator if the mechanical energy released in the explosion could be transferred to accelerated particles with an overall efficiency of a few percent.

8.   Theory

The currently favored theory for the origin of the Galactic cosmic rays is that they are produced by an efficient form of Fermi acceleration (the diffusive shock acceleration mechanism described in the accompanying article) operating at the strong shocks bounding supernova remnants. This theory can claim some major successes:

- It satisfactorily accounts for all aspects of the chemical composition in a rather natural manner (and is the only quantitative model to do so) [10, 11]

- The power and energy density of the Galactic cosmic rays are naturally explained if the shock acceleration process operates at the high efficiencies expected.

- The source spectrum below about 10^{14} eV is predicted to be close to E^{-2} but slightly concave and to fall off rapidly at higher energies. This is in reasonable agreement with the harder propagation models below 10^{14} eV although there is a clear problem with the "knee".

In addition, we now have strong evidence for electron acceleration in at least some shell-type supernova remnants to the 10^{14} eV region from nonthermal X-ray and radio synchrotron observations, the prime example being the remnant of SN1006 [25], confirmed by inverse Compton gamma-ray observations in the TeV region [26]. Very interestingly, recent Chandra observations of E0102.2-7219, a remnant located in the small Magellanic cloud and therefore at a known distance, strongly suggest that there is substantial "missing energy" which may be going into cosmic ray production [27]. However it must be said that the situation is still very confused:

- There is still no unambiguous direct evidence for the acceleration of ions as distinct from electrons in supernova remnants.

- The origin of the "knee" and of the particles above the "knee" remains totally obscure.

- There is a worrying disagreement between the propagation theories and the acceleration theories as to the preferred source spectrum.

One is tempted to conclude by recalling that back in the early days of the subject Lord Rutherford remarked that the solution to the problem of the origin of cosmic rays was "more work and less talk" [28], advice which still has merit today.

Acknowledgments

This article was prepared while visiting the Service d'Astrophysique of CEA Saclay with the support of the TMR programme of the EU under contract FMRX-CT98-0168. The hospitality of M Tagger and L Koch is gratefully acknowledged.

References

[1] V. F. Hess, Z. Phys. 13 (1912) 1084
[2] V. S. Berezinskii et al., "Astrophysics of Cosmic Rays", North-Holland Amsterdam Holland, 1990

[3] P. Auger, J. Phys. Radium 10 (1939) 39

[4] M. Boezio et al., Astrophys. J. 518 (1999) 457

[5] W. Menn et al., Astrophys. J. 533 (2000) 281

[6] M. Amenomori et al., Astrophys. J. 533 (2000) 281; 461 (1996) 408

[7] A.D. Erlykin and A.W. Wolfendale, Astr. Astrophys. 350 (1999) L1

[8] W. R. Webber, Astrophys. J. 506 (1998) 329

[9] J.-P. Meyer, L. O'C. Drury and D. C. Ellison, Space Science Reviews 86 (1998) 179

[10] D. C. Ellison, L. O'C. Drury and J.-P. Meyer, Astrophys. J. 487 (1997) 197

[11] J.-P. Meyer, L. O'C. Drury and D. C. Ellison, Astrophys. J. 487 (1997) 182

[12] A. W. Strong and I. V. Moskalenko, Astrophys. J. 509 (1998) 212

[13] V. S. Ptuskin et al., Astr. Astrophys. 321 (1997) 434

[14] D. J. Cutler et al., Astrophys. J. 248 (1981) 1166

[15] D. J. Cutler and D. E. Groom, Astrophys. J. 376 (1991) 322

[16] M. Takeda et al., Astrophys. J. 522 (1999) 225

[17] D. J. Bird et al., Astrophys. J. 511 (1999) 739

[18] N. Hayashida et al., Astropart. Phys. 10 (1999) 303

[19] V. S. Ptuskin and A. Soutoul, Space Science Reviews 86 (1998) 225

[20] J. J. Connell, Astrophys. J. 501 (1998) 59

[21] J. J. Connell, M. A. Duvernois and J. A. Simpson, Astrophys. J. 509 (1998) 97

[22] Wiedenbeck et al., Astrophys. J. 523 (1999) L61

[23] P. Sreekumar et al., Phys. Rev. Lett. 70 (1993) 127

[24] A. W. Strong and J. R. Mattox, Astr. Astrophys. 308 (1996) L21

[25] K. Koyama et al., Nature 378 (1995) 255

[26] T. Tanimori et al., Astrophys. J. 497 (1998) L25

[27] J. P. Hughes, C. E. Rakowski and A. Decourchelle, astro-ph/0007032; Astrophys. J. in press

[28] H. Geiger et al., Proc. Roy. Soc. A132 (1931) 331

Phenomenology of Extensive Air Showers

Iliana Magdalena Brancus

National Institute for Physics and Nuclear Engineering, P.O. Box MG-6 Bucharest
Romania
iliana@muon2.nipne.ro

Keywords: Cosmic rays, extensive air showers, EAS lateral and time structure

Abstract Higher energy primary cosmic rays, when penetrating from the outer space into the Earth's atmosphere initiate so-called Extensive Air Showers (EAS) by multiple particle production in the cascading interactions of the primary particle. This lecture presents the phenomenological features of the EAS development, the formation of the different shower components (electrons, photons, muons, hadrons) and considers various EAS parameters: total intensities (sizes), lateral and arrival time distributions etc. as observables of experimental studies. The most important interaction ingredients which drive the EAS development and lead to the EAS appearance at observation level are discussed, in particular in view of a basic understanding and realistic Monte Carlo simulations of the EAS features. The presentation will be illustrated with various recent results of the KASCADE experiment.

1. Introduction

Penetrating into the Earth's atmosphere, cosmic rays produce secondary radiation by interactions with the air-target and subsequent decays in subnucleonic particles, hadrons, pions, muons, electrons, photons etc. This phenomenon has been discovered by the Austrian physicist Victor Hess [1], in his balloon flights in 1912, by measurements of the height-dependence of ionization of air. It is first reported in a conference held in Karlsruhe in 1912.

If the energy of a primary particle is as high as 10^{14} eV or higher, the total number of particles is multiplied by cascading interaction processes in particle-air collisions with a large lateral spread, about several hundred meters and observed by particle coincidences as a phenomenon called Extensive Air Shower (EAS). The discovery of EAS ("Grandes

D. N. Poenaru et al. (eds.), Nuclei Far from Stability and Astrophysics, 361–372.

Gerbes") is attributed to the French physicist Pierre Auger who communicated the observation in June 1938. Nearly simultaneously the German Werner Kohlhörster has published the same observation as "Gekoppelte Höhenstrahlung" [2].

Primary cosmic rays consist mostly of charged particles (most abundantly of high-energy protons, alpha-particles, but also heavier nuclei $C, O, Si, ... Fe$), which are accelerated in our Galaxy and deflected by the interstellar magnetic field. The energy spectrum of the cosmic rays comprises more than 12 orders of magnitude, up to 10^{20} eV, following overall power laws, decreasing first $\propto E^{-2.7}$, with a distinct change of the spectral index around 10^{15} eV, called the "knee" [3]. A further change, called the "ankle" is observed at ultrahigh energies.

2.   Extensive Air Shower Genealogy and Components

After the collision of e.g. a high energy proton of the energy E_0 with air nuclei, in average after a free mean path (interaction) length λ_N, the proton (or neutron) survives with the average energy $(1 - K)E_0$. The rest of the energy KE_0 is transferred to the secondary particles (multiparticle production). Most of the secondaries are pions, but also kaons, antiprotons and exotic particles are produced with the multiplicities n_i (see Fig. 1). Neutral pions π^0 immediately decay into two gamma rays. The survival nucleon produces further secondary particles in collisions with air nuclei, thus forming a cascade of interaction processes called the hadronic or the nuclear active component. The most important parameters driving the EAS development are the average quantities λ_N (inverse proportional to the inelastic cross section), the "inelasticity" K and the mean multiplicity $< n >$ of particle production, and additionally the transverse momentum $< p_t >$ distributions [3].

With decreasing energies the charged pions and kaons increasingly decay in muons (and neutrinos) before doing further collisions and form the muon component of positive and negative muons ($\mu^\pm$), called the penetrating component, since muons are only weakly absorbed in matter and experience less multiple scattering. Muons are lost by decay processes (in electrons and neutrinos) and reduce their energy by ionization losses. The gamma rays originating from the immediate decay of the π^0 initiate electromagnetic showers and form the intensive electromagnetic component [4].

These main EAS components are usually observed by large detector arrays like in the KASCADE experiment whose layout is able to register all three main EAS components [5].

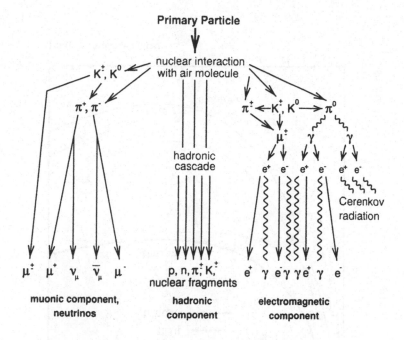

Figure 1 The progeny of EAS development

3. Development of Extensive Air Showers

3.1. Extensive Air Shower Longitudinal Development

The total number of particles that belong to the respective components (shower sizes) are denoted N_h, N_μ and N_e. Fig. 2 displays the average longitudinal development of the various components of 10^{15} eV showers of vertical incidence, calculated by Monte Carlo simulations of EAS, performed with the simulation code CORSIKA [7].

One notices the faster development of the electromagnetic component and the increased muon intensity of the heavy ion induced showers as compared to proton induced showers. These features are important with respect to the mass discrimination of the primaries [8].

The actual size of an EAS depends on the primary energy E_0 and on the atmospheric depth of the observation level [8]. But there are considerable fluctuations, especially due to the fluctuation of the first interaction point X_{in}. Hence for example an EAS observed at a particular observation level X with a certain size N_e may originate from an early X_{in} with large E_0 or from a deeper X_{in} with smaller E_0. There are two further components of EAS, which allow EAS observations, Cerenkov

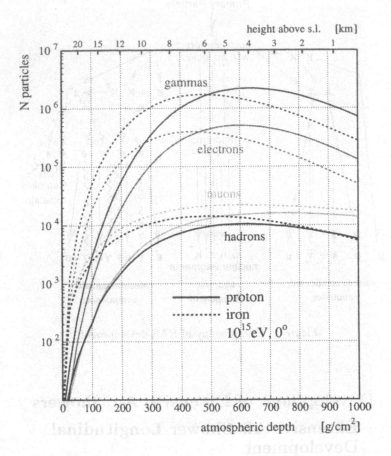

Figure 2   The longitudinal development of 1 *PeV* proton and *Fe* induced EAS from Monte Carlo simulations [6]

light and air fluorescence. Their observation gains increasing experimental importance under favorable observation conditions (see Ref. [9]).

3.2.   Electromagnetic Cascades and the Elongation Rate

The most intensive EAS component is the electromagnetic component. At high energies the dominant electromagnetic interaction processes are the production of electron-positron pairs by photons and the bremsstrahlung production by the electrons. The energy loss of relativistic particles by bremsstrahlung (similarly by pair production) follows an exponential form,

$$-dE/dX = E/X_0 \quad or \quad E(X) \propto exp(-X/X_0) \quad (1)$$

This exponential attenuation defines the characteristic radiation length X_0, i.e. the length over which the particle loses a fraction $(1 - 1/e)$ of its energy by electromagnetic processes.

In a schematic model of a pure electromagnetic shower, a high energy photon may generate an electron-positron pair in a distance $R = X_0 ln2$, each of which in turn generates high energy photons by bremsstrahlung, which further generate again electron-positron pairs and so on, building up an avalanche. Each step of the cascade approximately happens when the probability is 1/2. The cascade is going to be stopped when the energy of the particles reaches the critical energy E_c, i.e when the bremsstrahlung and pair production processes get dominated by ionization losses, Compton scattering and photoabsorption. This defines the maximum N_{max} of the particle production by $E_c = E_0/2^{N_{max}}$. In the schematic consideration the shower maximum is reached at $X_{max} = N_{max}X_0 ln2$. Thus, the maximum of the pure electromagnetic cascade is characterized by: $N_{max} = ln(E_0/E_c)/ln2$ and $X_{max} = ln(E_0/E_c)X_0$. We note the logarithmic dependence of X_{max} from the energy. Therefore a useful EAS quantity is the so-called electromagnetic elongation rate D_e [9, 10], defined for pure electromagnetic showers by

$$D_e = dX_{max}/dlnE_0 = X_0 \quad (2)$$

It characterizes the rate of change of the depth of the shower maximum with energy.

After the maximum X_{max}, the intensity drops down nearly exponentially, governed by the losses due to ionization and Compton scattering.

3.3.   The Shower Age

In the analytical description (electromagnetic cascade theory) of the longitudinal development of pure electromagnetic showers, a formal parameter enters, indicating the actual development of the shower: the age parameter s [9, 10]. The age parameter s varies from $s = 0$ (at the first interaction X_{in}), $s = 1$ at the depth X_{max} of the shower maximum, and reaches its maximum $s = 2$, when the shower size N_e is less than one particle. The same parameter s enters in the description of the form of the lateral distribution $f(s, R/r_0)$ of the electron-photon component derived by Nishimura and Kamata and independently by Greisen: NKG form (see Ref. [9, 10]). Tolerating the approximations of the electromagnetic cascade theory, for a pure electromagnetic shower the age s

at the observation level can be determined, by using the NKG form for adjusting the lateral distribution by fitting the parameter s.

EAS induced by hadronic particles develop as a superposition of electromagnetic cascades, which start with each hadronic interaction producing π^0's. Nevertheless the concepts of the theory of pure electromagnetic showers are often applied to hadronic EAS, defining in this way an age-parameter via a NKG description of the lateral distribution.

4.   Lateral Distributions of the EAS Components

The lateral distributions are the result of the transverse momentum distribution, with which the EAS particles are produced, of the distribution of the production heights of the arriving particles and their multiple scattering and attenuation in the atmosphere. There are various functional forms of phenomenological character in use. The starting point is the original Nishimura-Kamata-Greisen form (NKG-form) [10] exactly derived for pure electromagnetic showers, but also successfully used for the charged particle (the electron-photon), the muonic and hadronic components of hadronic showers. The lateral spread is measured in units of the so-called Molière radius r_0, which results from the multiple scattering theory. In case of the medium air and for the electron-photon component $r_0 = 9.6 \ g/cm^2$ or $79 \ m$ at sea level.

The muon and hadron components can be also parameterized by the NKG form, using different values of the scaling radius r_0. In addition there are various other forms in use in order to improve the phenomenological description of the lateral distributions. As mentioned previously, the other parameter entering in the NKG form is the age parameter s. Fig. 3 displays an example of a single shower as measured in KASCADE experiment [11], fitted by the NKG form.

Mean lateral distributions are obtained by fitting the mean distributions with one of the forms in vogue. It should be noted that the form depends also from the energy threshold of particle detection.

5.   Muon Arrival Time Distributions

Due to the relatively weak interaction and reduced multiple scattering muons of higher energy travel practically undisturbed through the atmosphere with nearly the velocity of light. Therefore, the distributions of the arrival times or of the angle of incidence of EAS muons, observed in the detector array, reflect largely the distributions of the production heights and are a source of information about the longitudinal development [12–15].

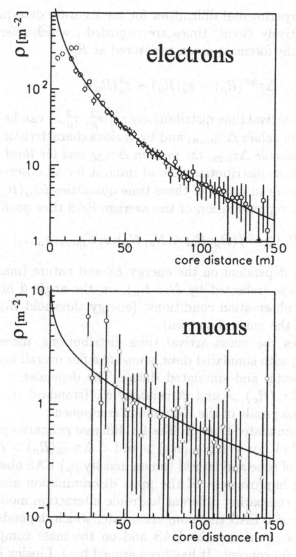

Figure 3 Particle densities of a single EAS as measured by KASCADE array and fitted by the NKG form [11].

Investigations of the relative muon arrival times τ_μ^1, τ_μ^2, τ_μ^3,... at a particular radial distance R_μ from the shower axis must refer to a zero-time, usually [12] to the arrival time τ_c of the shower center. We call such quantities *"global"* times :

$$\Delta\tau^{glob} = \tau_\mu(R_\mu) - \tau_c \qquad (3)$$
$$\Delta\tau_1^{glob} = \tau_\mu^1(R_\mu) - \tau_c \qquad (4)$$

Often there are experimental difficulties for an accurate determination of τ_c, and alternatively "*local*" times are regarded , which refer to the arrival $\tau_\mu^1(R_\mu)$ of the foremost muon registered at R_μ:

$$\Delta\tau^{loc}(R_\mu) = \tau_\mu(R_\mu) - \tau_\mu^1(R_\mu). \tag{5}$$

The single relative arrival time distributions:τ_μ^1, τ_μ^2, τ_μ^3,... can be characterized by the *mean values* $\Delta\tau_{mean}$, and by various characteristic quantities like *the first quartile* $\Delta\tau_{0.25}$, *the median* $\Delta\tau_{0.50}$ *and the third quartile* $\Delta\tau_{0.75}$ [15, 16], whose distributions are of interest for an observed EAS event sample. The mean values of these time quantities $\Delta\tau_\alpha(R_\mu)$, varying with R_μ, are a representation of the average EAS time profile:

$$< \Delta\tau_\alpha(R_\mu) >= f(R_\mu; E_0; A(N_e, N_\mu); \theta; E_{thres}; n_\mu \ldots) \tag{6}$$

which is generally dependent on the energy E_0 and nature (mass A) of the primary particle (reflected by N_e, N_μ), on the angle θ of shower incidence and on observation conditions, (energy threshold E_{thres} and multiplicity n_μ of the muon detection).

Fig. 4 compares the muon arrival time distributions, measured in KASCADE array, with simulated data, showing a fair overall agreement between experimental and simulated data. The dependence of their *mean values* $< \Delta\tau_\alpha(R_\mu) >$ *and dispersions* σ_α (standard deviations) represents the time profile of the EAS muon component.

Fig. 5 displays simulated time profiles for different primaries p and Fe of the energy $10^{15}\ eV$ for $< \Delta\tau_{0.50}(R_\mu) >$ and $< \Delta\tau_{0.50}(R_\mu) > /\rho_\mu(R_\mu)$. The combination of time and lateral (muon density ρ_μ) EAS observables leads to a visible improvement of the mass discrimination and is also recommended for comparing different hadronic interaction models [17].

In case of hadronic EAS the elongation rate, which depends on the hadronic interactions driving the EAS and on the mass composition, remains still a useful concept. It has been argued by J. Linsley [18] that muon arrival times, more precisely their dependence from the energy E_0 and the variation with the angle of EAS incidence can be directly related with the change of the elongation rate and with fluctuations of the atmospheric depth of the EAS maximum [19]. Fig. 6 shows results of EAS simulations, studying the suggested relation: the variation of the $< \Delta\tau_{mean}^{glob} >$ with $sec\,\theta$ and $log_{10}E_0$ for muon detection thresholds of $E_{thr} \geq 250\ MeV$ and $E_{thr} \geq 2\ GeV$ at sea level (KASCADE) and high-mountain altitude (ANI on Mt. Aragats, 3400 m a.s.l.).

The theoretical analysis of Linsley's theorem shows that the relation is rather complex, dependent from the atmospheric depth of the observation level, from the type of the primary particle and from the primary

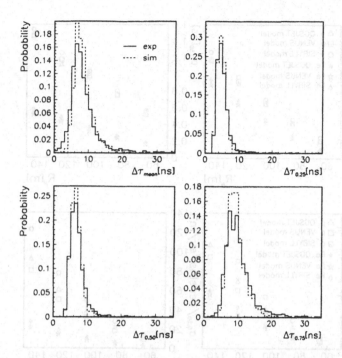

Figure 4 The distributions of different time quantities for a particular EAS sample observed with $4.25 < log_{10}N_{\mu}^{tr} < 4.45$, in the range $5^0 < \theta \leq 30^\circ$ for the EAS angle of incidence and $90\ m < R_{\mu} < 100\ m$. The truncated muon number N_{μ}^{tr} is an observable of the KASCADE experiment and turns out to be a good energy identifier [8]. The simulations adopt a mass composition $H : O : Fe = 4 : 1 : 2$.

energy. That means that a direct evaluation e.g. in terms of the elongation rate changes is hardly possible and needs to be calibrated, based on Monte Carlo simulations for each particular case [20].

6.   Concluding Remarks

The observation of EAS provides a unique opportunity to detect ultra high-energy particles and to study their interactions. This possibility is due to the fact that EAS are "extensive", with an intensity of millions of particles, coherently spread out over an effective area of $1\ km^2$. The main observables which are measured by earth-bound arrays are:

- For EAS of different angles of incidence, the total intensity, the lateral distribution and the time structure of the various different components.

- In case of the hadronic component, arrays with a calorimetric device study also the energy distributions.

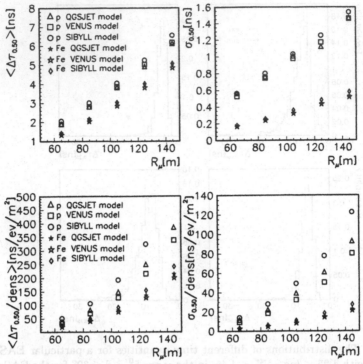

Figure 5 Simulated time profiles $< \Delta \tau_{0.50}(R_\mu) >$ and $< \Delta \tau_{0.50}(R_\mu) > /\rho_\mu(R_\mu)$ of the EAS muons for two primaries and for three hadronic interaction models, calculated for $E_0 = 10^{15}$ eV [17]

- The longitudinal development is signaled by muon arrival time distributions.

The physical significance of EAS observations has two interrelated aspects:

- One aspect is the astrophysical one, to infer from EAS observations the primary energy spectrum and mass composition as basis for theoretical explanations of the origin, acceleration mechanisms and propagation through the space.

- The other aspect arises from the fact that the nuclear collisions driving the EAS cascade provide a source of information on the features of high-energy hadronic interactions and a playground for studies in energy regions beyond the limits of man-made accelerators.

The interrelation of these two aspects originates from the fact that the cosmic accelerator does not provide a well specified beam for studies,

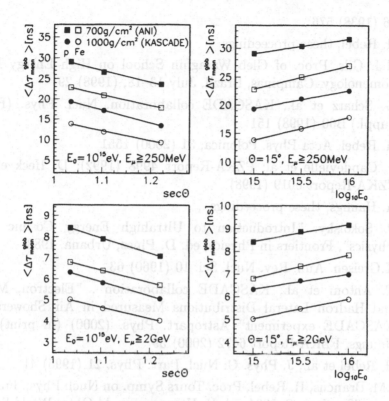

Figure 6 The variation of $< \Delta\tau_{mean}^{glob} >$ with angle-of-EAS incidence θ for $E_0 = 10^{15}$ eV and with $log_{10}E_0$ for $\theta = 15°$ [20].

with an a-priori unknown energy spectrum and a varying mass composition. On the other side to specify energy and mass of the colliding primary particle, we need just a better knowledge of the interaction properties.

Acknowledgments

The experimental results used for illustrations are obtained by the KASCADE collaboration. The substantial help and various communications of Prof.Dr. H. Rebel, Prof.Dr. H.J. Gils, Dr. R. Haeusler, Dr. A. Haungs, Dr. D. Heck, Dr. H.J. Mathes, Dr. M. Roth, Dr. B. Vulpescu, Dr. J. Wentz, DP F. Badea, DP A. Bercuci and DM J. Oehlschläger are gratefully acknowledged.

References

[1] V. Hess, Z. Phys. 12 (1911) 998; Z. Phys. 13 (1912) 1084

[2] P. Auger, R. Maze, T. Grivet-Meyer, Compt. Rend. Seance Acad. Si. 206 (1938) 354; W. Kohlhörster, I. Matthes, E. Weber, Naturwiss.

26 (1938) 576

[3] H. Rebel, these proceedings

[4] H.J. Gils, Proc. of Gleb Wataghin School on High Energy Phenomenology, Campinas, Brazil, July 13–18, (1998) 29

[5] G. Schatz et al., KASCADE collaboration, Nucl. Phys. (Proc. Suppl.) B60 (1998) 151

[6] H. Rebel, Acta Phys. Polonica, 31 (2000) 1551

[7] J. Capdevielle et al: FZKA-Report 4998 (1992); D. Heck et al, FZKA-Report 6019 (1998)

[8] A. Haungs, these proceedings

[9] P. Sokolsky: "Introduction to Ultrahigh Energy Cosmic Ray Physics", Frontiers in Physics, ed. D. Pines, Urbana, 1988.

[10] K.Greisen. Ann. Rev. Nucl. Sci. 10 (1960) 63

[11] T. Antoni et al., KASCADE collaboration , "Electron, Muon and Hadron Lateral Distributions Measured in Air Showers by KASCADE experiment" Astropart. Phys. (2000) (in print); A. Haungs, FZKA-Report 6472 (2000) 59

[12] H. Rebel et al., J. Phys. G: Nucl. Part. Phys. 21 (1995) 41

[13] I.M. Brancus, H. Rebel, Proc. Tours Symp. on Nucl. Phys., France, Aug. 30 – Sept .2, 1994, ed. H. Utsunomiya, M.Ohta, World Scientific, p.78

[14] H.Rebel, FZKA-Report 6472 (2000) 115

[15] T. Antoni et al., KASCADE collaboration, "Time Structure of the EAS Muon Component measured by KASCADE Experiment", Astropart. Phys., (in print)

[16] I.M. Brancus et al., FZKA-Report 6151 (1998)

[17] I.M. Brancus, "Simulation Studies of the Information Content of Muon Arrival Time Observations for High Energy Extensive Airshowers" Internal Report KASCADE 09/2000-02-51.02.03 (2000) Forschungszentrum Karlsruhe

[18] J. Linsley, 15th ICRC, Plovdiv, Bulgaria, 12 (1977) 89

[19] R. Walker A.A. Watson, J. Phys. G.: Nucl. Phys. 7 (1981) 1297; P.R. Blake et al., J. Phys. G.: Nucl.Part.Phys. 16 (1990) 755

[20] A.F. Badea et al., Astropart. Phys. (2000) (in print)

EAS Signatures of the Mass Composition of High-Energy Cosmic Rays

A. Haungs

KASCADE collaboration

Forschungszentrum Karlsruhe, Institut für Kernphysik
Postfach 3640, D-76021 Karlsruhe, Germany
haungs@ik3.fzk.de

Keywords: Cosmic rays, extensive air shower, elemental composition

Abstract   The elemental composition and the energy spectra of the different kinds of primaries of high energy cosmic rays are only accessible by indirect measurements, i.e. by analyzing the appearance of extended air showers (EAS). This lecture gives an overview on mass sensitive observables in EAS measurements with earth-bound detector setups. Various different methods for the reconstruction of the chemical composition from observations in the KASCADE and other experiments are compared.

1.   Introduction

Since the celebrated balloon flights of Victor Hess 1911 and 1912 cosmic radiation (CR) is known as a high energy particle rain entering the Earth's atmosphere [1].   Kohlhörster [2] and independently Auger [3] reported 1938 about particle coincidences in measurements with earth-bound detectors, concluding the existence of so-called extended air showers (EAS) generated as particle cascades in the atmosphere by the incoming primary high-energy cosmic rays.   The determination of the energy spectrum and the chemical composition of primary cosmic rays has been the goal of numerous subsequent experiments. Balloon and satellite borne experiments are able to measure directly the primary CR, but the all particle flux decreases very rapidly with the energy, allowing such measurements only up to energies of $10-100\,TeV$. Indirect measurements by the observation of the EAS, identify a steepening of the power law slope of energy spectrum at around $3-5\,PeV$, known as the "knee" of the cosmic ray spectrum. Since the discovery of the knee [4], 40 years ago, the physical origin of this kink in the spectrum could not yet be conclusively clarified. Nevertheless

D. N. Poenaru et al. (eds.), Nuclei Far from Stability and Astrophysics, 373–384.

there are many theoretical ideas and conjectures with the attempt to explain the knee feature (see L. O'C. Drury, these proceedings). The nature and composition of the source, the acceleration of the particles and the transport in the interstellar or intergalactic medium are the parameters of the models. The difference in the model approaches are expressed in the particular variation of the primaries abundance with energy, i.e. of the elemental composition of the cosmic rays. This lecture addresses the question: How and to which extent we are able to estimate the primary mass of the CR from EAS measurements? Current high-energy cosmic ray experiments are focused to this question, especially in the energy range around the knee.

Cosmic ray induced EAS consist of three measurable particle components (see I.M. Brancus, these proceedings): electrons, muons, and hadrons. The largest component, the electrons, produce in the atmosphere Cerenkov light, which is an additional observable phenomenon in earth-bound CR experiments. Air showers of primaries of different masses develop differently in the atmosphere, leading to varying numbers and distributions of the secondary particles of the shower cascade. Mainly the first high-energy hadronic interactions of the cosmic ray particle with an air molecule are responsible for the differences. But the steeply falling energy spectrum of the primaries, the large fluctuations in the shower development and the relatively small sampling area of the experimental detector arrays (compared to the lateral extension of the total air shower covering few km^2) smears out the average differences. In addition, all methods for inference of the primary mass are based on comparisons with Monte-Carlo simulations of the air shower development, invoking models for the hadronic interactions. Uncertainties of the ingredients of such simulations complicate the determination of the chemical composition by indirect measurements (see H. Rebel, these proceedings).

2.   Mass Sensitive Signatures in EAS

The air shower cascade is driven by the hadronic interactions of the primary particle and by the interactions of the secondary particles. Therefore a sufficient understanding of the interaction processes plays a crucial role to identify the primary mass. According to the Glauber theory [5] the cross sections of nucleus-nucleus interactions get larger with increasing target and projectile mass. For example, the inelastic cross section σ_{inel}^{A-Air} of iron is at 1 PeV approximately 6 times larger than for protons of equal energy. Hence a heavy ion induced air shower starts earlier in average and develops faster in the atmosphere with increasing primary mass. Approximately a primary nucleus of mass A and energy E_0 can be regarded as a superposition of A independent nucleons of energy E_0/A [6].

Consequently in showers induced by heavy primaries more secondary particles are produced but with smaller energies. That means the electron-gamma component gets faster absorbed and has less electrons at observation level (after the shower maximum), while the penetrating muon component is richer. This is due to the fact that the muons interact weakly and their decay time is large as compared to pions (Fig. 1). The ratio of muon number to electron number in EAS at observation level proves to be the strongest mass sensitive observable. The superposition model predicts for all additive observables power law dependences with the mass, which get approximately confirmed by Monte Carlo simulation (Fig. 2). Statistical arguments let expect that the fluctuations of the sum of A independent showers are smaller than of a shower generated by a single proton of higher energy (Fig. 2). This effect is smeared out by the limits of the superposition assumption, but can still guide the mass dis-

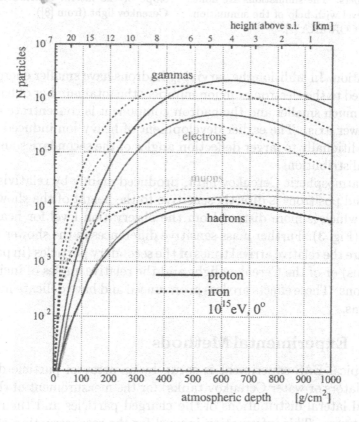

Figure 1   Longitudinal development of the various particle components of proton and iron induced air showers.

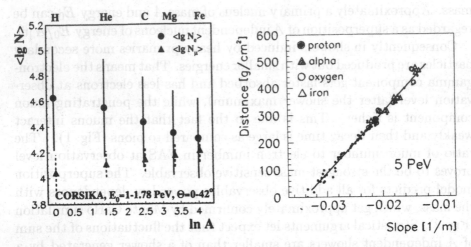

Figure 2 Dependence of the particle numbers of EAS on the primary mass. The error bars show the fluctuations in the numbers. The simulations are done for sea level with help of the simulation program CORSIKA [7].

Figure 3 The distance of the shower maximum from observation level for different primary masses as function of the slope of the lateral distribution of the Cerenkov light (from [8]).

crimination. In addition the surviving hadrons have smaller energies, but compared to the electron or muon numbers the total number of hadrons in EAS is much smaller and the hadron component is concentrated around the shower axis. The earlier development of heavy ion induced showers lead additionally to larger deflection angles of the secondaries and flatter lateral distributions.

The atmospheric Cerenkov light, produced mainly by relativistic electrons and positrons is directly sensitive to the height of the shower maximum, which is more distant from the observation level for heavier primaries (Fig. 3). Further mass sensitive differences in the shower development, are the relative arrival times of the secondary particles (in particular of muons) or of the Cerenkov light and the relative angles of incidence of the muons. These effects are less pronounced and have delicate measuring problems.

3. Experimental Methods

A typical EAS experiment is setup by an array of particle detectors (scintillators or water Cerenkov tanks) for the measurement of the intensity and lateral distributions of the charged particles and their relative arrival times. This information is used for the reconstruction of the energy of the incoming particle, the angle-of-incidence, and the position of the shower axis. Additionally most of the experiments operate large shielded or underground muon counters for the estimation of local or total muon numbers. Only few experiments use calorimeters for studies of the hadronic part of EAS. The EAS Cerenkov component is measured by photomultiplier arrays looking at the clear night sky or by large mirror telescopes with imaging cameras. The telescopes are most effective for primary energies somewhat lower than the knee region and are prevailingly used for high-energy gamma-ray astronomy. For energies in the *EeV* range the intensity of the air fluorescence is sufficiently large to enable measurements. Some experiments try to examine the structure of the shower core, i.e. the region around the shower axis, where at lower observation levels large particle densities exist. Experiments at very high altitudes measure EAS (especially proton induced ones) in a very early stage of the development, where only a few, high-energetic particles in the *TeV* range exist. These particles can be observed by devices using emulsion or X-ray films intersected by lead plates.

The results of EAS experiments determining the elemental composition of the primary cosmic rays are traditionally given by values of the mean logarithmic mass versus the primary energy. In the following, some experimental methods will be described and the results are compiled.

3.1.   Charged Particle Measurements

The correlation of the electron size with the EAS muon content displays the strongest mass sensitivity and is observed in many experiments. The MSU (Moscow State University) EAS experiment consists of an scintillator array and a muon underground detector ($\approx 35 \ m^2$ Geiger counters). Whereas the total number of charged particles (indicator for the primary energy) is reconstructed from the array data, the measured muon density in the underground detector is used as mass sensitive observable. With the help of simulated lateral distributions the total number of muons in the shower is reconstructed. This value and its deviation (fluctuation) from the mean value is compared with distributions obtained by simulations for differently adopted compositions [9]. The best fit values of this analysis are displayed in Fig. 8.

A different approach to infer the mass information carried by the muon component is applied by the HEGRA-CRT experiment [10]. Muon tracking detectors, set up inside the HEGRA scintillator array, measure the deflection angles of muons with respect to the shower axis. The muons of iron initiated showers are produced in average higher in the atmosphere than muons produced in proton induced EAS. This leads to different average deflection angles at the same distance to the shower center. Comparisons with simulated distributions lead to the mean logarithmic mass shown in Fig. 8.

A more sophisticated method for the reconstruction of the elemental composition is used in the CASA-MIA experiment [11]. The large CASA-MIA scintillator array includes 16 underground muon detectors, each of $\approx 190 \ m^2$ acrylic scintillators. Using the measured densities of electrons and muons at certain distances from the shower core and additionally the slope parameter of the electron lateral distribution, single showers are classified as belonging to "light ion" or "heavy ion" induced EAS . The obtained relative abundances of the various mass groups are converted in a mean mass value (see Fig. 8).

The Chacaltaya experiment setup in the altitude of 5200 m a.s.l. operates, in addition to the usual scintillator array, a so called "burst detector" ($\approx 7 \, m^2$) to use the electron-hadron correlation as mass discrimination parameter. It is built up by scintillator counters below 15 cm lead intersected by X-ray films. With such a device the number of hadrons above 1 TeV can be determined [12]. However as the development of the hadronic component is strongly interrelated with the electromagnetic component, the mass discrimination is modest. The procedure requires the shower core to be inside the relatively small burst detector what limits strongly the number of events, leading to only one average value of the mean mass (Fig. 8).

3.2.   Cerenkov Light Measurements

The mass sensitive observable of experiments registering the EAS Cerenkov light is the atmospheric height of the shower maximum. It is reconstructed from the lateral distribution of the Cerenkov light registered by the PMT arrays. The obtained values of X_{max} for four different experiments are displayed in Fig. 4. All the experiments use scintillator arrays for the determination of the core location and the angle-of-incidence of the EAS. Two of these experiments use the array information additionally for the reconstruction of the primary energy (HEGRA-AIROBICC [13] and SPASE-VULCAN [14]), the other two use simply the intensity of the Cerenkov light for the energy estimation (CACTI [15] and DICE [16]). AIROBICC, VULCAN, and CACTI are arrays of PMTs, DICE consists of two separated mirror telescopes. In Fig. 4 the values of X_{max} are compared with expectations for the extreme masses of proton and iron to display the tendency of the variation of the composition with primary energy.

The BLANCA experiment (an array of 144 PMTs, installed within the scintillator array CASA) reconstructs the height of shower maximum ana-

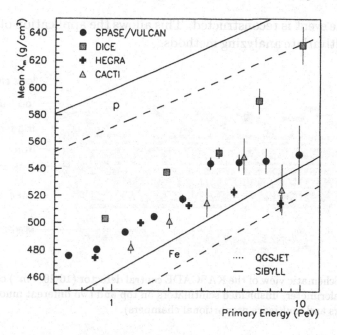

Figure 4 Mean height of shower maximum versus the primary energy measured at different Cerenkov experiments (references see text) compared with proton and iron induced EAS simulations based on different models (QGSJET [17], SIBYLL [18]).

lyzing the slope of the lateral Cerenkov light distribution, and additionally a transformation of X_{max} into mean masses [19] is performed (Fig. 8). The conversion parameters are deduced from detailed EAS simulations based on different high-energy interaction models. This leads to a rough estimate of the systematic error due to the unknown details of the interactions and results in ≈ 0.2 in $ln(A)$.

3.3. The KASCADE Experiment

KASCADE (KArlsruhe Shower Core and Array DEtector) [20] is an experiment especially dedicated to measurements of extensive air showers around the knee. From the very beginning the philosophy of the experiment was to build a modern multi detector setup to measure the different air shower components simultaneously on event-by-event basis, and in addition to promote the development of analyzing tools for the inference of primary energy and mass composition . The developed Monte Carlo code CORSIKA [7] is a program for detailed three-dimensional simulations of the EAS development. The hadronic interactions are optionally described by different interaction models. KASCADE measures the electromagnetic, the muonic (for different threshold energies) and the hadronic component of air showers simultaneously, and a large set of observables for each single event is reconstructed. This allows the application of sophisticated multivariate analyzing methods.

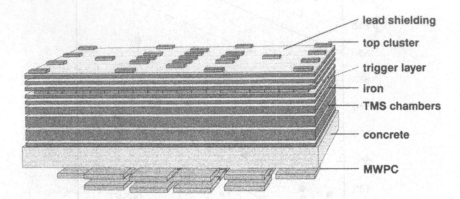

Figure 5   Schematic view of the KASCADE central detector ($16 \cdot 20 \cdot 4 \ m^3$) consisting of a hadron calorimeter, unshielded scintillators on top and two different muon detectors (scintillators and multiwire proportional chambers).

The KASCADE detector system is a large array of unshielded and shielded scintillators for the measurement of the electron and muon lateral distributions with a central detector of different kind of detectors for the hadron and muon detection (Fig. 5): among others an 8-layer iron sampling calorimeter and a system of large area position-sensitive multiwire proportional chambers (MWPC).

The concept of KASCADE implies also to perform various independent analyses of different sets of mass sensitive observables, applying different methods and considering different interaction models. Comparisons of the analyses is expected to help to understand systematic features, which may lead to the large spread of results of the various experiments (see Fig. 8).

One of the procedures analyzes showers with the core inside of the central detector area and exploits the mass sensitive particle structure of the shower core. Muons and secondaries produced by hadrons in the calorimeter exhibit a hit pattern in the MWPC which is analyzed in terms of multifractal moments. One of the resulting parameters is D_6 (Fig. 6), which signals somehow the uniformness of the pattern: Iron induced showers have flatter lateral distributions and the hadrons have smaller energies; hence the pattern is much more uniform [21]. Two of the multifractal parameters are combined with the number of muons in the shower core and the shower size as input in a neural net analysis to estimate the relative abundance of various masses resulting in values of the mean logarithmic mass (Fig. 8) [22].

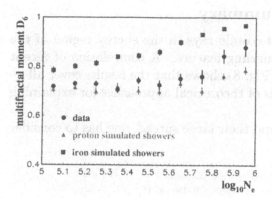

Figure 6 The mass sensitivity of the multifractal parameter D_6, reflecting the structure of the shower core, versus the total number of electrons in the EAS. The data are compared with the distributions expected for pure proton and iron induced showers.

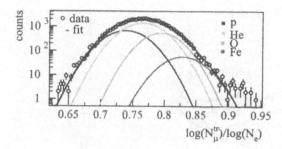

Figure 7 The measured distribution of the muon to electron number ratio in a certain energy range is fitted by a sum of simulated Gauss-functions predicted for different primaries.

In a first step a parametric method is applied to the analysis of the ratio of the electron and muon number, estimated per single shower. The measured distribution of these ratios in a certain energy range (Fig. 7) is assumed to be a superposition of parameterized Gaussian distributions for different primary masses with mean values and widths expected from the simulations [25]. In addition to minimize the a-priori constraints a nonparametric multiparameter analysis of the KASCADE data, based on the Bayesian decision rule, has been performed (see M. Roth, these proceedings). The limited number of Monte Carlo simulations do limit the number of observables which can be used for the multivariate analyses. Hence smaller sets of different observables (electron size, muon size, fractal, and hadron parameters) are studied and the results are averaged for presenting an actual result of this analysis (Fig. 8). Examples of hadronic observables are the number of reconstructed hadrons in the calorimeter ($E_h^{tresh} = 100\ GeV$), their energy sum, or the energy of the most energetic hadron ("leading particle"). The analyses have been performed independently on basis of two different high-energy interaction models (QGSJET [17] and VENUS [24]).

4.   Discussion and Summary

The chemical composition of cosmic rays in the energy region of the knee ($\approx 3-5\ PeV$) is still a puzzling feature. A compilation of recent results of various experiments (Fig. 8) shows that the results cover all — also controversial — predictions of theoretical approaches for explaining the knee.

When comparing the results and their large spread, one has to consider

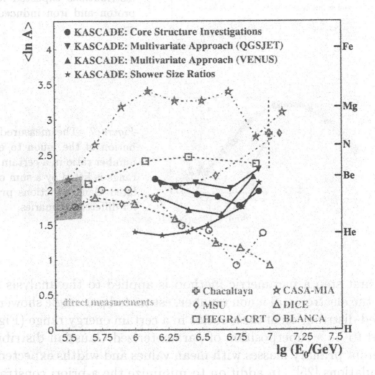

Figure 8 Mean logarithmic mass versus the primary energy measured by different experiments (references see text). Most of the analyses of the data are based on MC simulations using the QGSJET high-energy interaction model [7]. The error bars are omitted for sake of clarity.

following questions which may provide sources of the discrepancies:
— which EAS observables are used for the inference, i.e. from which shower components are the results derived?
— which high-energy hadronic interaction model is the generator for the Monte Carlo simulations used in the analyses?
— how is the energy of the EAS estimated?
— which analyzing techniques have been applied to compare data with simulations (parametric methods or nonparametric approaches)?

— to which extent are the studied observables sensitive to the mass A, to $ln(A)$, or directly to A?

— how many element groups are considered for the conversion of the results in $ln(A)$, i.e. how is the procedure determining $\langle ln(A) \rangle$?

Most of the results show a slight tendency to a heavier mean mass above the knee region. Especially all different analyses of the KASCADE data show this tendency independently from method, the kind of observables or the invoked hadronic interaction model. However, the absolute scale of the resulting mean mass differs distinctly. A comparison of the various experiments is very difficult, as they have on the one hand still large statistical errors, and on the other hand the systematic uncertainties are unknown or not comparable. However, with a large set of observables, KASCADE should be able to examine the sources for the systematic differences. In fact a significant dependence on the particular choice of the analyzed observables is found. The use of hadronic observables leads to a result rather different from results taking only muonic parameters into account. Also the pre-chosen hadronic interaction model underlying the analyses has a noticeable influence. The dependence on the analyzing method seems to be comparatively less dramatic.

As origin of the different answers of the analysis when different EAS observables are regarded, we identify mainly the internal inconsistencies of the hadronic interaction models generating the Monte-Carlo simulations. In fact the appearance of large discrepancies in the elemental composition using observables from different shower components can be interpreted as a strong hint that the energy balance of the particle components of the shower development is not well taken into account in the models [26]. Hence any progress towards the astrophysical solution of the knee enigma and of the question of the mass composition needs a substantial progress in the understanding of the high energy hadronic interaction.

Acknowledgments

The author wishes to thank the colleagues of the KASCADE and of other experiments for various discussions and contributions to this lecture, in particular Dr. M. Roth, Dr. A. Lindner, Prof. H. Rebel, and Prof. G.B. Khristiansen[1].

References

[1] V. Hess, Phys. Z. 13 (1912) 1084

[2] W. Kohlhörster, I. Matthes, E. Weber, Naturw. 26 (1938) 576

---

[1]Prof. Khristiansen, who discovered the knee, recently died after long illness. I remember nice and fruitful discussions with him at his visits in Karlsruhe in the beginning of KASCADE.

[3] P. Auger, R. Maze, T. Grivet-Meyer, Compt. Rend. Hebd. Seanc. Acad. Si. 206 (1938) 354

[4] G.V. Kulikov and G.B. Khristiansen, Sov. Phys. JETP, Volume 35(8), Number 3, MARCH 1959

[5] R.J. Glauber and G. Matthiae, Nucl. Phys. B21 (1970) 135

[6] G. Schatz et al., J. Phys. G: Nucl. Part. Phys. 20 (1994) 1267

[7] D. Heck et al., FZKA 6019, Forschungszentrum Karlsruhe 1998

[8] A. Lindner, Astropart. Phys. 8 (1998) 235

[9] Y.A. Fomin et al., J. Phys. G: Nucl. Part. Phys. 22 (1996) 1839

[10] K. Bernlöhr, Astropart. Phys. 5 (1996) 139

[11] M.A.K. Glasmacher, Astropat. Phys. 12 (1999) 1

[12] C. Aguirre et al., Phys. Rev. D62 (2000) 032003

[13] A. Röhring et al., Proc. 26th Int. Cosmic Ray Conf., Salt Lake City (1999) OG 1.2.09

[14] J.E. Dickinson et al., Proc. 26th Int. Cosmic Ray Conf., Salt Lake City (1999) OG 1.2.05

[15] S. Paling et al., Proc. 25th Int. Cosmic Ray Conf., Durban (1997) Vol. 5, p. 253

[16] S.P. Swordy et al., Proc. 26th Int. Cosmic Ray Conf., Salt Lake City (1999) OG 1.2.07

[17] N.N. Kalmykov, S.S. Ostapchenko, A.I. Pavlov, Nucl. Phys. B (Proc. Suppl.) 52B (1997) 17

[18] J. Engel et al., Phys. Rev. D46 (1992) 5013

[19] J.W. Fowler, Astropart. Phys. (2000), submitted; preprint astro-ph/0003190

[20] H.O. Klages et al. — KASCADE collaboration, Nucl. Phys. B (Proc. Suppl.) 52B (1997) 92

[21] A. Haungs et al., Nucl. Instr. and Meth. A372 (1996) 515

[22] A. Haungs et al. — KASCADE collaboration, Proc. 26th Int. Cosmic Ray Conf., Salt Lake City (1999) HE 2.2.39

[23] M. Roth et al. — KASCADE collaboration, Proc. 26th Int. Cosmic Ray Conf., Salt Lake City (1999) HE 2.2.40

[24] K. Werner, Phys. Rep. 232 (1993) 87

[25] J. Weber et al. — KASCADE collaboration, Proc. 26th Int. Cosmic Ray Conf., Salt Lake City (1999) HE 2.2.42

[26] T. Antoni et al. — KASCADE collaboration, J. Phys. G: Nucl. Part. Phys. 25 (1999) 2161

Cosmic Ray Studies at High Mountain Altitude Laboratories

Oscar Saavedra

Dipartimento di Fisica Generale and INFN, Torino
Universita' degli Studi di Torino, Torino, Italy
saavedra@to.infn.it

Keywords: Cosmic rays, astroparticle physics, high energy interactions

Abstract   Cosmic ray investigations in high mountain altitude laboratories, their experimental aspects and scientific aims are presented. The main features of high-energy nuclear interaction phenomena, with particular emphasis to experimental observations with the emulsion chamber technique, are described. Finally, future prospects of detector installations in high mountain altitude for cosmic ray investigations are considered.

1.   Introduction

This lecture discusses some important scientific achievements that have been derived from cosmic ray experiments in high mountain altitude laboratories and which should attract the interest of young scientists for this exciting field of science. There appear still many unsolved important problems with attractive new perspectives, in particular from experimental point of view for studies in high altitude sites. Cosmic Ray physics was born in 1912 when V.F. Hess [1], after launching 7 balloons investigating the penetrating radiation from the outer space, opened a fantastic new window for the research of the Universe through observations of cosmic radiation. Additionally a great impact resulted from the observation of the Extensive Air Showers (EAS), independently discovered by Kohlhörster [2] and by Auger [3] in 1938. The discovery has been developed to powerful tool to study indirectly primary cosmic radiation, up to the extreme energy ranges of 10^{20} eV, which are not accessible to direct measurements. About 10 years later, Lattes, Occhialini and Powell [4] exposing nuclear emulsions plates on Pic du Midi and Chacaltaya discovered the π-meson. This discovery prompted the study of elementary particles physics and the strong interaction features.

D. N. Poenaru et al. (eds.), Nuclei Far from Stability and Astrophysics, 385–396.
© 2001 *Kluwer Academic Publishers. Printed in the Netherlands.*

In the following years the emulsion technique discovered a variety of new "elementary" particles revealing important properties of strong interaction. Although our-days-knowledge about the primary cosmic radiation is much more detailed, there are still lively discussions about the nature and origin of cosmic rays (A. Wolfendale, these proceedings) in particular about the still enigmatic "knee" in the cosmic ray energy spectrum (A. Haungs, these proceedings, a feature discovered by Khristiansen et al. in 1959 [6].

Why are high altitudes sites so important for observations of cosmic rays? Two main reasons are obvious: (1) The EAS cascades produced by primary cosmic rays in the air with lower energies can be observed there, while usually died out in larger atmospheric depths. (2) The shower cascades are observed in their early stages of development, with minimal fluctuations (Fig. 1)

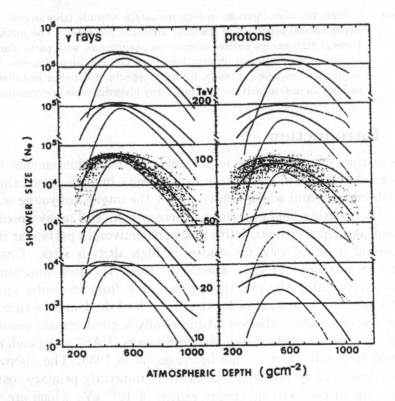

Figure 1 Transition curves of air showers induced by gamma rays and protons for several primary energies.

In the following, some experiments (INCA, SLIM and Emulsion Chambers) which are currently operated in the Chacaltaya Laboratory, will be sketched, showing the advantage of the high altitude studies for various observations of the cosmic radiation. Finally a possible experiment, using a variety of different techniques within a larger international collaboration, is described.

2.   The Chacaltaya Laboratory

The Chacaltaya Cosmic Ray Research Laboratory, one of the most famous cosmic ray laboratory in the world, on Mt. Chacaltaya, near La Paz, Bolivia, is located in an altitude of 5220 m a.s.l., corresponding to an atmospheric depth of 530 gr/cm^2. Note that this atmospheric thickness is equivalent to only 6.6 nuclear mean free paths and 14.1 radiation lengths. The geographic position of Mt. Chacaltaya is at 16° S of lat. 291.8° E long. and $-4°$ of geomagnetic latitude. The geographic location in the Southern Hemisphere is also interesting in view of γ-astronomy, in particular since instrumental installations are rather scarce there as compared with the Northern hemisphere. The Chacaltaya laboratory is the highest continuously operated research station on the globe, and it provides unique opportunities for cosmic ray research. As already indicated at energies above 10^{14} eV, the flux of primary cosmic rays is so low that direct observations by balloon or satellite-borne instruments (with areas of only a few square meters) are not practical. For example, the integral primary cosmic ray flux of energies above 10^{16} eV is only one particle per $(m^2 \, sr \, yr)$. Consequently, in order to improve the sensitivity of indirect studies for energies of and above 10^{15} eV (one PeV), it is favorable to install dedicated detector systems in altitudes as high as possible.

3.   INCA Experiment

INCA is the acronym of INvestigation of Cosmic Anomalies experiment dedicate to search for Gamma Ray Bursts (GRBs). GRBs are the most conspicuous phenomenon in the γ-ray wavelengths observed so far and have been one of the most exciting discoveries in the high-energy astrophysics. Their appearance lasts timely between milliseconds and minutes; then the phenomenon vanishes, and the trail goes cold. The BATSE experiment (see for example [7] on board of CGRO satellite observed more than 2000 bursts, coming from random directions on the sky and suggesting that such bursts originate from cosmological distances. What is inducing these sudden flares of radiation? How are they able to outshine in a short period every thing else in the Universe at γ-ray energies, then vanishing without any trace? These are few of many questions about GRBs. The measured γ-ray energy spectrum with BATSE extends up to few MeV showing that the spectrum becomes steeper at about one MeV. The EGRET experiment on board the CGRO observed 6 GRBs with energies $\geq 1\ GeV$, including the emission of a 18 GeV photon as the GRB940217 burst. The detection of a high energy component would constrain the models of the emission mechanism and the range of the source distances.

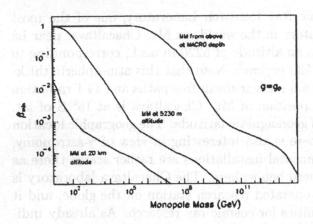

Figure 2 Accessible regions in the plane (mass,β) for monopoles coming from above for an experiment at altitudes of 20 km, 5230 m and for an underground detector at the Gran Sasso Lab.

The detection of GeV γ-ray by ground based experiment can be performed using the single particle technique i.e. observing the secondary particles generated by the γ-ray entering the atmosphere. Details of this technique are given by Vernetto [8]. The number of secondary particles in the EM cascade produced by a primary γ-ray is larger in the higher altitudes, we

consider here. As an example, the mean number of particles generated by a 16 GeV γ-ray reaching 5200 m is ≈ 1 while at 2000 m a.s.l. is only ≈ 0.03 particles. The INCA experiment (operated by a collaboration between Italy, Bolivia and Japan) is running at Chacaltaya to search GRBs by using the single particle detection technique. The experimental layout consists of 12 scintillation detectors (4 m^2 each) of a total area of 48 m^2 and the experiment is running since December 1996. Comparing the results of INCA with those of the EAS-TOP experiment in terms of the upper limits of the fluence (erg/cm^2) vs. the zenith angle of GRB. Aglietta et al. [9], Cabrera et al. [10], it turns out that obviously the sensitivity at Chacaltaya is better, even if the INCA area (only 48 m^2) is considerably smaller than at EAS-TOP (350 m^2).

4.   SLIM Experiment

SLIM is the acronym of Search for LIght magnetic Monopoles. The existence of magnetic monopoles (MMs) in cosmic radiation is a currently discussed question and one of the main topics of present-days non-accelerator particle physics. Grand Unified Theories of electroweak and strong interactions predict the existence of superheavy MMs with masses larger than 10^{16} GeV. However such MMs cannot be produced by energies of existing artificial accelerators. Recent papers [11] conjecture that relatively low mass MMs could be a source of the highest energetic cosmic rays since the basic magnetic charge should be very large and the relatively light MMs could be accelerated to energies of the order of 10^{20} eV. Moreover, since nuclearites (strangelets, aggregates of u, d and s quarks in equal proportions) and Q-balls [12] lose a relatively large amount of energy for $\beta \geq 4 \cdot 10^{-5}$ their detection appears to be possible by SLIM experiment.

Fig. 2 and Fig. 3 show the accessible regions in the plane (mass, β) for MMs and nuclearites at MACRO depth, at Chacaltaya and in 20 km height. Exposures in the high altitude of Chacaltaya would allow detection of the above mentioned particles even if they had strong interaction cross section, which would prevent them from reaching the earth surface. In fact, according to M. Rybczynski et al.

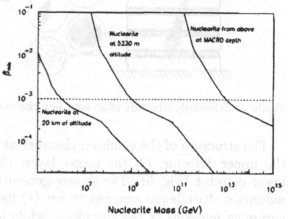

Figure 3 Accessible regions in the plane (mass,β) for nuclearites coming from above for an experiment at altitudes of 20 km, 5230 m and for an underground detector in the Gran Sasso Lab.

[13] the flux at Chacaltaya is estimated to be $\approx 6 / 100$ $m^2 y$.

The SLIM experiment is operated in the Chacaltaya laboratory since January 2000. It comprises track-etch detectors of 24 cm × 24 cm, each setup by 3 layers of CR-39, 3 layers of polycarbonate and an aluminum absorber being 1 mm thick. The total area of Chacaltaya detector is 400 m^2 and it is planned to be exposed for 4 years. Within 4 years operation a sensitivity level of 10^{-14} $eV / cm^2 \, s \, sr$ should be reached

5. Emulsion Chamber Experiment

The basic structure of the emulsion chamber is a multi-layered sandwich of nuclear emulsion plates and lead plates, sketched in Fig. 4. A γ-ray or an electron induces an electromagnetic shower cascade in the Pb plates which is recorded as tracks in the inter-laid emulsion plates in the various stages of the cascade development. The evaluation of the observed tracks with a microscope at various depth thickness of lead implies an energy measurement of the cascade produced by the incident γ-ray. The energy threshold for detection depends on the observation procedures, the structure of the chamber and exposure conditions.

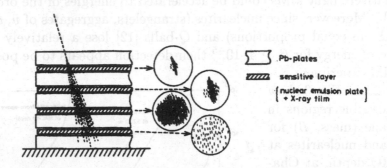

Figure 4   Schematic structure of an emulsion chamber as electron shower detector.

The structure of the emulsion chamber at Chacaltaya comprises: (1) the upper detector, (2) the target layer, (3) the air gap and (4) the lower detector (Fig. 5). The shower-generating cosmic ray particles at mountain altitude are assumed to be: (1) hadrons — some of them are surviving nucleons or cascade pions, while hadrons of other type have negligible contributions to too short lifetimes or small yield, respectively. (2) electrons and γ's from decaying π°'s and their cascade particles produced in the atmosphere. By the experimental arrangement a bundle of high energy electrons, γ's and hadrons, which originate from nuclear and electromagnetic cascade processes initiated high in atmosphere by a high energy single primary cosmic particle. Such a coherent bundles are called a "cosmic-ray family". The tracks observed in the emulsion plates under a microscope have a nickname of "jets". They are classified according to their origin: (1) A-jets when they are produced in the atmosphere above the chamber, (2) C-jets when they are produced in the target material, carbon, and (3) *Pb*-jets when they are generated in the lead material of the upper or lower detector. The reasons for the two-storey structure is: (1) the upper chamber works as a shield against atmospheric γ's and electrons coming into the target and the lower detector, (2) the target material is made of low *Z* material that is nearly transparent for γ-rays produced in C-jets and (3) the air gap server as spacer, resolving the γ-rays of C-jets. The design of the detector facilitates the classification of the shower-generating particles into hadrons and γ-rays. Showers developing in the upper chamber are associated to γ-rays, those in the lower chamber are ascribed to a hadronic origin.

This association is based on the different characteristics of the interactions of γ-rays (radiation length $X_o = 0.57$ cm) and hadrons (collision m.f.p. $\lambda = 18.5$ cm). Because of the larger penetration of hadrons the showers due to nuclear interactions can happen at any place in the chamber, while showers appearing near the top of the chamber are due to γ-rays [5].

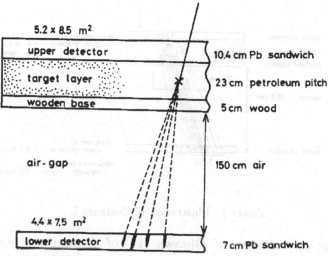

Fig. 2. Chacaltaya emulsion chamber of two-storey structure.

Figure 5   Chacaltaya emulsion chamber of two-storey structure.

5.1.   Centauro Events

A new class of events, called Centauros, has been identified by the Brazil-Japan collaboration [5] with a first example observed in the chamber No. 15 (Fig. 6). This type of events exhibit characteristics being very different from normal events: the appearance in the upper half of chamber does not fit to the appearance in lower part, like the highly asymmetric Centaur in the Greek mythology, with the top half of a man and the legs of a horse. The showers display in the lower detector local nuclear interactions of hadrons while the showers in the upper detector can be either due to the atmospheric (e,γ) or to the nuclear interactions in the upper detector itself.

The characteristics of this Centauro I event are: (1) high primary energy: 1.650 *TeV*, (2) an estimated production height of about 50 *m* above the chamber, (3) a large hadron multiplicity of ≈ 100, (4) zero-multiplicity of π°'s),that means that the parent interaction produces a large number of hadrons but none or only very few of them do rapidly decay in γ ($\pi°$'s), (5) large values of the mean transverse momentum p_T of the secondaries (~ 1.8 GeV/c.)

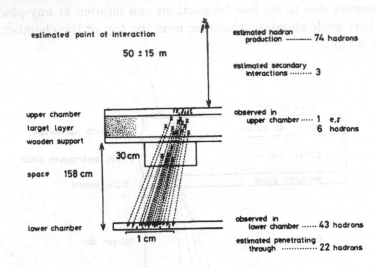

estimated point of interaction

50 ± 15 m

estimated hadron production ---------- 74 hadrons

estimated secondary interactions ········· 3

upper chamber
target layer
wooden support

observed in upper chamber ····· 1 e,γ
6 hadrons

30 cm

space   158 cm

lower chamber

observed in lower chamber ······· 43 hadrons

1 cm

estimated penetrating through ·············· 22 hadrons

Figure 6   Illustration of Centauro I.

Following an estimate of Ohsawa [14] of Centauro events, taking into account the total exposure time of two-storied chambers: $t = 3.49 \cdot 10^2$ $(m^2 \cdot yr)$, the experimental probability to observe a Centauro I event equals $6.3 \cdot 10^{-3}$), while simulations for such event as rare fluctuations of normal events lead to a probability of $1.0 \cdot 10^{-5}$. This distinct difference between the estimated probabilities indicates that Centauro events do not originate from eventual fluctuations of the involved interaction processes. Fig. 7 shows a diagram of N_h vs Q_h, where N_h is the number of hadron-induced showers (with energies exceeding 1 TeV) and Q_h is the ratio between the total energy of the hadron-induced showers and the total reconstructed energy. The diagram shows the distribution of the observed Centauro events (I-V) and the contours of the distribution of 5119 simulated events.

The observations of high- energy phenomena by emulsion chamber techniques seem to indicate that the characteristic features of the hadron interaction may get changed drastically from that what we know at lower energies. There are many conjectures about the origin of the Centauro phenomena, either as arising from exotic primaries [15, 16] or from exotic interactions like as Centauro fire-ball, [5, 17, 18]. In addition recent attempts ascribe Centauro events to extreme fluctuations in high energy hadronic interactions. Simulating the emulsion chamber experiment at Chacaltaya G. Schatz and J. Oehlschläger [19] have found that large fluc-

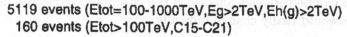

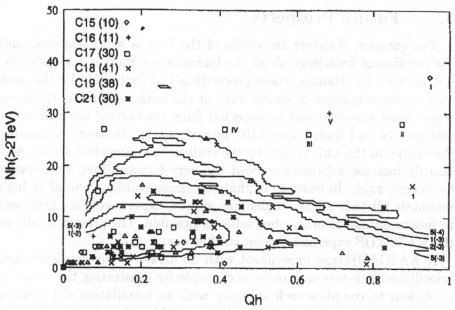

Figure 7 Diagram of N_h and Q_h (see text) for all the events observed at Chacaltaya. The contours are the simulated events.

tuations in the ratio of hadrons to electromagnetic particles may occur. In contrast, M. Tamada [20], on basis of CORSIKA simulations, finds that the current models are unable to explain consistently the extraordinary characteristics of the high-energy cosmic ray events observed in high mountain altitudes.

Considering further results from other different experiments with quite different conditions, analyzing various different EAS observables and using various experimental techniques (like discrepancies reported by the KASCADE collaboration [21], see H.Rebel, these proceedings, and results observed at Chacaltaya [22, 23]) the findings may indicate that at primary energies of $10^{15} - 10^{16}$ eV the "inelasticity" of the nuclear interaction mechanism may be larger than theoretically expected. The present discussion shows that investigations of the knee phenomena by large detector arrays, measuring particle numbers, have an intimate relation to the emulsion chamber experiments observing γ families. Obviously the interpretation of data in terms of the primary energy spectrum and mass composition of cosmic rays with energies higher than the knee region will be seriously affected if the characteristic of the hadron inter-action would really change radically.

6.   Future Prospects

The question of nature and origin of the knee is still unclarified, and the insufficient knowledge about the hadronic interaction is a basic rea-son for that (A. Haungs, these proceedings). What could be the next step for investigations of cosmic rays in the knee region? At these en-ergies some answers could be expected from the current accelerator in-vestigations and from future LHC investigations. However, increasing the energy in the c.m. system by the collider experiments does not nec-essarily increase information about the very forward direction relevant for cosmic rays. In order to exploit the considerable potential of high mountain altitude studies under this aspect, experiments are proposed to measure there as many observable as possible. The present results of the KASCADE experiment point to this way.

A KASCADE-type experiment with an array of densely distributed scintillator detectors may serve as example for illustrating the ideas. It is obvious to complete such an array with an installation of Cerenkov counters since the atmospheric conditions at Chacaltaya are optimal for Cerenkov light observations. Of course a hadron calorimeter has to in-tegrate the following design features:

— A sufficient thickness to cover hadron energies up to 100 TeV.
— A sufficiently large area (100 m^2) to collect a larger number of events.
— A carpet detector on the top (streamer tubes or RPC counters) to measure the fine structure of the electromagnetic component of the as-sociated EAS.
— A tracking system on the top to determine the arrival direction of survival hadrons.
— A fine grain tracking device like emulsion chambers or scintillator fibers.
Such a detector arrangement would enable studies of the following items:

— **The direct calibration of the proton spectrum up to 100 TeV.**
So far the energy region up to some 100 TeV has been investigated by balloon direct measurements, while at higher energies the indirect EAS measurements provide the only information. By the present proposal a direct measurement of the primary proton spectrum could be achieved at higher energies. In fact, the survival proton spectrum of protons that arrive at the Chacaltaya level experience an attenuation given by $N = N_o \cdot e^{-530/\lambda(E)}$, where $\lambda(E)$ is the nuclear interaction m.f.p. The

energy of such events can be measured by the calorimeter, with an identification by the absence of particles in the carpet detector or in the EAS
array. In addition, during the moonless nights the Cerenkov detectors
could help to select the primary protons. With the foreseen area of the
calorimeter 80 events/year for $E_p (\geq E) = 100 \, TeV$ could be registered.
The proposed direct measurements up to (or higher) $100 \, TeV$ are requested for the calibration of the indirect EAS measurements carried
out with the same installation at Chacaltaya.

— **Pure events.** Pure events are events arising from about 2 collision
m.f.p. above the observation level and which have not experienced further fluctuations of cascade development. Two nuclear interaction m.f.p.
≈ 4 radiation length (r.l.) corresponds only $\approx 2 \, km$ above Chacaltaya.
Therefore, when a very collimated hadronic jet is produced it initiates
an electromagnetic cascade by decaying π°'s, observed in a very early
stage with a very steep lateral distribution. The lateral distribution of
electromagnetic cascade remains well concentrated in the $100 \, m^2$ carpet
detector. The high energy hadrons would appear well inside the $100 \, m^2$
calorimeter area. Thus an interesting possibility is provided to study jet
production in the very forward direction process, very difficult or even
impossible to be studied at colliders. Few events/year of this type are
expected at $1 \, PeV$ and ~ 100 events/year at $200 \, TeV$. Such events
originate from primary cosmic ray protons since only protons can reach
the observation level of Chacaltaya.

— **The detection of multihadrons.** Multihadrons events (≥ 2) registered by the calorimeter would enable investigations of the primary composition in a semi-direct way. In fact, Chacaltaya is only at $530/\lambda(E)$
interaction m.f.p. and 14 r.l. from the atmospheric top. Therefore a
study of events with a simultaneous measurements of the hadron energy
and multiplicity, with the associated EAS and the Cerenkov light arrays
seems feasible at Chacaltaya. In addition, Centauro-like events could be
possibly associated to an accompanying EAS.

7. Conclusions

This lecture emphasizes the advantages of cosmic ray measurements
in high altitudes, exploiting the unique conditions there in laboratories
like on Mt. Chacaltaya. With a suitable design the proposed experiment
could demonstrate these possibilities with interesting answers to current
problems of cosmic ray physics.

Acknowledgments

I am grateful to Prof. H. Rebel for many enlightening discussions on the subject and for providing me the opportunity to present the discussed aspects in a lecture at the Advanced Study Institute in Predeal.

References

[1] V. Hess, Phys. Z. 13 (1912) 1084

[2] W. Kohlhörster, I. Matthes, E. Weber, Naturwiss. 26 (1938) 576

[3] P. Auger, R. Maze, T. Grivet-Meyer, Compt. Rend. Hebd. Seanc. Acad. Si. 206 (1938) 354

[4] C. Lattes, G. Occhialini and C. Powell, Nature 160 (1947) 486

[5] C. Lattes, Y. Fujimoto, S. Hasegawa Phys. Rep. 65 (1980) 151

[6] G.V. Kulikov and G.B. Khristiansen, JETP 35 (1959) 635

[7] G.J. Fishman, Astron. Astrophys. Suppl. 138 (1999) 395

[8] S. Vernetto, Astropart. Phys. 13 (2000) 75

[9] M. Aglietta et al., Astron. Astrophys. Suppl. 138, (1999) 595

[10] R. Cabrera et al., Astron. Astrophys. Suppl. 138, (1999) 599

[11] E. Huguet and P. Peter hep-ph/9901370 (1999); C.O. Escobar and R.A. Vazques, Astropart. Phys. 10 (1999) 197

[12] A. Kusenko et al., CERN-TH/97-346, hep-ph/9712212 (1997)

[13] M. Rybczynski et al. Presented at Chacaltaya Meeting July 2000

[14] A. Ohsawa, ICRR Report-454-99-12 (1999)

[15] J.D. Bjorken et al. Phys. Rev. D20 (1979) 2353

[16] E. Witten, Phys. Rev. D30 (1984) 272

[17] S. Hasegawa, ICR Report151-87-5 (1987)

[18] J.D. Bjorken, Int. J. Mod. Phys., A7 (1992) 4189

[19] G. Schatz and J. Oehlschläger, Nucl. Phys. Proc. Suppl. 75A (1999) 281

[20] M. Tamada ICRR Report-454-99-12 (1999)

[21] KASCADE collaboration, Proc. 26th ICRC vol. 1 (1999) 131

[22] C. Aguirre et al., ICRR-report-460-2000-4, accepted in Phys. Rev. D.

[23] N. Kawasumi et al., Phys. Rev D53 (1996) 3634

What Do We Learn About Hadronic Interactions at Very High Energies From Extensive Air Shower Observations?

Heinigerd Rebel

KASCADE collaboration

Forschungszentrum Karlsruhe, Institut für Kernphysik

Postfach 3640, D-76021 Karlsruhe, Germany

rebel@ik3.fzk.de

Keywords: Cosmic rays, extensive air shower, hadronic interaction

Abstract   Extensive air showers (EAS) induced by the collisions of primary cosmic particles with the Earth's atmosphere are a playground for studies of particle interaction processes. The hadronic interaction is a subject of uncertainties and debates, especially in the ultrahigh energy region extending the energy limits of man made accelerators and the experimental knowledge from collider experiments. Since the EAS development is dominantly governed by soft processes, which are presently not accessible to a perturbative QCD treatment, one has to rely on QCD inspired phenomenological interaction models, in particular on string-models based on the Gribov-Regge theory like QGSJET, VENUS and SIBYLL. Recent results of the EAS experiment KASCADE are scrutinized in terms of such models, by comparison of the data with predictions of simulations using various models en vogue as generators in the Monte Carlo EAS simulation code CORSIKA.

1.   Introduction

In addition to the astrophysical items of origin, acceleration and propagation of primary cosmic rays, cosmic ray studies follow the historically well developed aspect of the interaction of high-energy particles with matter. Cosmic rays interacting with the atmosphere produce the full zoo of elementary particles and induce by cascading interactions extensive air showers (EAS) which we observe with large extended detector arrays distributed in the landscapes, recording the features of different

D. N. Poenaru et al. (eds.), Nuclei Far from Stability and Astrophysics, 397–407.
© 2001 *Kluwer Academic Publishers. Printed in the Netherlands.*

particle EAS components. The EAS cascade carries information about the hadronic interaction, though it has to be disentangled from the a-priori unknown nature and quality of the beam. When realizing the present limits of man made accelerators, it is immediately obvious why there appears a renaissance of interest in cosmic ray studies from the point of view of particle physics. EAS observations at energies above 10^{15} eV represent an almost unique chance to test theoretical achievements of very high energy nuclear physics.

This lecture reviews some relevant aspects of hadronic interactions affecting the EAS development, illustrated with recent results of EAS investigations of the KASCADE experiment [1], especially of studies of the hadronic EAS component using the iron sampling calorimeter of the KASCADE central detector [2].

2.   EAS Development and Hadronic Interactions

The basic ingredients for the understanding of EAS are the total cross sections of hadron air collisions and the differential cross sections for multiparticle production. Actually our interest in the total cross section is better specified by the inelastic part, since the elastic part does not drive the EAS development. Usually with ignoring coherence effects, the nucleon-nucleon cross section is considered to be more fundamental than the nucleus-nucleus cross section, which is believed to be obtained in terms of the first. Due to the short range of hadron interactions the proton will interact with only some, the so-called wounded nucleons of the target. The number could be estimated on basis of geometrical considerations, in which size and shape of the colliding nuclei enter. All this is mathematically formulated in the Glauber multiple scattering formalism, ending up with nucleon-nucleus cross sections.

Looking for the cross features of the particle production, the experiments show that the bulk of it consists of hadrons emitted with limited transverse momenta ($< P_t > \sim 0.3$ GeV/c) with respect to the direction of the incident nucleon. In these "soft" processes the momentum transfer is small. More rarely, but existing, are hard scattering processes with large P_t-production.

It is useful to remind that cosmic ray observations of particle phenomena are strongly weighted to sample the production in forward direction. The kinematic range of the rapidity distribution for the Fermilab proton collider for 1.8 TeV in the c.m. system is equivalent to a laboratory case of 1.7 PeV. Here the energy flow is peaking near the kinematical limit. That means, most of the energy is carried away longitudinally. This dominance of longitudinal energy transport has initiated the concept, suggested by Feynman: The inclusive cross sections are expressed by factorizing the longitudinal part with an universal transverse momentum distribution $G(P_t)$ and a function scaling with the dimensionless Feynman variable x_F, defined as the ratio of the longitudinal momentum to the maximum momentum. Though this concept, expressing finally the invariant cross sections by

$$E \cdot d^3\sigma/dp^3 \sim x_F \cdot d^3\sigma/dx_F dp_T \tag{1}$$

provides an orientation in extrapolating cross sections, is not correct in reality, and the question of scaling violation is a particular aspect in context of modeling ultrahigh-energy interactions.

3. Hadronic Interaction Models as Generators of Monte-Carlo Simulations

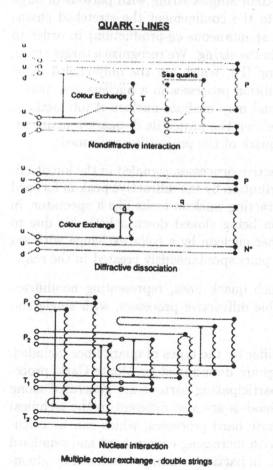

Microscopic hadronic interaction models, i.e. models based on parton-parton interactions are approaches, inspired by the QCD and considering the lowest order Feynman graphs involving the elementary constituents of hadrons (quarks and gluons). However, there are not yet exact ways to calculate the bulk of soft processes since for small momentum transfer the coupling constant α_s of the strong interaction is so large that perturbative QCD fails. Thus we have to rely on phenomenological models which incorporate concepts from scattering theory.

A class of successful models is based on the Gribov-Regge theory. In the language of this theory the interaction is mediated by exchange particles, so-called Reggeons. At high energies,

Figure 1 Parton interaction lines.

when non-resonant exchange is dominating, a special Reggeon without color, charge and angular momentum, the Pomeron, gets importance. In a parton model the Pomeron can be identified as a complex gluon network or generalized ladders i.e. a colorless, flavorless multiple (two and more) gluon exchange. For inelastic interactions such a Pomeron cylinder of gluon and quark loops is cut, thus enabling color exchange ("cut cylinder") and a re-arrangement of the quarks by a string formation. Fig. 1 recalls the principles by displaying some parton interaction diagrams.

- The interacting valence quarks of projectile and target rearrange by gluon exchange the color structure of the system (the arrow indicates the color exchange by opening the cylinder). As a consequence, constituents of the projectile and target (a fast quark and slow di-quark e.g.) for a color singlet string with partons of large relative momenta. Due to the confinement the stretched chains start to fragment (i.e. a spontaneous $q\bar{q}$-production) in order to consume the energy within the string. We recognize a target string (T) and a projectile string (P), which are the only chains in pp collisions. In multiple collision processes in a nucleus, sea quarks are additionally excited and may mediate nucleon-A interactions. While in the intermediate step the projectile di-quark remains inert, chains with the sea quark of the projectile are formed.

- Most important are diffractive processes, signaled in the longitudinal momentum (x_F) distribution by the diffractive peak in forward directions. Here the interacting nucleon looks like a spectator, in some kind of polarization being slowed down a little bit due to a soft excitation of another nucleon by a color exchange with sea quarks (quark-antiquark pairs spontaneously created in the sea).

- There is a number of such quark lines, representing nondiffractive, diffractive and double diffractive processes, with single and multiple color exchange.

The various string models differ by the types of quark lines included. For a given diagram the strings are determined by Monte Carlo procedures. The momenta of the participating partons are generated along the structure functions. The models are also different in the technical procedures, how they incorporate hard processes, which can be calculated by perturbative QCD. With increasing energy hard and semihard parton collisions get important, in particular minijets induced by gluon-gluon scattering.

In summary, the string models VENUS [3], QGSJET [4] and DPM-JET [5] which are specifically used as generators in Monte-Carlo EAS simulations are based on the Gribov-Regge theory. They describe soft particle interactions by exchange of one or multiple Pomerons. Inelastic reactions are simulated by cutting Pomerons, finally producing two color strings per Pomerons which subsequently fragment into color-neutral hadrons. All three models calculate detailed nucleus-nucleus collisions by tracking the participants nucleons both in target and projectile. The differences between the models are due to some technical details in the treatment and fragmentation of strings. An important difference is that QGSJET and DPMJET are both able to treat hard processes, whereas VENUS, in the present form, does not. VENUS on the other hand allows for secondary interactions of strings which are close to each other in space and time. That is not the case in QGSJET and DPMJET. SIBYLL [6] and HDPM [7] extrapolate experimental data to high energies guided by simple theoretical ideas. SIBYLL takes the production of minijets into account. These models are implemented in the Karlsruhe Monte Carlo simulation program CORSIKA [7, 8] to which we refer in the analyses of data. An extensive comparison of the various models and studies of their influence on the simulated shower development and EAS observables have been made in Ref.[9]. There are distinct differences in the average multiplicities and the multiplicity distributions generated by different models. Nevertheless the variations in the average longitudinal development, though visible, appear to be relatively small. It should be noted that when inspecting the development of single showers with identical initial parameters, instead of average quantities, we get impressed by the remarkable fluctuations and sometimes unusual EAS developments. A further aspect which affects the accuracy of the simulations are the tracking algorithms propagating the particles through the atmosphere. In devising the CORSIKA code great care has been taken of this aspect, since the outcome for arrival time and lateral distributions could be significantly influenced by the tracking procedures.

4.   EAS Analysis Scheme

The general scheme of the analysis of EAS observations involves Monte Carlo simulations constructing pseudo experimental data which can be compared with the real data. The king-way of the comparison is the application of advanced statistical techniques of multivariate analyses of nonparametric distributions [10].

The KASCADE experiment, whose general layout and multi-detector system have been discussed elsewhere [11] provides a number of various different observables. We consider, in particular, the hadronic observables [12], in dependence from shower parameters which characterize the registered EAS, in particular indicating the primary energy:

- The shower size N_e, i.e. the total electron number

- The muon content N_μ^{tr} which the number of muons obtained from an integration of the lateral distribution in the radial range from 40 to 200 m. It has been shown that this quantity is approximately an mass independent energy estimator for the KASCADE layout, conveniently used for a first energy classification of the showers [13].

5.   Test of Single EAS Observables

The mass composition of cosmic rays in the energy region above 0.5 PeV is poorly known. Hence the comparison of simulation results based on different interaction models has to consider two extreme cases of the primary mass: protons and iron nuclei, and the criteria of our judgment about a model is directed to the question, if the data are compatible in the limits of the predicted extremes of protons and iron nuclei. First, the dependence of the average number of hadrons N_H with an energy $E_H > 100$ GeV from the shower size is shown and

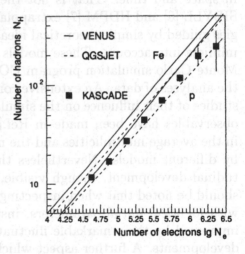

Figure 2 Hadron number N_H - shower size N_e correlation.

compared with the predictions of the VENUS and QGSJET model. The energy range covers the range from 0.2 PeV to 20 PeV. The result shows some preference for the QGSJET model, and such an indication is corroborated by other tests.

There is another feature obvious. When shower observables are classified along the electromagnetic shower sizes N_e, a proton rich composition is displayed. This effect is understood by the fact that at the same energy protons produce larger electromagnetic sizes than iron induced showers, i.e. with the same shower size iron primaries have higher energies, where the steeply falling primary induces the dominance of protons in the sample.

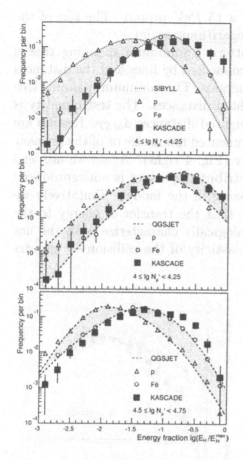

Figure 3 Distribution of the energy fraction of the EAS hadrons.

Another example considers the frequency distributions of the energy of each single hadron E_H with respect to the energy of the most energetic hadron E_H^{max}. The data are compared with predictions of SIBYLL and QGSJET for iron and proton induced showers.

- For a primary proton one expects that the leading particle is accompanied by a swarm of hadrons of lower energies. For a primary iron nuclei the energy distribution appears narrowed.

- The two upper curves display the case for a primary energy below the knee (about 3 PeV). The deficiencies of SIBYLL are obvious and have been also evidenced by other tests, especially with the muon content [12]. SIBYLL seems to produce a wrong EAS muon intensity, and it is fair to mention that just this observation has prompted the authors to start a revision of the SIBYLL model.

- At energies well above the knee (about 12 PeV) also the QGSJET exhibits discrepancies, at least in the energy distribution of the hadrons of the shower core. Other observables like lateral distribution and the total number of hadrons, however appear more compatible with the model.

How to interpret this results? Tentatively we may understand that in the simulations E_H^{max}, the energy of the leading hadrons is too large. Lowering E_H^{max} would lead to a redistribution of the E/E_H^{max} distribution shifting the simulation curves in direction of the data.

A further test quantity is related to the spatial granularity of hadronic core of the EAS. The graph (Fig. 4 left) shows the spatial distribution of hadrons for a shower induced by a 15 PeV proton. The size of the points represents the energy (on a logarithmic scale).

For a characterization of the pattern a minimum spanning tree is constructed. All hadron points are connected by lines and the distances are weighted by the inverse sum of energies. The minimum spanning tree minimizes the total sum of all weighted distances. The test quantity is the frequency distribution of the weighted distances d_{MST}. Results are shown for two different bins of the truncated muon size or of the primary energy (2 and 12 PeV), respectively (Fig. 4 right). Again we are lead to the impression that either the distribution pattern is not reproduced or the high-energy hadrons are missing in the model. Tentatively we may deduce from these indications, that the transfer of energy to the secondaries — what we phenomenologically characterize with the not very well defined concept of the inelasticity of the collision — appears to be underestimated.

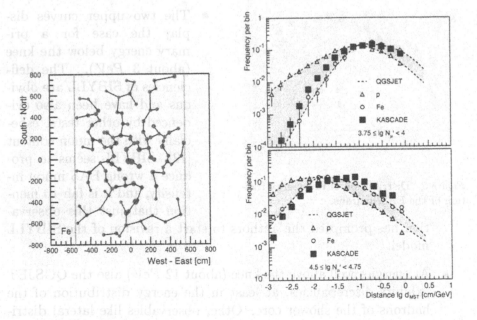

Figure 4 Left: Example of a hadronic core observed in the calorimeter (top view). Right: Frequency distributions of the distances of the minimum-spanning-tree.

6.   Correlation Tests

There are methods of advanced statistical inference and pattern recognition (by Bayes decision making and neural networks) to compare the measured distributions with the model distributions, generated by Monte Carlo simulations and to classify the event to a certain class, i.e to certain energy and mass value of the primary [14].

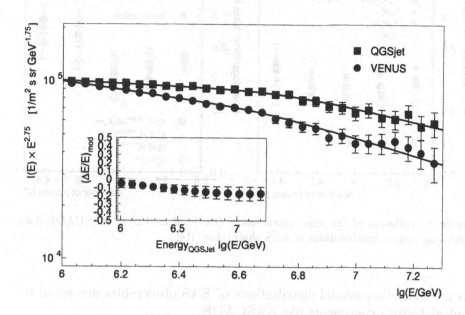

Figure 5 The primary energy spectrum around the knee resulting from KASCADE data analyses on basis of two different hadronic interaction models [14].

The result of such a classification obviously depends from the type of the hadronic interaction model.Thus, in general,the application of the pattern recognition procedures based on different interaction models, lead to different shapes of the primary spectrum [14]. This model dependence is illustrated in Fig. 5 showing the resulting primary energy spectra inferred from the same data set. However we may try more than just compare the results of different models. We are able to explore the correlations of the observables within in a certain specific model and look for the influence of the particular observables and their correlations.

The Fig. 6 displays the results for the mass composition variation, based on the QGSJET model and expressed by the mean logarithmic *A*. Various different combinations of EAS parameters have been regarded. We see, different combinations, though with the same tendencies, lead to differing results, and these differences reveal that the model does not reproduce all the internal correlations, which the nature requires. Moreover we are able to look systematically for the role and the mass sensitivity of the correlations, as feedback for the model builders. Such an approach is not yet worked out, especially in the direction to specify the margin of model uncertainty. Let us realize: The experimental ba-

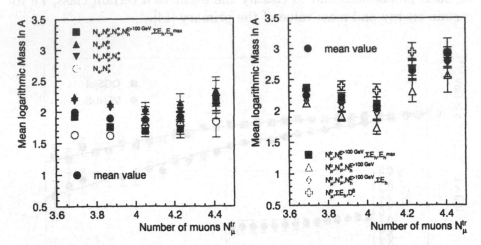

Figure 6 Variation of the mass composition ⟨lnA⟩ inferred from KASCADE data analyzing various combinations of EAS observables [14].

sis are multidimensional distributions of EAS observables measured by multidetector experiments like KASCADE.

7.   Concluding Remarks

From the investigation of a series EAS observables and comparisons with different hadronic interaction models, en vogue for ultrahigh energy collisions, we conclude with following messages:

In general there are tentative indications that the inelasticity in the fragmentation region is underestimated especially with increasing energy. All models are in a process of refinements. Actually somehow triggered by the experimental indications, there is a common enterprise of VENUS and QGSJET towards a combined model descriptions: NEXUS [15]. That is a unified approach combining coherently the Gribov-Regge theory and perturbative QCD. Faced with the experimental endeavor to set up giant arrays for astrophysical observations at extremely high energies, the Monte Carlo simulations need certainly a safer ground of model generators. Hence our efforts in KASCADE are directed to extend the array and to refine the present studies with results towards primary energies of 10^{17} eV.

Acknowledgments

The experimental results are based on results of the KASCADE collaboration. I acknowledge the help of A. Haungs and M. Roth.

References

[1] H.O. Klages et al. — KASCADE collaboration, Nucl. Phys.B (Proc. Suppl.) 52B (1997) 92

[2] J. Engler et al., Nucl. Instr. Meth. A427 (1999) 528

[3] K. Werner, Phys. Rep. 232 (1993) 87

[4] N.N. Kalmykov and S.S. Ostapchenko, Phys. At. Nucl. 56 (1993) 346

[5] J. Ranft, Phys. Rev. D51 (1995) 64

[6] R.S. Fletcher el al., Phys. Rev. D50 (1994) 5710

[7] J.N. Capdevielle et al., KfK Report 4998, Kernforschungszentrum Karlsruhe (1992).

[8] D. Heck et al., FZKA Report 6019, Forschungszentrum Karlsruhe (1998)

[9] J. Knapp, D. Heck, G. Schatz, FZKA Report 5828, Forschungszentrum Karlsruhe (1996)

[10] A. Chilingarian, "ANI — Nonparametric Statistical Analysis of High Energy Physics and Astrophysics Experiments", Users Guide, 1998;
M.Roth, FZKA Report 6262, Forschungszentrum Karlsruhe (1998)

[11] A. Haungs, these proceedings

[12] T. Antoni et al. — KASCADE collaboration, J. Phys. G: Part. Phys. 25 (1999) 2161

[13] R. Glasstetter for the KASCADE collaboration, *Proc. 25th ICRC (Durban)* 6 (1997) 157

[14] M. Roth, these proceedings

[15] H.J. Drescher et al., *preprint hep-ph/9903296* (March 1999)

Application of Neural Networks in Astroparticle Physics

Markus Roth

KASCADE collaboration

Institut für Kernphysik, Forschungszentrum Karlsruhe

P.O. Box 3640, D-76021 Karlsruhe, Germany

roth@ik3.fzk.de

Keywords: Neural networks, Bayes classifiers, cosmic rays, mass composition

Abstract   Based on a well defined statistical inference procedure for multiparameter experiments an introduction and some results of the applied Bayesian and congruent Neural Network (NN) algorithms are given. The chosen Bayesian methods provide the opportunity to study physical data on event-by-event basis. These non-parametric approaches evade to analyze data describing explicitly the physical process by a functional form, which is often not known, but rely on Monte Carlo (MC) simulations of the physical quantities. Some results of the experiment KASCADE are given to illustrate the application of the algorithms.

1. Introduction

Data analysis means transferring information from data (feature) space to map (physics) space. Thus, the theory of the experiment can simply be regarded as an mathematical operator mapping from physics to feature space. The, so called, *forward problem* consists of applying this operator. The *inverse problem* is related with the inverse operator that might take us from the feature to the physics space. The inverse operator might not exist at all, since it usually operates on a space larger than the data space. Nevertheless, the inverse problem has to be solved in a consistent way.

If observed data contain only one or two variables, simplified methods are available, showing or emphasizing some global properties or relations between observed and physical quantities, but neglecting correlations of different observables. The simplification often implies the use of parameterizations of the average behavior, which biases the results and limits

409

D. N. Poenaru et al. (eds.), Nuclei Far from Stability and Astrophysics, 409–420.
© 2001 *Kluwer Academic Publishers. Printed in the Netherlands.*

the accuracy due to neglecting the fluctuations. When multivariate data are examined, it is nearly impossible to comprehend the feature space structure. For the analysis of multivariate parameter distributions and accounting for the influence of the fluctuations more sophisticated methods are needed.

The Bayesian and Neural Network approaches respond to these substantial necessity. The main advantage of the involved algorithms is the use of non-parametric procedures by introducing the constraints of *a priori* chosen parameterizations. The methods facilitate an event-by-event analysis. For classification tasks we define for each given event x a *posterior* probability $p(A_j|x)$, giving the likelihood of belonging to class A_j. Monte Carlo events set up a multidimensional probability space, which is taken to calculate such *posterior* probabilities.

A detailed investigation of the energy spectrum and mass composition of cosmic rays in the energy range of $10^{15} - 10^{16}$ eV is a typical example of application in astrophysics, based on the analysis of extensive air shower (EAS) events, which are registered by the KASCADE [1, 2, 3] detector installation.

2.   Classifiers in Principle

Pattern recognition techniques are efficient tools to determine the correct association of a given sample to a certain category or class. From the measurements or simulations of a physical phenomenon, a set of quantities (observables) is obtained which define an observation vector x. This observation vector serves as the input to a procedure based on decision rules, by which a sample is assigned to one of the given classes. Thus it is assumed, that an observation vector is a random vector whose conditional density function depends on its class A_i.

In the following we consider *non-parametric* techniques like Bayes classifiers and artificial Neural Networks [4]. The term non-parametric indicates, that the representations of the distributions (like probability density functions of Bayes classifiers or weights of Neural Networks) get no more specified by a-priori chosen functional forms. But they are constructed through the analysis process by the given data distributions themselves.

For most of the applications a feedforward network is used. In feedforward networks neurons are arranged in layers without back-loops. There are no connections between neurons of the same layer. The basic element of a Neural Network is an artificial neuron described by

$$y_i = g\left(\sum_{j=1}^n w_{ij}x_j + S_i\right),$$

where w_{ij} are the connection weights from neuron j to neuron i. The input activities x_j are multiplied with the connection weights, accumulated and transfered by a nonlinear activation function to the output activities y_i (Fig. 1). The threshold S_i represents the center of the critical region of each neuron, where the activation value changes drastically. If all neurons of one layer are regarded (Fig. 2), the function of the complete layer can be described as a matrix/vector multiplication

$$\vec{y} = g\left(\vec{W} \cdot \vec{x} + \vec{S}\right),$$

where $\vec{x}$ is the input vector and $\vec{W}$ the weight matrix, which keeps all connection weights between two related layers. The sigmoidal function

$$g(x) = \frac{1}{1+e^{-x}}$$

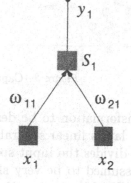

Figure 1 Model of an artificial neuron.

is mostly used as the activation function in feedforward networks. Feedforward networks are very powerful when using one or more hidden layers. The structure or topology of the network determines the class of geometry for pattern recognition, for function approximation or for

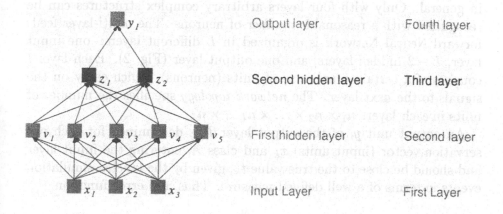

Output layer	Fourth layer
Second hidden layer	Third layer
First hidden layer	Second layer
Input Layer	First Layer

Figure 2 Model of a feedforward Network.

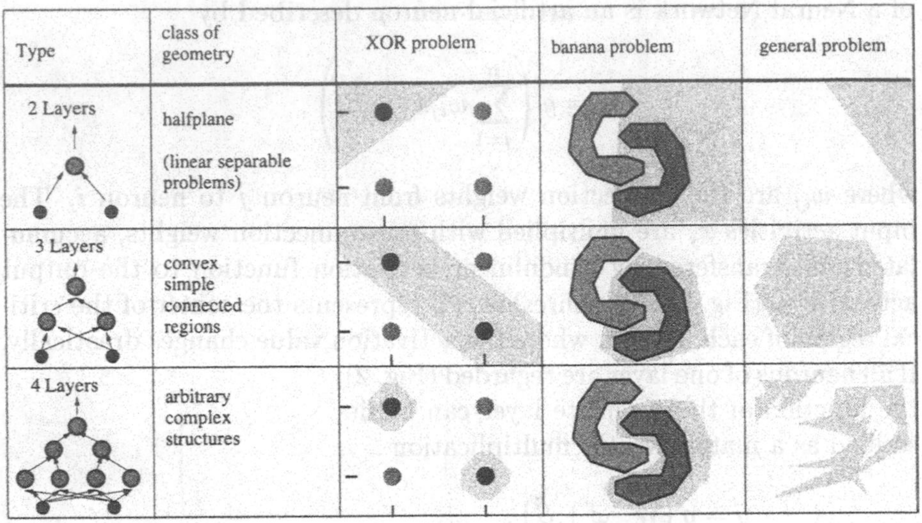

Type	class of geometry	XOR problem	banana problem	general problem
2 Layers	halfplane (linear separable problems)			
3 Layers	convex simple connected regions			
4 Layers	arbitrary complex structures			

Figure 3 Capacity of multi-layer feedforward networks.

transformation to be described by the Neural Network (Fig. 3). With two layers linear separable problems can be solved. Every output neuron divides the input space into two regions. If the sigmoidal function is assumed to be very sharp the function of an output neuron may be visualized by a separation line (in a 2-dimensional input space) or a hyperplane (in an n-dimensional input space). Non-linear problems cannot be solved by this type of network. In Fig. 3 we can see that a three-layer network is able to solve all convex tasks. Even non-convex objects may be separated by one hidden layer networks if the problem is given in a pixel or boolean representation, but many neurons are required in general. Only with four layers arbitrary complex structures can be recognized with a reasonable number of neurons. The multi-layer feedforward Neural Network is organized in L different layers: one input layer, $L-2$ hidden layers, and one output layer (Fig. 2). Each layer l consists of a certain number n_l of units (neurons), which carry on the signals to the next layer. The *network topology* specifies the number of units in each layer: $n_1 \times n_2 \times \ldots \times n_{L-1} \times n_L$.

An output unit y_i of the output layer L is determined for each observation vector (input units) x_k and class A_i entering the input layer and should be close to the true value t_k, given by the labeled simulation events in terms of a well defined measure. Thus, the error function

$$E(W) = \frac{1}{2N} \sum_{i=1}^{N} (y_k - t_k)^2 \tag{1}$$

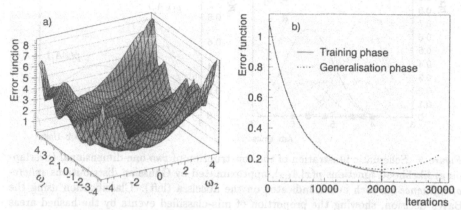

Figure 4 A snapshot at a certain iteration of the network error $E(W)$ as a function of neuron weights is shown in the left panel. The network error $E(W)$ vs. number of iterations is given on the right hand side. During the training phase the error decreases monotonically, but an independent test sample gives a rise after a certain number of iterations.

has to be minimized (Fig. 4). The most common algorithm for network training, i.e. minimizing $E(W)$, is an iterative adjustment of the weights w_{ij} and thresholds S_i by the *back-propagation* algorithm [6]. There exist different other algorithms or extended versions of this basic back-propagation, which try to circumvent problems in finding the global minimum or sticking in a local minimum. Additional problems arise, if the training process leads to an overtraining of the network by adopting the properties of the training samples, but cannot give satisfactory results, when it is applied to another validation set. Thus, in a generalization phase one has to control the quality of the network with an independent labeled set of samples (Fig. 4).

In general, the output y_i is a continuous function. Hence not only the classification can be done applying Neural Networks, but also parameter estimation (regression) is possible. The Bayes classifier is a powerful algorithm. The performance is generally excellent and asymptotically *Bayes optimal*, so that the expected *Bayes error* (see below) is less than or equal to that of any other technique [7]. The estimated probability densities converge asymptotically to the true density with increasing sample size [8].

The method is based on the *Bayes Theorem* [9]

$$p(A_i|x) = \frac{p(x|A_i) \times P(A_i)}{p(x)} \Leftrightarrow \text{posterior} = \frac{\text{likelihood} \times \text{prior}}{\text{normalization}} \quad (2)$$

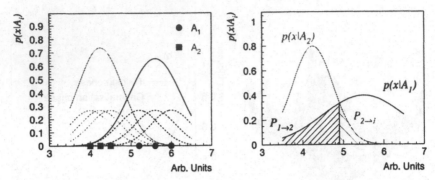

Figure 5 Schematic illustration of the construction of two one-dimensional (overlapping) likelihood functions $p(x|A_{1,2})$, approximated by Gaussian distributions *sphere-of-influence* for each event, indicated on the abscissa (left). Classification using the Bayes decision, showing the proportion of mis-classified events by the hashed areas (right).

with $p(x) = \sum_{j=1}^{N} p(x|A_j)P(A_j)$, which holds, if the different N hypotheses A_i (i.e. classes) are mutually exclusive and exhaustive. By a prior and a normalization factor the theorem connects the likelihood for an event x of a given class A_i with the probability of a class A_i, being associated to a given event x. The prior gives the *a priori* knowledge of the relative abundance of each class and is major basis of debates on Bayesian inference procedures. It is nearly always the best to follow the advice given by Bayes himself [9], generally known as *Bayes' Postulate* (occasionally also referred to as *Principle of Equidistribution of Ignorance*): So far there exists no further knowledge, the prior probabilities should be assumed to be equal $P(A_i) = \frac{1}{N}$. In the fortuitous case that the likelihood functions $p(x|A_i)$ are known for all populations, the Bayes optimal decision rule is to classify x into class A_i, if

$$p(A_i|x) > p(A_j|x) \tag{3}$$

for all classes $A_j \neq A_i$, as illustrated (with the mis-classification probabilities) in Fig. 5 (right). To construct an estimate $\hat{p}(x|A_i)$ of the likelihood $p(x|A_i)$ of class A_i, the k-th simulated event x_{ki} is assumed to have a *sphere-of-influence*, where it contributes to the probabilities (see Fig. 5 left). There are various procedures to specify this contributions, whose superpositions lead to continuous likelihood functions, replacing the frequency distributions of discrete simulated events N_i of each class A_i in the n-dimensional *observation* space. A standard choice of such *spheres* are multivariate normal distributions, because they are simply well behaved, easily computed and have been shown in practice to perform well.

The *Bayes error* ϵ represents the total sum (integral) of mis-classified events and is given in case of two classes by the simple relation (hashed area in Fig. 5 right)

$$\epsilon = \int \min\{p(A_1|x), p(A_2|x)\} \cdot p(x)dx. \tag{4}$$

To account for the mis-classification, the rates $P_{ij} = P_{A_i \to A_j}$, i.e. the probability of an event $x \in A_i$ being classified in the class A_j, are estimated by the *leave-one-out* method (also called *jack-knifing*). Each simulated event is held back once, while the others are used to estimate the association of this particular event. By a *bootstrap* method different subsets of each simulated class are used to perform the *leave-one-out* method to give an asymptotically unbiased estimate of the variance of the $\hat{P}_{ij}$ [7]. Thus, in simple case the true number of events $n_i^\star$ can be deduced from the classified events n_j by a matrix inversion

$$\sum_j \hat{P}_{ij}^{-1} n_j = n_i^\star \qquad \text{with} \qquad \hat{P}_{ij} = \hat{P}_{A_i \to A_j}. \tag{5}$$

But other unfolding methods should be applied, if the problem has a lot of degrees of freedom, because the inverse $\hat{P}_{ij}^{-1}$ might become singular (see e.g. [10]).

It can be shown that the NN behaves like a Bayesian Classifier, i.e. the output nodes produce Bayesian *a posteriori* probabilities [11]. In previous publications the consistency and equivalence of Neural Network and Bayes classifier results in EAS analysis have been demonstrated [12, 13]. Hence an adequate choice of the particular decision rule and of the appropriate algorithm is just a matter of the actual conditions (e.g. computing time and memory workload).

3. Classifiers in Practice

As an example the analysis of extensive air shower data are given. The goal of the KASCADE experiment is to determine the primary cosmic ray (CR) spectrum and elemental composition in the knee region ($10^{14} - 10^{16} eV$) by simultaneously measuring a large number of EAS observables for each individual event and with high quality. This 'redundancy' uncovers systematic biases in the interpretation of the data, thereby enabling also tests of the high-energy hadronic interaction models plugged into EAS simulations.

To familiarize the reader with the classification methods the elemental composition is analyzed in the following. The observables N_e and N_μ^{tr} (definition see [2, 3]) generated by the Monte Carlo program

CORSIKA [14] are used to build the likelihood distributions in Fig. 6 (hadronic interaction model: QGSJet [15]). If $\hat{p}(x|A_{\mathrm{Fe}}) > \hat{p}(x|A_i)$ ($i \in \{\mathrm{p}, \mathrm{O}\}$), then the distribution is colored in dark grey, in lieu thereof p and O are colored in middle grey and light grey. A rough separation can be recognized, but also a strong overlapping of the likelihood distributions.

In the most probable cases the different particle types are identified correctly, but the incorrectly classified fraction is used for correction further-on (Fig. 7). Due to the stronger fluctuations and weaker correlations with mass and/or energy, different other sets of observables result in lower true-classification rates P_{ii}. Therefore it is not possible to consider more than 3 classes, since this would require an analysis of further observables simultaneously, with a number of Monte Carlo simulations larger than actually available. Within the statistical errors there is no difference visible between NN and Bayes classification. Thus only the NN classification rates using a $2 \times 5 \times 2 \times 1$ network topology are displayed in Fig. 7.

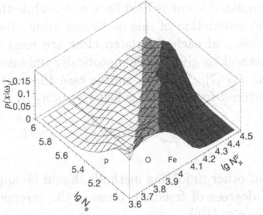

Figure 6 Superposition of three probability density distributions $\sum_{i=1}^{3} \hat{p}(x|A_i)/3$ deduced from QGSJet simulations using the observables N_e and N_μ^{tr}. Events in the dark shaded area mark the region classified as iron, middle grey as oxygen, and light grey as proton.

As a result, analyses of different sets of observables (definition see [2, 3]) are compiled in Fig. 8. In general the tendencies are the

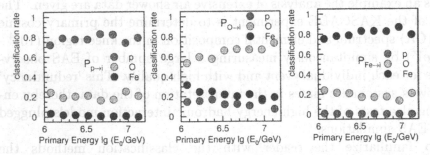

Figure 7 Classification rates for three classes (p,O and Fe). The used observables are N_μ^{tr} and N_e.

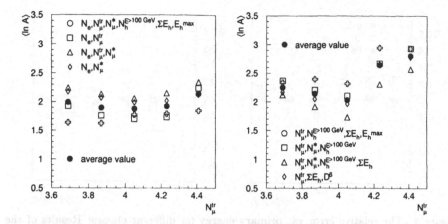

Figure 8 Mean logarithmic mass $\langle \ln A \rangle$ resulting from the analysis of different sets of observables vs. $\lg N_\mu^{tr}$. The sets displayed on the right do not include the observable N_e.

same. However, remarkably, all sets omitting the electron size N_e (right graph) result in a heavier composition and a more pronounced increase above the knee ($\lg N_\mu^{tr} = 4.15 \simeq E_{knee}$). As the electron size has the strongest mass correlation, as well as the smallest fluctuations, the mass compositions are predominantly determined by N_e and N_μ^{tr} (left). The compositions resulting from sets of weaker correlated observables lead to different compositions (right). If the degree of the correlations of different observables in simulations and measurement for each primary mass would be the same, the determined compositions (shown in Fig. 8) should be identical within the statistical errors. Hence the shown inconsistency may be arising from differently marked correlations among measured and simulated events. Thus, with such an amount of redundant information one can try to rule out, if a certain hadronic interaction model is able to describe the measured data and give consistent results or not.

Nevertheless, different models can also be compared and tested whether their predictions are in agreement or not. Thus, the energy estimation of the primary all-particle energy spectrum of the cosmic rays is considered.

For estimating the primary energy E the most important EAS parameters are again the shower size N_e and the truncated muon number N_μ^{tr}, where now N_μ^{tr} carries most of the information. For the training of the network (according to Equ. 1) two independent samples have been generated to allow a validation of the results. After training the network, the quality of the training is checked.

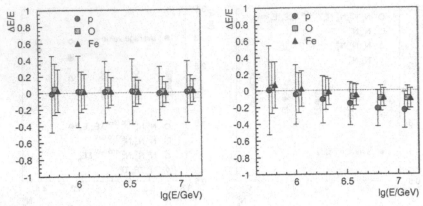

Figure 9 The relative error vs. primary energy for different classes: Results of the network trained with QGSJet samples (left) and the result of a network, which was trained with the QGSJet samples, but the analyzed data are VENUS samples (right).

Before applying, the response and the biases of the trained Neural Network have to be carefully scrutinized. For these studies, the two samples of quasi-data resulting from the Monte Carlo simulations using the QGSJet [15] and the VENUS [16] hadronic interaction models, respectively, were used. The following questions were putting with which uncertainties the original subsamples of different mass classes are recognized and well associated by the network, and how the recognition capability depends on the energy. With this aspect we consider the relative deviation of the reconstructed energy E_{est} from the true value E_{true} (which is known for the simulated events), more precisely, the distribution of $(\Delta E/E)_{est} = (E_{est} - E_{true})/E_{true}$, whose mean value and the standard deviation represent the bias and the energy resolution (relative error) of the reconstruction, respectively. Fig. 9 displays the relative error of the estimated energy of different primary particles. The relative error of a network, trained with QGSJet samples, is displayed (left). In general, the bias for the various classes is less than $3-5$ %, but the energy resolution (spread) proves to be strongly mass dependent. As expected, the iron class has the smallest energy spread. The network trained with VENUS samples leads to the same results. The result of a network, which has been trained with QGSJet samples, is also shown (right), but analyzing samples generated by the VENUS model. With increasing energy the QGSJet trained network underestimates systematically the true energy of the VENUS samples. Moreover, the bias appears to be mass dependent, implying that the degree of the correlations between energy and the analyzed shower observables is varying differently for the different primary particles and models. This is a caveat for the estimate

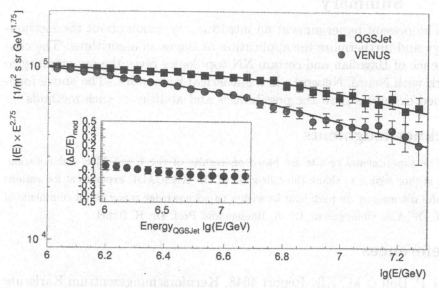

Figure 10 Energy spectra resulting from the analysis of data of the KASCADE experiment using two differently trained networks (by VENUS and QGSJet samples). The reconstructed energies are compared on event-by-event basis and their differences are given as relative error vs. the energy reconstructed on the basis of the QGSJet model in the inset.

of true energies of the measured event samples, that hidden mass dependent correlations lead to an all-particle spectrum, depending from the true mass composition.

Fig. 10 presents the reconstructed energy spectra resulting from the analysis using two different networks, trained with QGSJet and VENUS samples, respectively. Apparently, the spectrum using the VENUS trained network, results in an extremely steep spectrum as compared with the QGSJet findings. It should be emphasized, that the used network takes into account not only the absolute values of the observables N_e and N_μ^{tr}, but also their internal correlations. In order to specify the relative error, arising from the model dependence, mean value and spread of $\Delta E/E = (E_{\text{VENUS}} - E_{\text{QGSJet}})/E_{\text{QGSJet}}$ are additionally given (inset). The variation of this model error displays a change at higher energies, which could indicate a change of the composition. The results of the fitted spectra are discussed in [2, 3, 13]. A final conclusion about the quality of different models cannot be made, but an extrapolation to direct measurements at lower energies indicates, that the VENUS model might not be able to describe the EAS data.

4.   Summary

The present paper aims at an introductory lesson about the methodology and furthermore the application of Bayesian algorithms. The congruence of Bayesian and certain NN topologies gives the opportunity to work with Neural Networks in a statistical framework. The above mentioned examples show the possibilities and abilities of such methods.

Acknowledgments

The experimental results are based on results of the KASCADE collaboration. The author wishes to thank the colleagues of the KASCADE experiment for various helpful discussions. In particular he wishes to acknowledge many useful comments of Prof. Dr. A.A. Chilingarian, Dr. A. Haungs and Prof. Dr. H. Rebel.

References

[1] P. Doll et al., KfK-Report 4648, Kernforschungszentrum Karlsruhe (1990)

[2] H. Rebel, these proceedings, Predeal (2000)

[3] A. Haungs, these proceedings, Predeal (2000)

[4] A.A. Chilingarian, Comput. Phys. Commun. 54 (1989) 381

[5] A.A. Chilingarian, *ANI users guide*, unpublished

[6] D. Rummelhart, J. McClelland, *Parallel Distributed Processing*, MIT Press, Cambridge (1986)

[7] K. Fukunaka, *Introduction to Statistical Pattern Recognition*, Academic Press (1972)

[8] E. Parzen, Annals of Mathematical Statistics 33 (1962) 1065

[9] T. Bayes, Phil. Trans. Roy. Soc. 53 (1763) 54; reprinted in Biometrika 45 (1958) 296

[10] V. Blobel, *Unfolding methods in high-energy physics experiments*, report CERN 85-02 (1985)

[11] H. Gish, Proc. IEEE Conf. on Acoustics Speech and Signal Processing, (1990) 1361

[12] A. Chilingarian et al., Nucl. Phys. B (Proc. Suppl.) 52B (1997) 237

[13] M. Roth, FZKA-Report 6262, Forschungszentrum Karlsruhe (1999)

[14] D. Heck et al., FZKA-Report 6019, Forschungszentrum Karlsruhe (1998)

[15] N.N. Kalmykov, S.S. Ostapchenko, Yad. Fiz. 56 (1993) 105

[16] K. Werner, Phys. Rep. 232 (1993) 87

List of Participants

Marilena Avrigeanu
National Institute of
Physics and Nuclear Engineering
POB MG-6
RO-76900 Bucharest-Magurele
Romania
mavrig@ifin.nipne.ro

Vlad Avrigeanu
National Institute of
Physics and Nuclear Engineering
POB MG-6
RO-76900 Bucharest-Magurele
Romania
vavrig@ifin.nipne.ro

Aurelian Florin Badea
National Institute of
Physics and Nuclear Engineering
POB MG-6
RO-76900 Bucharest-Magurele
Romania
badea@muon1.nipne.ro

Madalina Badea
National Institute of Plasma,
Laser and Radiation Physics
Atomistilor Str. 109
RO-76900 Bucharest-Magurele
Romania
mbadea@muon1.nipne.ro

Gerhard Baur
Forschungszentrum Jülich
Institut für Kernphysik
D-52425 Jülich
Germany
g.baur@fz-juelich.de

Alexandru Bercuci
National Institute of
Physics and Nuclear Engineering
POB MG-6
RO-76900 Bucharest-Magurele
Romania
alex@muon1.nipne.ro

Richard Russell Betts
Argonne National Laboratory
Physics Division
Argonne, IL 60439
USA
betts@uic.edu

Charles J. Beyer
Vanderbilt University
Department of Physics
POB 1807-B
Nashville, TN 37235
USA
cj@styx.phy.vanderbilt.edu

Iliana Magdalena Brancus
National Institute of
Physics and Nuclear Engineering
POB MG-6
RO-76900 Bucharest-Magurele
Romania
iliana@muon1.nipne.ro

Mark Caprio
WNSL Yale University
272 Whitney Ave
New Haven, CT 06520-8124
USA
mark.caprio@yale.edu

421

Florin Carstoiu
National Institute of
Physics and Nuclear Engineering
POB MG-6
RO-76900 Bucharest-Magurele
Romania
carstoiu@theor1.theory.nipne.ro

Richard F. Casten
Yale University
WNSL, Physics Department
POB 208124
272 Whitney Ave
New Haven, CT 06520-8124
USA
rick@riviera.physics.yale.edu

Gheorghe Cata-Danil
National Institute of
Physics and Nuclear Engineering
POB MG-6
RO-76900 Bucharest-Magurele
Romania
cata@ifin.nipne.ro

Caner Cicek
Canakkale Onsekiz Mart University
Department of Physics
17100 Canakkale
Turkey
canerk@bornova.ege.edu.tr

Madalina Colci
National Institute of
Physics and Nuclear Engineering
POB MG-6
RO-76900 Bucharest-Magurele
Romania
mcolci@personal.ro

Cary N. Davids
Argonne National Laboratory
Physics Division
Argonne, IL 60439
USA
davids@anl.gov

Luke O'Connor Drury
Dublin Institute for Advanced Studies
School of Cosmic Physics
5 Merrion Square
Dublin 2
Ireland
ld@cp.dias.ie

Alexandru Enulescu
National Institute of
Physics and Nuclear Engineering
POB MG-6
RO-76900 Bucharest-Magurele
Romania
alen@ifin.nipne.ro

Daniela Fluerasu
National Institute of
Physics and Nuclear Engineering
POB MG-6
RO-76900 Bucharest-Magurele
Romania
dannyf@ifin.nipne.ro

Gilles de France
Ganil BP 5027
F-14076 Caen Cedex 5
France
defrance@ganil.fr

Werner Gast
Forschungszentrum Jülich
Institut für Kernphysik
D-52425 Jülich
Germany
w.gast@fz-juelich.de

Radu Alexandru Gherghescu
National Institute of
Physics and Nuclear Engineering
POB MG-6
RO-76900 Bucharest-Magurele
Romania
rgherg@ifin.nipne.ro

Nguyen Van Giai
Institut de Physique Nucléaire
Groupe de Physique Théorique
91406 Orsay Cedex
France
nguyen@ipno.in2p3.fr

Tudor Glodariu
National Institute of
Physics and Nuclear Engineering
POB MG-6
RO-76900 Bucharest-Magurele
Romania
glodariu@ifin.nipne.ro

Mehmet Cem Guclu
Istanbul Technical University
Department of Physics
Ayazaga Kampusu Maslak
80626 Istanbul
Turkey
guclu@sariyer.cc.itu.edu.tr

Rüdiger Haeusler
Forschungszentrum Karlsruhe
Institut für Kernphysik
POB 3640, D-76021 Karlsruhe
Germany
haeusler@ik3.fzk.de

Oliver Haug
Universität Tübingen
Institut für Theoretische Physik
Auf der Morgenstelle 14
D-72076 Tübingen
Germany
oliver.haug@uni-tuebingen.de

Andreas Haungs
Forschungszentrum Karlsruhe
Institut für Kernphysik
POB 3640
D-76021 Karlsruhe
Germany
haungs@ik3.fzk.de

Aurelian Isar
National Institute of
Physics and Nuclear Engineering
POB MG-6
RO-76900 Bucharest-Magurele
Romania
guest24@theo.physik.uni-giessen.de

Naoyuki Itagaki
Department of Physics
University of Tokyo
7-3-1 Hongo Bunkyo-ku
Tokyo 113-0033
Japan
itagaki@tkyntm.phys.s.u-tokyo.ac.jp

Anna Iwan
Department of
Experimental Physics of
Lodz University
Pomorska 149/153
90236 Lodz
Poland
ianna@krysia.uni.lodz.pl

Valeri Kalinin
V.G.Khlopin Radium Institute
194021 2nd Murinsky av. 28
Sankt Petersburg
Russia
kalinin@atom.nw.ru

Franz Käppeler
Forschungszentrum Karlsruhe
Institut für Kernphysik
POB 3640
D-76021 Karlsruhe
Germany
kaepp@ik3.fzk.de

Jürgen Kiener
Centre de Spectrométrie Nucléaire
et de Spectrométrie de Masse
IN2P3-CNRS Bat 104-108
91405 Campus Orsay
France
kiener@csnsm.in2p3.fr

Tzanka Todorova Kokalova
Hahn-Meitner-Institut Berlin
Glienicker Straße 100
D-14109 Berlin
Germany
kokalova@hmi.de

Pierre Leleux
Université Catholique de Louvain
B-1348 Louvain-la-Neuve
Belgium
leleux@fynu.ucl.ac.be

Alexander Lisetskiy
Universität Köln
Institut für Kernphysik
Zulpicher Str 77
D-50935 Cologne
Germany
lis@ikp.uni-koeln.de

Nikolay Petrov Minkov
Institute for Nuclear Research
and Nuclear Energy
72 Tzarigrad Road
1784 Sofia
Bulgaria
nminkov@inrne.bas.bg

Mihail Doloris Mirea
National Institute of
Physics and Nuclear Engineering
POB MG-6
RO-76900 Bucharest-Magurele
Romania
mirea@ifin.nipne.ro

Elvira Moya de Guerra
Instituto de
Estructura de la Materia
Consejo Superior de
Investigaciones Cientificas
Serrano 123
28006 Madrid
Spain
imtem22@pinar2.csic.es

Alex C. Mueller
Institut de Physique Nucléaire
F-91406 Orsay Cedex
France
mueller@ipno.in2p3.fr

Gottfried Heinz Münzenberg
Gesellschaft für
Schwerionenforschung
Planckstrasse 1, Postfach 11 05 52
D-64291 Darmstadt
Germany
g.muenzenberg@gsi.de

Karl Manfred Mutterer
Technische Universität Darmstadt
Institut für Kernphysik
Schlossgartenstrasse 9
64289 Darmstadt
Germany
mutterer@hrz1.hrz.tu-darmstadt.de

Yasuki Nagai
Research Center for
Nuclear Physics
Osaka University
10-1 Mihogaoka, Ibaraki
Osaka 567-0047
Japan
nagai@rcnp.osaka-u.ac.jp

Wolfram von Oertzen
Hahn-Meitner-Institut Berlin
Glieniker Str. 100
D-14109 Berlin
Germany
oertzen@hmi.de

Riza Ogul
University of Selcuk
Department of Physics
42079 Kampus Konya
Turkey
rogul@selcuk.edu.tr

Alexandra Olteanu
National Institute of
Physics and Nuclear Engineering
POB MG-6
RO-76900 Bucharest-Magurele
Romania
secretar@ifin.nipne.ro

Takaharu Otsuka
University of Tokyo
Department of Physics
7-3-1 Hongo, Bunkyo-ku
Tokyo 113
Japan
otsuka@phys.s.u-tokyo.ac.jp

Larisa Florina Pacearescu
National Institute of
Physics and Nuclear Engineering
POB MG-6
RO-76900 Bucharest-Magurele
Romania
larisa@theor1.theory.nipne.ro

Alexandra Petrovici
National Institute of
Physics and Nuclear Engineering
POB MG-6
RO-76900 Bucharest-Magurele
Romania
spetro@ifin.nipne.ro

Dorin N. Poenaru
National Institute of
Physics and Nuclear Engineering
POB MG-6
RO-76900 Bucharest-Magurele
Romania
poenaru@ifin.nipne.ro

Joann Prisciandaro
National Superconducting
Cyclotron Laboratory
Michigan State University
164 S. Shaw Lane
East Lansing, MI 48824 - 1321
USA
prisc@nscl.msu.edu

Adriana Rodica Raduta
National Institute of
Physics and Nuclear Engineering
POB MG-6
RO-76900 Bucharest-Magurele
Romania
araduta@ifin.nipne.ro

Alexandru Horia Raduta
National Institute of
Physics and Nuclear Engineering
POB MG-6
RO-76900 Bucharest-Magurele
Romania
hraduta@ifin.nipne.ro

Apolodor A. Raduta
National Institute of
Physics and Nuclear Engineering
POB MG-6
RO-76900 Bucharest-Magurele
Romania
raduta@ifin.nipne.ro

Akunuri V. Ramayya
Vanderbilt University
Department of
Physics and Astronomy
POB 1807, Station B
Stevenson Center 6301
Nashville, TN 37235
USA
ramayya1@ctrvax.vanderbilt.edu

Heinigerd Rebel
Forschungszentrum Karlsruhe
Institut für Kernphysik
POB 3640
D-76021 Karlsruhe
Germany
rebel@ik3.fzk.de

Peter Ring
Technische Universität München
Physik Department T30
James-Frank-Straße
85748 Garching
Germany
peter_ring@physik.tu-muenchen.de

Markus Roth
Forschungszentrum Karlsruhe
Institut für Kernphysik
POB 3640
D-76021 Karlsruhe
Germany
roth@ik3.fzk.de

Oscar Saavedra
Universita' di Torino
Dipartimento di Fisica Generale
"A. Avogadro"
Via P. Giuria 1, 10125 Torino
Italy
saavedra@to.infn.it

Maria Sanchez-Vega
Texax A & M University
Cyclotron Institute
College Station, TX 77843
USA
svega@comp.tamu.edu

Nicolae Sandulescu
National Institute of
Physics and Nuclear Engineering
POB MG-6
RO-76900 Bucharest-Magurele
Romania
nsandu@theor1.theory.nipne.ro

Pedro Sarriguren
Instituto de Estructura de la Materia
Consejo Superior de Investigaciones
Cientificas
Serrano 123
28006 Madrid
Spain
imtps61@pinar2.csic.es

Werner Scheid
Universität Giessen
Institut für Theoretische Physik
Heinrich-Buff Ring 16
D-35392 Giessen
Germany
werner.scheid@theo.physik.uni-giessen.de

Joachim Scholz
Forschungszentrum Karlsruhe
Institut für Kernphysik
POB 3640
D-76021 Karlsruhe
Germany
scholz@ik3.fzk.de

Gerhard Schrieder
Technische Hochschule Darmstadt
Institut für Kernphysik
D-64289 Darmstadt
Germany
schrieder@ikp.tu-darmstadt.de

Ion Silisteanu
National Institute of
Physics and Nuclear Engineering
POB MG-6
RO-76900 Bucharest-Magurele
Romania
silist@theor1.theory.nipne.ro

Elena Stefanova
INRNE, BAS
72 Tzarigrad Road
1784 Sofia
Bulgaria
elenas@inrne.bas.bg

Fotis Tervisidis
Democritus University of Thrace
Nuclear Engineering Laboratory
Greece
tervis@lib.auth.gr

Robert Tribble
Texas A & M University
College Station, TX 77843
USA
tribble@comp.tamu.edu

Ekaterini Tsoulou
Institute of Nuclear Physics
NCSR "Demokritos"
153 10 Aghia Paraskevi, Attiki
Greece
ktsoulou@mail.demokritos.gr

Carmen Tuca
National Institute of
Physics and Nuclear Engineering
POB MG-6
RO-76900 Bucharest-Magurele
Romania
secretar@ifin.nipne.ro

Hiroaki Utsunomiya
Konan University
Department of Physics
Okamoto 8-9-1
Higashinada Kobe 658-8501
Japan
hiro@konan-u.ac.jp

Bogdan Vulpescu
National Institute of
Physics and Nuclear Engineering
POB MG-6
RO-76900 Bucharest-Magurele
Romania
bogdan@muon1.nipne.ro

Mark Wallace
Cyclotron Laboratory
Michigan State University
East Lansing, MI 48824
USA
wallace@nscl.msu.edu

Jürgen Wentz
Forschungszentrum Karlsruhe
Institut für Kernphysik
POB 3640
D-76021 Karlsruhe
Germany
wentz@ik3.fzk.de

Sir Arnold Wolfendale
University of Durham
Department of Physics
South Road, Durham DH1 3LE
United Kingdom
a.w.wolfendale@durham.ac.uk

Ramon Wyss
Royal Institute of Technology
SE-100 44 Stockholm
Sweden
wyss@msi.se